Marcel Bessis
Professor, School of Medicine
Director, Institute for Cell Pathology (INSERM 48)
Hospital of Bicêtre, Paris, France

George Brecher
Professor and Chairman, Department of Laboratory Medicine
School of Medicine, University of California
San Francisco, California, U.S.A.

Marcel Bessis

Blood Smears
Reinterpreted

Translated by George Brecher

With 342 Figures, Some in Color

Springer International 1977

This book has been written and illustrated at the

Institute for Cell Pathology
Inserm Unit 48, Faculty of Medicine, Paris

Georgette Delpech has assured the completeness of
the references
Maurice Gaillard has drawn the illustrations
Geneviève Prenant has executed the microphotographs
Catherine Sémik has assisted at every stage of the production
of the final manuscript

ISBN 3-540-07206-3 Springer-Verlag Berlin Heidelberg New York
ISBN 0-387-07206-3 Springer-Verlag New York Heidelberg Berlin

French Edition: Réinterprétation des Frottis Sanguins
Masson/Springer 1976

Library of Congress Cataloging in Publication Data. Bessis, Marcel, 1917–. Blood smeare reinterpreted. Translation of Réinterprétation des frottis sanguins. Bibliography: p. Includes index. 1. Blood-Examination. I. Title. RB45.B3813   616.07'561   77-23901

Printed in Germany.

The use of general descriptive names, trade names, trade marks etc. in this publication, even if not specifically identified, does not indicate that such names, as understood by the Trade Marks and Merchandise Marks Act, may accordingly be used freely by anyone.

Type-setting, printing and bookbinding: Universitätsdruckerei H. Stürtz AG, Würzburg.
2121/3140-543210

Table of Contents

V

CHAPTER 3

Granulocytic Series

CHAPTER 4

Thrombocytic Series

CHAPTER 5

Lymphocytic Series

CHAPTER 6

Monohistiocytic Series

CHAPTER 7

Plasmocytic Series

Translator's Preface

The translation of *Blood Smears Reinterpreted* was begun when the French original was still undergoing revisions. I was accorded the opportunity to question any statement or turn of phrase that was unclear to me or appeared liable to misinterpretation. It is my hope that as a result, ambiguities—particularly those inherent in differences between American and European usage—have been removed and that I have at least approached the ultimate goal of any translation: to reflect the author's intention accurately while remaining as readable as the original.

Beyond the role of translator, I was encouraged to assume the role of critic. As a result, some pages or even single sentences were hotly debated, sometimes for hours, as Marcel Bessis insisted that any interpretations on which we could not agree should be so indicated. In fact our discussion invariably ended in agreement, though they led to changes of a sentence here or a word there and, on occasion, to the addition of a footnote or a brief paragraph.

Inevitably our discussions extended beyond the individual page to the structure of the book. Although thoroughly conversant and in tune with its purpose, I had to be reminded again and again that this was not to be a comprehensive treatise of morphologic hematology, the traditional titles and sequence of the chapter notwithstanding. The length of the chapter on red cells and the cursory treatment of mastocytes reflect no value judgement on the importance of red cell spicules on the one hand and mastocyte granules on the other.

From this personal experience it seems worthwhile to emphasize once more that the purpose of the book is to give examples of a method of making dead blood smears come alive. The selection of examples and the omission of entire subjects such as multiple myeloma or monocytic leukemia are, once again, not value judgements but the natural outcome of Marcel Bessis's particular interests and prior work on reinterpretation. His intent is for all of us to use this method in our daily work, to extend and improve it and to challenge the interpretations given as we develop new insights. We hope the reader will derive as much enjoyment from that exercise as from the reading of the book itself.

Author's Preface

I

This book attempts to bridge the gap between the hematology of smears and the cell biology of today. Most hematologists find it difficult to accept blood smears for what they are: flattened, brilliantly colored cadavers. Yet from certain appearances of these corpses a great deal can be inferred about the ultrastructure and pathophysiology of the living cells. This book is a collection of examples of this way of thinking.

II

"*A point of view is always false*" said Paul Valery. To take an example, there is the red blood cell of classical hematology, that of the phase contrast microscope, that of the electron and of the scanning microscope. Add to these appearances the red cell of the biochemist, the immunologist, the isotopist, the rheologist, even the mathematician ... All too often different points of view are not confronted and the individual specialist remains confined within the limits of his own technique. Yet, only the integration of all these data can bring out the reality hidden beneath the diverse appearances.

III

One retains best what one comprehends. I shall take the features of the plasma cells as an example: plasma cells have a dark blue (but occasionally bluish purple) cytoplasm with an eccentric nucleus and a pale paranuclear area; they are seen in great numbers in chronic infections. The student is expected to commit these purely descriptive features to memory. Yet, these pecularities can all be related to function. Plasma cells secrete antibodies: *this explains their increase in infectious states.* The machinery that produces these antibodies requires a very large number of ribosomes—*hence the dark blue cytoplasm.* The antibodies pass through the Golgi body which is located in the centrosome: *the result is a pale paranuclear area*, for the Golgi body is not stainable by Giemsa. The centrosome is the most rigid part of the cell: *this explains the nucleus being pushed toward the periphery.* Finally, it is known that that in certain pathologic cases, antibodies may accumulate in the cytoplasm. Since antibodies stain red with Giemsa, *the variation in coloration from blue to red* (and the occasional bluish purple) *become comprehensible.*

X

IV

No doubt examination of the blood cells based on their appearance in smears will be replaced eventually by methods based on their pathophysiology and these techniques will be automated. To accelerate this evolution, to guide it, to maintain continuity with the past, we must discriminate between what blood smears can still teach us and what we must not expect from them. That is one of the aims of this book.

V

This book does not pretend to be comprehensive. A good many details are yet to be reinterpreted and the significance of others is yet to be discovered. No doubt my reinterpretations contain errors and I have overlooked clues to reinterpretation. The reader will correct the errors and discover the meaning of new clues.

Some of my reinterpretations are hypothetical. I have usually indicated them and pointed out problems that remain to be solved. The aim of this book is not to teach new facts but to give examples of reinterpretation which will induce the reader to so reinterpret himself every cell he observes in smears. "The role of the teacher is not to teach new facts but to inspire new ideas" as my mentor Policard used to say.

VI

One last remark: this exercise in reinterpretation applies not only to cells as they appear on smears, but also to cells viewed with phase contrast, electron microscopy, etc. Ultimately, it applies to any visual image: the significance of the image lies in the mind of the viewer.

The Uses of Morphology

An Image Is Not Reality

Every image is a collection of signals which the initiated transforms into symbols. For example, when we look at a forest, we add hidden images to the images which are clearly apparent. Our memory tells us that the forest contains trees, branches, and leaves. Someone who has never walked through a forest could not know that the forest is shady and cool. Thus the image is not the reality. The reality is invisible: it can only be imagined, based on the totality of our experiences which the image conjures up in our mind.

When we observe a cell in a smear, we see only an image. The real cell is hidden behind that image. We may add images of the cell from phase or electron microscopy, cytochemistry or any technics past or future; the real cell cannot be seen, it can only be imagined.

The Image Is Ambiguous

Gombrich recalls a famous mosaic at the entrance of a house in Pompei, depicting a black dog with bared teeth, carrying the inscription: "Cave Canem". Without that legend, one might have thought that the sign of the black dog indicated a pet shop or the "Black Dog Cabaret"...

The images of the cells in smears have no legends; one cannot understand them except by recognizing them. For a hundred years, hematologists have "seen" L.E. cells, Sézary cells, stomatocytes, echinocytes, etc. These images remained unnoticed and were passed over by thousands of hematologists until the image could recall to their minds an earlier description and preferably an explanation of the phenomenon. Thus the meaning of an image for each of us depends on the recollection of our past experiences, our education, and the code each of us applies to the signals we encounter.

The painter Escher noted that most of us understand a picture more readily through the intermediary of a text than through the picture itself. The reason is that a picture can be interpreted in a variety of ways. The text (or one's inherited or learned program) specifies a single meaning and excludes all others.

An Image Does Not Convey a Precise Message

I have often quoted the Chinese proverb that a picture is worth a thousand words. Indeed a picture is worth a thousand words of description and communicates that description in an instant. However, it cannot convey a precise message.

It is not the picture, it is the legend that conveys the precise message. The image supports it, at times well, at times poorly. If one inverts the legends of color plates in a book, it is sometimes easy, sometimes difficult, and occasionally impossible to establish the author's intended message.

"Morphologic evidence" (drawing by Siné)

Even less can a picture prove anything. A proof can only be derived from an experiment. A sequence of morphologic appearances implies transition, even without a legend. Yet such a sequence does not prove a relationship. This applies equally to the derivation of a cell line from its stem cell or the derivation of a particular leukemia from a given normal line.

The Image is Irreplaceable

It has recently become fashionable in some quarters to consider morphology little more than an outmoded (if agreeable) pastime, at a level of scientific interest about that of a butterfly collection. The notion is not novel. Paul Valéry ascribed it to Descartes (1639): "The notion of substituting numbers for a picture, of subsuming all knowledge into a comparison of sizes and the depreciation of any relationship that cannot be expressed arithmetically has had the greatest consequences in all fields of Science." One of the consequences is the wish of some investigators to study cytology without the aid of their eyes.

The proponents of this idea forget an inherent property of living cells: shape and structure are linked to function. They depend on the genetic code and accidental events during the lifetime of the cell. The trained observer can thus extrapolate from the appearance of a living cell to its internal organization and reconstruct the stages of its development. But that is not all. The visual image evokes an emotion and stimulates ideas. The emotion created by the Mona Lisa cannot readily be evoked by a table of spatial distributions of densities of specific wavelengths...

The images of cells and other "microscopic creatures" will thus continue to play a major role in the study of life processes.

True, the images are ambiguous and cannot communicate ideas: they occasion them...

Nomenclature

Nomenclature is one of the plagues of hematologic exposition. Throughout the years, hematologic nomenclature has inevitably reflected prevailing concepts of the origin or function of blood cells and at times conflicting theories have led to heated debates. As advancing knowledge has pinpointed errors of previous interpretations, terms embodying these earlier notions have become unsuitable or meaningless. To make matters worse European and American "schools" have occasionally used the same term to convey different facts or ideas. Some terms are simply ill chosen and carry implications other than those intended. Some are ungrammatical or consist of a mixture of Latin and Greek roots, thereby offending scholars of language.

Any attempt to correct these faults of current nomenclature must proceed with caution. As Horace warned in 40 B.C.: "Usage is the supreme law of language." It has remained so since Horace's time. Consequently, even absurd terms should not be changed without good and sufficient reason. In the following paragraphs I suggest guidelines for changes of nomenclature. I am, however, under no illusion: even well-reasoned changes will be resisted. Hence, when change appears impractical, a clear redefinition of older terms is suggested.

When Should One Coin New Terms?

1) When new phenomena are discovered, e.g., new structures are revealed by electron microscopy

2) When new or better understanding of known phenomena requires a new vocabulary to convey the new concepts

Thus the three terms—*determination, amplification,* and *maturation*—facilitate the explanation of our present concepts of differentiation (see p. 117).

Similarly, the use of echinocyte, acanthocyte, and keratocyte has removed some of the earlier confusion surrounding certain poikilocytes and spiculed cells (see p. 64).

When Should Terms be Abandoned or Redefined?

When terms are sanctioned by long continued usage, it may be unwise to attempt to replace them by new ones. At a minimum, however, we should draw attention to errors incorporated in terms that are based on disproven theories, as these may retard or prevent the birth of new ideas.

1) Examples of words based on erroneous theories or inadequate observations are "dedifferentiation" when applied to the development of leukemia (see p. 212); "spherocyte" when applied to red cells that are not truly spheres (see p. 96); and "classification" when applied to categorizations (see pp. 187, 188). Words based on only partially verified ideas are "virocyte" (see p. 74), "heparinocyte" (see p. 184) and "immunocyte" (see p. 145).

2) Redefinition of terms becomes essential when the present definition is inadequate. For example, as presently defined, a "monoblast" is a cell which cannot be recognized with certainty.

3) Redefinition is particularly desirable when the same expression is used differently in different countries. For example, "granulocyte" (see p. 114), "myeloblast" (see p. 108) and "erythroblast" (see pp. 30, 32) have different meanings in Europe and America.

4) Abolition of some redundant terms appears desirable when a large number of synonyms are in use. We can probably dispense readily with the use of "thesaurocyte", which is seldom used, "rubicyte" which is both rarely used and linguistically improperly formed from one Latin and one Greek root, and "Biermer's anemia" which is outdated.

5) When the significance of a term has been weakened. For example, since the study of cells originally called "cytology" has become prominently associated with diagnostic "exfoliative cytology", the new term "cellular biology" has been created to convey the meaning of cytology in its original sense. Similarly, "hematologic morphology" is nowadays thought of as a science of yesterday no longer contributing new insights into physiology and pathology. The depreciation of the term is indicated by the proposal to call "neo-morphology" those recent studies, particularly of living cells, which have clearly generated new ideas. Yet, what is the purpose of morphology? Certainly not to collect pictures, but to understand them. Morphology needs to be considered in that sense, not to be renamed.

How Should New Terms be Formed?

Having convinced oneself that a new term is truly necessary to describe a new observation or to improve the understanding of a subject, one should make sure that:

1) The new term is usable in any language. This is best assured by following the traditional practice of medical nomenclature, i.e., the derivation of new terms from the Greek. While English has replaced Latin and Greek as the new international language, the usage of common English words such as "burr" to designate a particular red cell may be confusing even to the scientist familiar with colloquial English. The combination "burr cell" is not immediately recognizable, the specific term: "echinocyte" is. The presumed advantage of voking a familiar picture in the reader's mind may also be questioned: while Webster defines burr first as the "prickly envelope of a fruit", it also gives a "thin ridge or roughness left by a cutting tool" and

"a punched out piece of metal", "a washer". It is only after one has seen an echinocyte that one can associate the term "burr cell" with the image it is meant to evoke.

2) The term should permit the formation of derived and composite terms to facilitate wide usage, easy indexing and retrieval of relevant information from the literature. Terms derived from the Greek again readily serve those aims. Echinocyte gives rise to echinocytogenic and to echino-acanthocyte, echino-spherocyte and other terms that would require one or more sentences for description if these combinations were not available to match the composite appearances.

3) Linguists among hematologists will be pleased if new terms follow standard linguistic rules, e.g., do not mix Latin and Greek roots. We probably have little chance of replacing the faulty "schistocyte" by the correct "schizocyte" in American and British usage, but perhaps we can avoid committing similar linguistic "crimes" in the future.

Nomenclature Adopted in This Book

My aim has been to use a rational nomenclature without shocking the traditionalist. New or relatively new terms have been introduced sparingly. I have always noted the reasons for their adoption. The erroneous deviation of other terms such as "reticulocyte" is merely mentioned (see p. 44). I am content to consider as provisional the few new terms I have used and to await their acceptance or rejection by hematologists at large.*

---

* I published a large treatise (Living Blood Cells, 1973) in which some of these rules are not respected. The reader who consults this treatise will thus have to verify the meaning of some words used in it.

Chapter 1

General Anatomy and Physiology of Blood Cells

This chapter reviews the principal features of:

1) Cell organelles as seen in the electron microscope, which has revealed their structure and occasionally their very existence.

2) The morphologic expression of cell functions such as endocytosis, exocytosis and locomotion, as revealed by examination of living cells by phase contrast microscopy.

3) The production, destruction and behavior of blood cells as revealed by a variety of experimental techniques such as isotopic labeling, transplantation and others.

This general introduction will avoid repetition during consideration of each cell line. It will explain the symbols used in the diagrams for the representation of the different organelles. It deals with what one cannot see but what one must know about the life of a cell when examining a stained blood smear.

1 — Inventory of Cell Organelles[1-3]

To demonstrate the difference between the "real" cell and the cell as it appears in a blood smear, we will use the lymphocyte as an example.

In a smear (Figs. 1 and 2), all one can distinguish is a round nucleus with blocks of chromatin, a bluish cytoplasm and a small, clear juxtanuclear zone. Yet this cell contains a large number of organelles (Fig. 3):

— a nucleus surrounded by a nuclear envelope, containing chromatin, nuclear sap and nucleoli;
— mitochondria;
— a Golgi complex;
— centrioles;
— microtubules;
— microfilaments;
— ribosomes;
— endoplasmic reticulum;
— lysosomes and/or specific granulations;
— vacuoles and inclusions;
— a cell membrane;
— a cytoplasmic matrix in which all these organelles are bathed.

These organelles—and perhaps others that we do not yet recognize—enable the lymphocyte to regulate all its functions of metabolism, multiplication, maturation, secretion and locomotion. It is tempting to attribute to each organelle a particular role, paralleling the role of organs in higher organisms.

The nucleus contains the "blueprint" of the cell and controls synthetic activity specific for the cell.

The mitochondria supply the necessary energy, particularly by synthesizing ATP.

The ribosomes and Golgi complex manufacture the proteins.

The microtubules represent the skeleton.

The microfilaments represent the muscles of the cell.

The lysosomes are equivalent to the organs of digestion.

The vacuoles control some of the mechanisms of entry and exit of various materials from the cell.

The inclusions are stockpiles of such materials.

The plasma membrane protects the protoplasm from the surrounding environment, regulates the exchanges with the environment and receives signals which govern the behavior of the cell.

This functional anatomy is correct in general, but is full of inaccuracies and deficiencies in details: the model of the lymphocyte depicted here recalls, on a scale 100,000 times smaller, the naive anatomic designs of surgeons in the middle ages who believed that the bile emptied directly into the bloodstream and that the aorta carried air ...

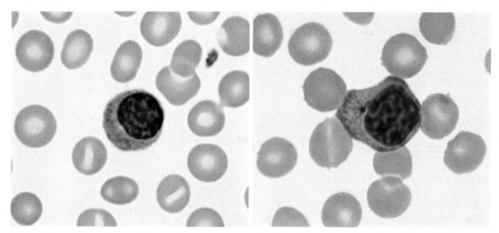

Fig. 1 – *A small lymphocyte*. The centrosome, unstained, can be seen to indent the nucleus

Fig. 2 – *A large lymphocyte*. In this cell, the area
occupied by the centrosome is stained violet

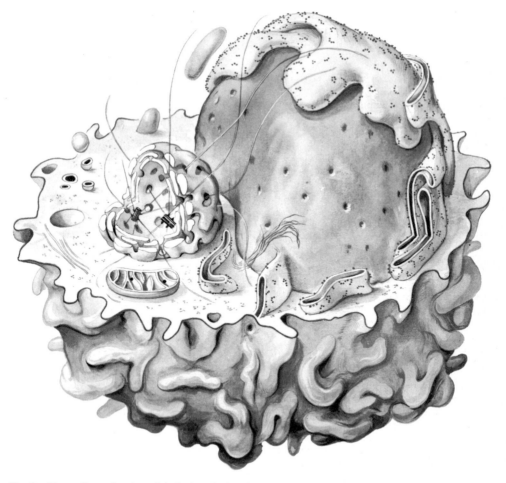

Fig. 3 – *Three-dimensional model of a lymphocyte.* Compare this reconstruction with a section examined
by electron microscopy (diagram p. 5, microphotographs pp. 231 to 235)

3

2 — The Nucleus

The nucleus, in its intermitotic resting stage, consists of four elements: the chromatin, the nucleoplasm (or nuclear sap), the nucleoli, and the nuclear envelope.

Chromatin

The appearance of the chromatin varies depending on the cell line and the state of maturation.

The younger the cell, the larger the nucleus and the more finely distributed the chromatin. The youngest cells have a large nucleus, in which the chromatin forms long, fine, interwoven or tangled filaments. This "dispersed" chromatin is believed to reflect an intense metabolic activity.

The older the cell or the more inactive, the more condensed the clumps of chromatin appear. These clumps form the "chromocenters*". It is generally believed that these areas represent portions of the chromatin that are not engaged in the active control of the ongoing cellular metabolism.

The chromocenters are always attached to the inside of the nuclear membrane (except where pores exist), and to the nucleolus (perinucleolar chromatin).

Sex chromatin. In the female sex, the nucleus contains a small mass of condensed chromatin, shaped like a triangle or lentil. In the neutrophilic granulocytes, this sex chromatin forms a nuclear appendage (see p. 120).

Nucleoplasm

Its amount varies from one cell type to another. In the same cell line, it becomes progressively reduced with aging. A diminution of the nuclear volume results. In pathologic states, the nucleoplasm can become very abundant (nuclear edema) or can disappear altogether (pyknosis of the nucleus, see p. 22).

In smears, the chromocenters are readily visible as are the sex chromosome and the nucleoplasm. The electron microscope has so far added little to our understanding of these structures.

Nucleolus

It is composed of the nucleolonema (a bunched filament or aggregate of particles very rich in RNA) and protein fibrils, a small quantity of DNA in the form of an intranuclear reticulum.

In smears, nucleoli are generally readily seen. The intensity of their blue color depends on the quantity of RNA which they contain. Sometimes they do not stain and are only recognized by a border of more heavily staining chromatin. Sometimes, however, they are completely hidden by perinucleolar chromocenters.

Nuclear Envelope

The nucleus is surrounded by an extension of the endoplasmic reticulum which forms a nuclear membrane around it. The perinuclear space or cisterna is normally 50 nm in width, but varies greatly with the functional state of the cell. The perinuclear space is continuous with the lumen of the endoplasmic reticulum, as shown in Figure 1 (9).

In the perinuclear cisterna exist circular openings of approximately 50 nm (called nuclear pores), through which the cytoplasm and the nucleoplasm communicate (see p. 224). Normally, these nuclear pores are closed by a diaphragm. It should be recalled that the nuclear pores always face the nucleoplasm, never the chromocenters. In some cells, pores occupy as much as 10% of the surface of the nuclear envelope. In general, the number of pores diminishes as the cell matures.

The junction between the nuclear envelope and the endoplasmic reticulum does not hinder the movement of the nucleus in the living cell. In some cells, one can observe continuous rotations of the nucleus around its own axis.

In smears, one can infer the presence of a nuclear envelope only in pathologic states in which the perinuclear space is dilated. In some cells one can identify the nuclear pores between the chromocenters located at the circumference of the nucleus (see p. 30, Figs. 1 and 2).

Cell Cycle and Mitosis

The cell cycle is generally subdivided into four segments: G_1, S, G_2 and M (see Fig. 2). Prolonged periods of quiescence of the nucleus have been designated as G_0.

In smears, one can distinguish cells in S and G, although some hematologists have claimed that the nucleus becomes lighter and larger as the S phase progresses. It would be extremely helpful if a ready distinction of the different phases of the cycle was possible. Today, this can only be done by autoradiography (after isotopic labeling) for identification of the S phase, or by spectrophotometry for identification of $G_1 + G_2$.

---

* Or "karyosomes."

4

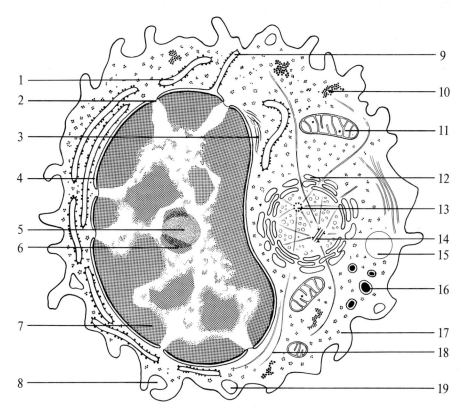

Fig. 1 – *Section of a lymphocyte examined in the electron microscope:* (1) rough endoplasmic reticulum (RER); (2) nuclear pore; (3) microfilaments; (4) nuclear envelope; (5) nucleolus; (6) perinuclear chromatin; (7) chromocenter; (8) beginning of pinocytosis; (9) RER communicating with nuclear envelope; (10) aggregate of glycogen; (11) mitochondrion; (12) Golgi complex; (13) centriole (cut transversely); (14) centriole (cut longitudinally); (15) contractile vacuole; (16) lysosome; (17) polyribosome; (18) microtubule; (19) end stage of pinocytosis

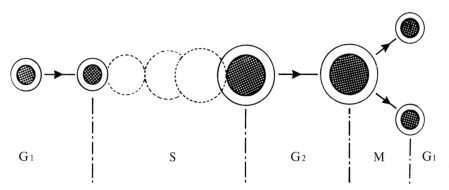

G_1 S G_2 M G_1

Fig. 2 – *Cell cycle.* G_1 postmitotic (intermitotic) phase; S phase of DNA synthesis; G_2 premitotic phase; M mitosis. For a given cell type, S, G_2 and M are constant

3 – Cytoplasmic Organelles

Centrosome and Golgi Complex

The centrosome consists of the Golgi complex and the centrioles which are situated in its center. This particular relationship between centrioles and the Golgi complex is not universal in all cells: it is characteristic of blood cells.

In the living state, the centrosome is recognizable as a clear space in the cytoplasm. When cellular movements are recorded by cinematography and time lapse exposure, the centrosome is seen to be the most rigid portion of the cell. The nucleus molds itself around it. The centrosome undergoes rhythmic oscillations which appear to facilitate the circulation of materials and organelles (see p. 16) and which cease only at low temperatures or in agonal stages of the life of the cell (see p. 22). The oscillations deform the nucleus rhythmically.

The Golgi complex is formed by several assemblies of closely packed cisternae (dictyosomes) which are grouped in a circular fashion to form a hollow sphere in which the two centrioles are located. Each of the assemblies consists of two to eight flattened sacs (see p. 224). The size and the appearance of the sacs reflect the metabolic activity of the cell. During heightened activity, the sacs separate and a large number of vesicles detach themselves from the sacs and carry away the materials manufactured by the Golgi complex such as gamma globulins, specific granules, lysosomes and others.

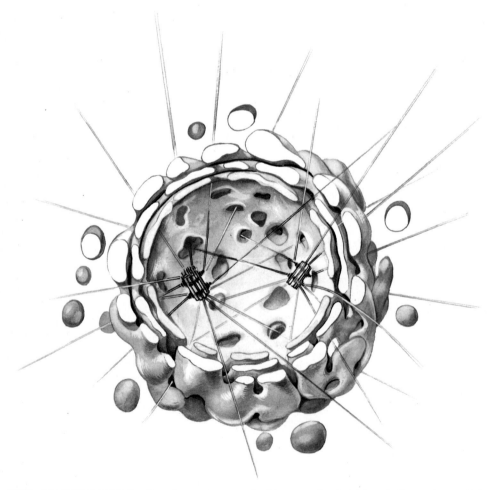

Three-dimensional reconstruction of a centrosome. In this case, the Golgi is a hollow sphere which includes the two centrioles. Their axes are at right angles to each other. Microtubules are attached to their satellites

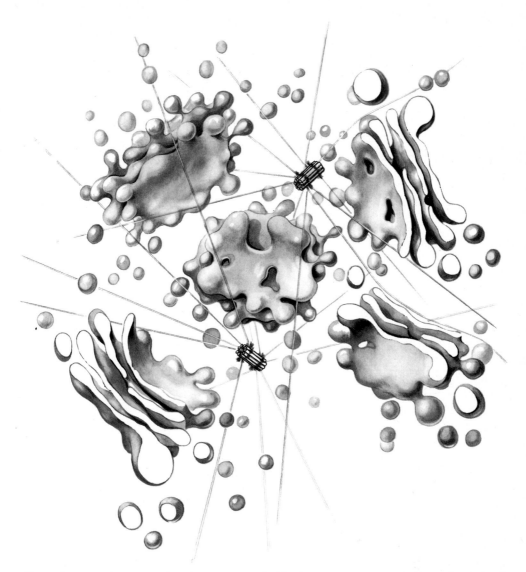

Three-dimensional reconstruction of a centrosome. In this, the most common arrangement, the stacks of cisternae are separated from each other (compare with microphotograph on p. 232)

The Golgi complex is the place where different products of cellular synthesis are combined to form the finished product. It is the end stage of the assembly line for different constituents of cellular activity where complex molecules and enzymes are packaged into granules.

In smears, the centrosome appears as a clear space in the cytoplasm which impinges upon the nucleus. The nucleus molds itself around this more resistant mass and often assumes a horseshoe shape or even a bilobed appearance. Single indentation of the nucleus and horseshoe shaped nuclei seen in smears are usually due to the presence of the centrosome.

3 – Cytoplasmic Organelles

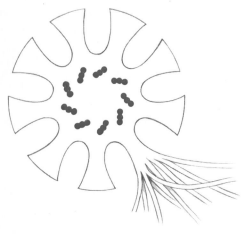

Fig. 1. – *Diagram of centriole*

Centrioles

Two centrioles are always present in intermitotic cells. They appear as small cylinders of 150 nm in diameter and 300 to 500 nm in length. They are always oriented in such a way that their long axes form a right angle with each other (see figs. on p. 6 and 7).

The wall of the cylinder is formed by nine sets of three tubules each. The axes formed by the three tubules are rotated against each other like the blades of a ship's screw (Fig. 2).

In blood cells, two sets of pericentriolar satellites surround the centriole tubules near each end. Microtubules are attached to the satellites (see p. 225).

The centrioles have important functions during mitosis and in cellular movement. They replicate themselves at the end of each cell division.

In polyploid blood cells such as megakaryocytes the number of centrioles corresponds to the ploidy of the cells.

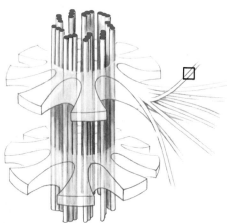

Fig. 2 – *Reconstruction of a centriole and its satellites*

Microtubules

Microtubules are structures approximately 200 nm in diameter and of variable lengths (see p. 233).

The wall of each microtubule has a thickness of approximately 4 nm and is formed of 13 filaments which consist of rows of spheres of the same diameter.

It has been hypothesized that the microtubules form a skeleton which maintains the shape of the cell. They do disappear as a result of cold or colchicine and reappear spontaneously on warming or washing the cell.

During mitosis, the microtubules form the fibers of the spindle. In telophase, the fibers of the spindle are concentrated at the line of division of the two cells and for a time the two daughter cells remain united by a bridge of spindle fibers (see p. 60). In normoblasts, this bridge of spindle fibers is frequently very long. Some are fixed at one end to a satellite of the centriole and at the other to the cell membrane.

In smears, the centriole and the microtubules cannot be seen directly. However, some techniques allow one to stain them reddish violet and to localize them quite clearly (see p. 3, Fig. 2 and p. 60, Cabot rings).

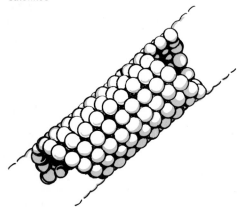

Fig. 3 – *Structure of a microtubule*

Ribosomes

These are particles approximately 20 nm in diameter which are present in all blood cells except the mature red cells. They consist primarily of RNA and proteins. They are comprised of two separate units of different density: 60S and 40S. They are present in the nucleus, the nucleolus and predominantly in the cytoplasm. They receive messenger and transfer RNA and play a cardinal role in the synthesis of proteins.

In the cytoplasm, ribosomes may be single (monoribosomes), but most frequently they form groups of 3 to 10 units and sometimes many more (polyribosomes, see p. 234). In that case the ribosomes are interconnected by a fine filament which represents messenger RNA.

The ribosomes can be attached to the sacs of endoplasmic reticulum (see below).

In smears, the greater or lesser basophilia (the more or less deep blue of Giemsa-stained smears) corresponds to the greater or lesser number of cytoplasmic ribosomes.

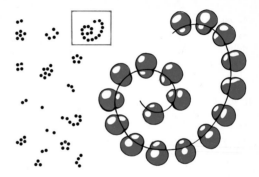

Ribosomes, polyribosomes and messenger RNA

Endoplasmic Reticulum

The endoplasmic reticulum is a system of flattened sacs and canaliculi which are more or less interconnected (see p. 231). The reticulum can be smooth or granular, i.e. covered by ribosomes, also called rough endoplasmic reticulum or RER.

Proteins produced by the ribosomes may accumulate or circulate through the endoplasmic reticulum to gain access to other regions of the cells, most frequently the Golgi complex (see p. 174).

Some cells (e.g., plasma cells) are filled with well-developed RER. Others (e.g. myeloblasts) contain only the first stages of its development.

In smears, the smooth ER cannot be seen. The RER can be recognized occasionally by the laminated appearance of the blue cytoplasm, described by the early cytologists as ergastoplasm (see p. 176).

Microfilaments

These are fibrils or groups of fibrils of approximately 5 nm diameter, often present in the vicinity of the nucleus or adjacent to the nuclear envelope (see p. 235). These fibrils belong to the group of actomyosins and play a role in intercellular and cellular movements (see p. 16).

In smears, they cannot be seen.

3 — Cytoplasmic Organelles

Mitochondria

These are round, oval or rod-shaped organelles, 3 to 14 nm in length and 2 to 10 nm in width. The total quantity of mitochondria of a given cell type is relatively constant but their number is quite different in different cell lines and in different stages of maturation.

The mitochondria are bordered by two membranes of the same thickness, one external and the other internal, which form the mitochondrial clefts or cristae. They contain enzymes which furnish the cell with the energy for its metabolic processes (ATP). They are more numerous in more active cells. The interior of the mitochondria contains proteins, phosphorylase, ribosomes, and also DNA.

In smears, mitochondria cannot be seen. One can infer their presence from their negative images in the heavily stained cytoplasm of some cells (see p. 29, Fig. 1 and p. 177, Fig. 1).

Reconstruction and diagram of a mitochondrion

Lysosomes

These are granulations which contain proteolytic enzymes. The enzymes become active when the lysosome membrane ruptures, usually at the edge of a phagocytic vacuole into which the lysosome empties its content (see p. 15).

The lysosomal enzymes include ribonuclease, desoxyribonuclease, and a number of hydrolytic enzymes. Esterase and acid phosphatase reactions are used to identify and characterize lysosomes in both the optical and electron microscopes. Lysosomes fused with phagosomes or autophagosomes (see p. 14) are called secondary lysosomes.

In smears, lysosomes appear commonly as azurophilic granules. Most of them, however, are not visible except after special enzyme stains.

Granules

Granules are membrane bound cytoplasmic organelles which contain diverse chemical substances stored in concentrated form.

Leukocyte granules are either primary lysosomes (e.g. azurophilic granules) or contain various enzymes and substances which are usually characteristic of a given cell line (e.g. histamine in granules of basophils).

In smears, the various granules stain in different ways. It should be noted, however, that the granules may not be discernible because they have lost the ability to stain with Giemsa or Wright's (see pp. 106 and 122) or because they are too small to be seen with the light microscope.

Vacuoles

One may distinguish:
1) Contractile vacuoles, remotely related to the contractile vacuoles of protozoa.
2) Lipid vacuoles. These are rare in normal cells but they are frequent in pathologic states or during in vitro aging of cells.
3) Multivesicular bodies. Their function is not clear. They probably originate from pinocytic vesicles.

In smears all of these vacuoles appear, provided they are large enough in size, as empty spaces in the stained cytoplasm of blood cells.

Inclusions

These are secretions temporarily stored by the cell, possibly awaiting later use; they can contain proteins, hemosiderin, mucopolysaccharides, and other products. In pathologic states, such inclusions can fill the entire cytoplasm.

In smears, special fixation and stains may be necessary to identify these inclusions. Alcohol fixation used for the Giemsa stain generally dissolves them, in which case they appear as vacuoles.

Cytoplasmic Matrix

A ground substance exists between the various organelles. It is composed of water and soluble substances which disappear during fixation and preparation. The matrix plays an important role in cell physiology and pathology: it constitutes the microenvironment which joins the organelles and makes their interaction possible.

Cell Envelope

Electron microscopy demonstrates the existence of a limiting cell membrane. It is formed, like all biologic membranes, by two layers of phospholipids into which proteins are inserted. The thickness of the membrane is approximately 800 nm. The function of the two-layered "unit membrane" can be understood only in conjunction with the immediately subjacent layer of cytoplasm (which contains enzymes and contractile proteins) and the exterior coat (which contains carriers of cell identity such as surface receptors and blood groups). For this reason, it may be preferable to speak of the cell envelope to designate the entire moiety involved in the interaction between the cell and its environment.

The technique of cryofracture has recently allowed us to separate the two layers of the cell membrane and to observe their separate surfaces. They reveal a large number of particles distributed either randomly or, in some cases, in an orderly array. They probably represent intramembranous proteins. In spite of this technical advance, we are still far from understanding the molecular structure of the membrane of different cells, or the localization of enzymes, receptors and antigens which they contain.

Physiology of the cell envelope. Time lapse cinematography and electron microscopy reveal microvilli and invaginations. The cell surface is never at rest (see p. 16). It has three basic functions:
1) It regulates the interchange with the environment by selective permeability, endocytosis, exocytosis and locomotion.
2) It detects hormonal or other signals which influence the function of the cell.
3) It carries the signs of cellular identity (blood groups, histocompatibility loci, and various receptors) which allow their recognition by other cells (see p. 24).

In smears, only major modifications of the membrane can be appreciated. It is impossible to see the microvilli or microinvaginations in the light microscope. Moreover, villosities and cytoplasmic extensions (e.g. in tricholeukocytes, see p. 200) may disappear as cells are spread out during the preparation of a smear.

Renewal of organelles. Almost all the organelles of a cell can be renewed. This renewal takes place in one of two ways: the entire organelles or the molecules which make up the organelles (or some of them) may be replaced. Here are some examples:

Microtubules become disorganized and disappear in response to a variety of influences, particularly after mitosis. As soon as this process is completed, the monomers of tubulin which are now dispersed in the cytoplasm begin to polymer-

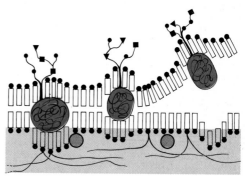

Diagram of molecular composition of the red cell membrane. At right, the two leaves of the membrane have been separated by the technique of freeze fracture. The large red particles represent proteins and glycoproteins, the small ones various enzymes. The filaments just below the membrane are spectrin. The terminal chains projecting above the cell membrane are glycolipids and polysaccharides (the major constituents of blood group substances).

ize and reform microtubules (colchicine blocks this polymerization).

Ribosomes are used up in the process of performing their function and are replaced by new ribosomes derived from the nucleus.

The cell membrane and all intracellular membranes are in a state of continual renewal. The portions of the membrane which are lost in the process of excretion or in the formation of granulations or vacuoles are continuously replaced by newly synthesized membrane.

This renewal of organelles cannot be suspected from the examination of the smears. Only certain special techniques of autoradiography, immunofluorescence, etc., allow its demonstration.

4 – Endocytosis

Definition

Endocytosis comprises all the different ways by which extra-cellular substances can gain access to the interior of the cell. Two types of processes are involved. The first is concerned with ions and small molecules which can traverse the cell membrane by a complex mechanism which allows a high degree of selectivity. It has no direct morphologic counterpart and does not concern us here. The second mechanism comprises phagocytosis, pinocytosis, rhopheocytosis: the cell membrane forms extensions and/or invaginations which introduce solid particles or droplets of liquid into the cytoplasm of the cell.

Although the vacuole formed in this fashion is surrounded by cytoplasm, its content is still extracellular as long as it remains separated from the cytoplasm by a membrane. The incorporation of the content of the vacuole into the cytoplasm is a separate and secondary event (see p. 14).

Phagocytosis[1, 2]

It introduces solid particles into the cytoplasm. The first necessary condition of phagocytosis is the adherence of the particle to the surface of the cell; otherwise the particle will be repulsed by the phagocyte. If the particle adheres, the cytoplasm of the cell envelops the particle, carries it with it and incorporates it into the interior of the cell. The third phase comprises the digestion and disappearance of the vacuole and at times the rejection of the phagocytosed particle or end products (see p. 14).

Phagocytosis may involve a variety of objects: dead cells, cells covered by antibodies, pathologically altered cells, circulating cryoglobulins or lipids, inert particles such as carbon, silica, etc.

Pinocytosis[3]

It introduces a droplet from the surrounding fluid into the cytoplasm. The cytoplasm extrudes veils which fuse so as to enclose some of the surrounding medium in a vacuole. Within a few minutes, the vacuole moves away from the periphery, the content of the vacuole is concentrated, reduced in size and finally disappears. The entrapment of a part of the ambient environment with all its components, the large macromolecular complexes as well as the small ions, is entirely different from the selective absorption of molecules through the specialized cell membrane receptors.

Pinocytosis plays an important physiologic role: the cell can absorb through this mechanism a considerable amount of liquid in a short time. For example, a histiocyte can form within an hour 50 to 120 vacuoles of a diameter of 1 to 6 µm and absorb more than a third of its own volume in liquids which the cell filters, metabolizes and of

which it returns the residue to the environment. Each of these steps may be affected separately by pathologic processes.

Micropinocytosis: Electron microscopy has revealed the existence of a phenomenon which is identical to pinocytosis but on a scale 10- to 100-fold smaller (Fig. 3).

Rhopheocytosis[4]

It is useful to distinguish the process of rhopheocytosis (Greek "to aspirate") from micropinocytosis. Micropinocytosis is initiated by the extrusion of veils. Rhopheocytosis appears to be initiated by the adherence of particles to the cell membranes. The particles may be as different as silica, carbon or gamma globulins. A characteristic example is rhopheocytosis of ferritin by erythroblasts (see p. 28). Ferritin molecules adhere to the cell membrane; invagination of the membrane with the adherent ferritin molecules leads to the formation of a vacuole. The vacuole, carried away by cytoplasmic streaming, moves away from the cell membrane.

Particles ingested by multiple rhopheocytic vesicles can be brought together by the action of cytoplasmic streaming and fuse into larger conglomerations of particles which may remain inside the cell or be extruded.

One may also observe the process of rhopheocytosis without any particle attaching to the membrane. Either attachment of a particle is not an absolute requirement for the process or present techniques do not permit us to visualize the particular particle which adheres and initiates invagination and vacuole formation.

Smears

In smears, all stages of phagocytosis may be observed (see pp. 118 and 164). Pinocytosis is difficult to differentiate from the emission of cytoplasmic veils. The vacuoles of micropinocytosis and rhopheocytosis are too small to be seen. However, the particles which enter the cytoplasm in this fashion may be visualized by the optical microscope whenever they coalesce into larger conglomerates (see pp. 34 and 56).

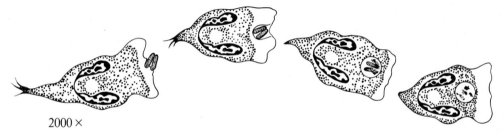

2000 ×

Fig. 1 — *Phagocytosis of bacteria by a granulocyte (PMN)* (phase contrast)

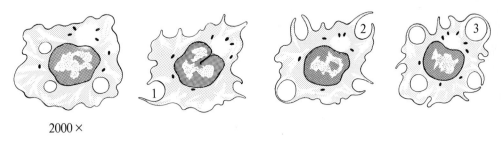

2000 ×

Fig. 2 — *Pinocytosis of plasma by a monocyte* (phase contrast)

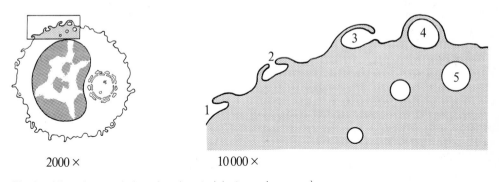

2000 ×   10 000 ×

Fig. 3 — *Micropinocytosis by a lymphocyte* (electron microscopy)

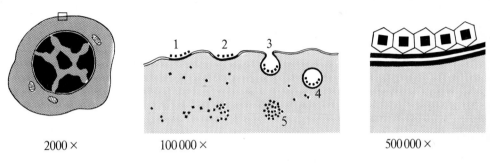

2000 ×   100 000 ×   500 000 ×

Fig. 4 — *Rhopheocytosis of ferritin by an erythroblast* (electron microscopy)

13

5 – Digestion and Exocytosis

Intracytoplasmic Digestion [1, 2]

After the process of phagocytosis or pinocytosis, the ingested material remains for a time within the digestive vacuole or phagosome: it remains separated from the cellular cytoplasm by a membrane which originated from the cell membrane. The content of the vacuole is thus located intracellularly but is not incorporated into the cytoplasm.

The mechanism which permits the actual digestion of the material is as follows. The phagosome comes in contact with several lysosomes, their membranes fuse, the lysosomes empty their content of enzymes into the phagosome and the phagocytosed particle is digested by the enzymes. The debris of this process is usually completely degraded and the vacuole contracts and disappears. Any indigestible residue remains in a sac called the residual body which stays within the cell or can be extruded. When the digestion vacuole fuses with a lysosome, it is referred to as a secondary lysosome. These secondary lysosomes sometimes contain indigestible material (for example, dextran and silica) which accumulates in the cell.

In some diseases, substances which are normally digested by the cell such as glycogen and lipids (see pp. 168 and 170) may also accumulate because of abnormalities of lysosomes (deficiency of certain enzymes). In other diseases, bacteria may be ingested, but not killed. They proliferate in the phagocyte and may be extruded alive into the environment.

Autophagosomes

Portions of the cell which have accidentally undergone some proteolytic degradation can become separated from the cytoplasm by a fine membrane. Lysosomes join and digest the contents of the autophagosome which can disappear entirely or leave behind a residual body (Fig. 2), which in turn can be extruded or may remain within the cell without causing any malfunction. Some of these residual bodies contain myelin forms, iron, calcium, etc.

Autophagosomes allow the elimination of pathologically altered organelles. They play, therefore, an important role in organelle renewal.

Intracytoplasmic Incorporation

In the course of digestion, some small molecules can pass the limiting membrane of the phagosome and become incorporated in the cytoplasm. In the case of rhopheocytosis, the vacuoles gradually disappear and molecules (such as ferritin, see p. 40 and Fig. 3) become intracytoplasmic.

Exocytosis

The term indicates the exit of any kind of material from a cell. It comprises very different phenomena such as the secretion of substances produced by the cell, the extrusion of remnants of the endocytic processes and clasmatosis.

Clasmatosis (or clasmacytosis) [3, 4]

The cytoplasm at the periphery of the cell becomes fragmented and the fragments are liberated into the environment without destruction of the cell. One must not confuse the clasmatosis with artifacts or pathologic changes which lead to the fragmentation of the entire cell. Microcinematography has shown that clasmatosis can take place without destruction of the cell.

The physiologic role of clasmatosis is ill-defined. It appears, however, that it could deliver plasma proteins into the environment. The fragments of cytoplasm correspond to the "hyaline bodies" of earlier authors. One may encounter them normally in smears of almost all aspirates of hemopoietic organs. In some pathologic conditions, particularly leukemia, they can become very numerous and can be mistaken for normal or abnormal platelets or basophilic erythrocytes (see p. 147, Fig. 1).

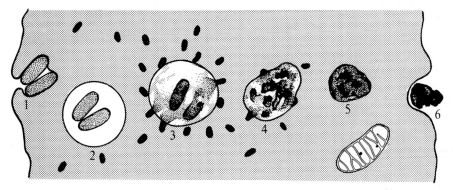

Fig. 1 – *Digestion and exocytosis of two bacteria:* (1) adsorption; (2) formation of a digestive vacuole; (3) lysosomes discharge their contents into the digestive vacuole; (4) formation of a secondary lysosome; (5) condensation of the secondary lysosome; (6) exocytosis

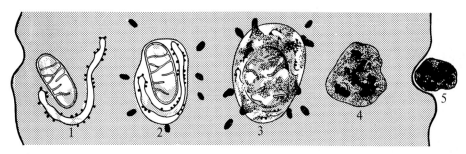

Fig. 2 – *Formation of an autophagosome:* (1) damaged territory including a mitochondrion and RER; (2) formation of digestive vacuole; (3) discharge of lysosomes; (4) condensation of secondary lysosome; (5) exocytosis

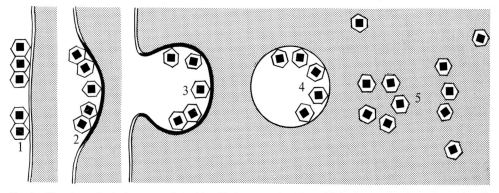

Fig. 3 – *Rhopheocytosis.* Incorporation of ferritin in an erythroblast: (1) ferritin molecules adhere to cell membrane; (2) and (3) invagination; (4) the vacuole closes; (5) the membrane of the vacuole dissolves and the ferritin molecules are free in the cytoplasm

6 — Cell Movements

Movement is one of the fundamental characteristics of life. Yet a great majority of cytologic studies deal with cadavers. Each cell line requires separate study which will be found in the corresponding chapters. Here, I will only indicate the general phenomena which one must always keep in mind when examining a cell in a smear.[1, 2, 3]

Locomotion

Leukocytes have this property, erythrocytes and platelets do not. Leukocytes crawl on a supporting surface (see pp. 116, 150 and 160). They cannot swim. In a leukocyte in motion one can distinguish, in the direction of its motion, the anterior part (protopod) which emits ruffles or veils, the middle portion which includes the nucleus, the centrosome, mitochondria and granulations, and the posterior portion (uropod) which can be greatly elongated and terminate in fine filaments which adhere to the supporting surface. The uropod is much more viscous than the rest of the cell. It adheres more tenaciously to other cells or to particles which it encounters. When a cell changes its direction, a new protopod forms on the side of the cell. The original uropod disappears into the body of the cell, and a newly formed uropod appears.

Direction of Locomotion

The locomotion of a cell can take place randomly or may be directional. Its direction depends on the distribution and gradients of chemical substances, electric charges or other potential stimuli present in the environment. The contact of a cell with another may inhibit further movement of the cell (contact inhibition).

The processes classified as chemotaxis, necrotaxis, galvanotaxis, phototaxis, and thigmotaxis have been particularly well studied in blood cells.

Movements of the Cell Surface

The surface of leukocytes and platelets is in continuous movement. Some shape changes of the surface such as protopods, blebs, veils, and other projections are large enough to be seen in the light microscope. Microvilli and microinvaginations are demonstrable only with the electron microscope. The movements of the cell surface depend on the nature of the cell and the state of the microenvironment in which the cell happens to find itself.

Intracellular Movements[4]

Examination of the living cell reveals currents in the interior of the cytoplasm (streaming) which carry along granules and mitochondria. Their essential role is to transport raw materials to the organelles and to carry away the metabolic products resulting from the activity of these organelles. Another movement which can be easily seen in leukocytes is the oscillation of the centrosome.

Intracellular Circulations[5]

In addition to the visible movement, some microscopic modes of transportation exist which are known only by their effects. Diffusion and thermal agitation are insufficient to explain how specific molecules unite for the normal functioning of the cellular machinery. It is necessary to postulate the existence of an organized transport of molecules in the protoplasm.

Today, the circulation of molecules in the cell appears to us of inextricable complexity. So did the blood circulation to Harvey who wrote, 300 years ago, on the first page of *De Motu Cordis et Sanguinis:*

"When I first gave my mind to vivisection as a means of discovering the movements and uses of the heart... I found the task so truly arduous and so full of difficulties that I was almost tempted to think with Frascatorius that the movement of the heart was only to be comprehended by God".

Smears

In smears few signs of cell movements remain. The cells are killed and deformed in the process of spreading. Moreover, they are commonly taken from blood in which cells are spherical and generally do not have visible surface extensions.

7 — Hemopoiesis

The mechanism which causes a pluripotential stem cell to generate (or develop into) stem cells committed to produce only red cells, only white cells or only platelets escapes us completely. Needless to say, the question is one of the most important problems of hematology. Unfortunately, the examination of blood smears contributes very little.

Experimental work in progress such as culture of cells in vitro inclines one to think that the problem is near its solution. I will only indicate, therefore, the principal hypotheses concerning stem cells (p. 18) and the general characteristics of maturation of blood cells (p. 20).

It will be useful to define first certain terms used, and sometimes used differently, by hematologists, embryologists, cytologists and oncologists.

Differentiation[1]

All cells of an organism have the same genetic information. Differentiation is a phenomenon which leads to the appearance of specific characteristics by progressive restriction of other potentialities of the genome. One must not confuse differentiation with the criteria by which we identify the process of differentiation, such as the appearance of granules or hemoglobin. It is evident that before the appearance of these granules, precursors of such granules must exist. With more and more refined techniques one might hope to identify the first moment at which the initiation of differentiation can be characterized.

Commitment

This is the instance when two cells, derived from the same precursor, take a separate route. The committed cells may have an undifferentiated appearance. They are, nevertheless, irreversibly committed and can only evolve in a single direction. If the circumstances are not favorable, maturation may not proceed. However, the cell cannot go back ("dedifferentiate"), and take another direction*.

It is as if the cell at the moment of commitment receives a program which it must follow. Commitment assigns the program, maturation executes it.

Commitment is not only irreversible, it is also exclusive. In the normal state one has never observed the existence of a cellular chimera, such as a cell that would produce at the same time hemoglobin and neutrophilic granules.

Maturation

This is the totality of phenomena which begins with the moment of commitment and ends when the cell has all its characteristics. The different stages of maturation of a cell are easily recognized in smears (see p. 20).

Amplification

Amplification is defined as the number of cell progeny from a single committed cell (i.e., a cell that can develop only into one specific line). Amplification normally takes place concomitantly with maturation.

Amplification varies in different cell lines. In the red cell series, one committed cell ordinarily gives rise to 16 fully differentiated cells (see p. 20).

Modulation

At each stage of maturation, the cell may assume different morphologic aspects if the environment changes. For instance, a histiocyte in the circulation is very different from the same histiocyte in culture or in the tissues of the organism. Nevertheless, it remains the same cell. When it returns from one environment to another, it again assumes the appearance and functions appropriate to that environment. However, under pathologic circumstances or after experimental manipulation it is sometimes extremely difficult to recognize modulated cells for what they are.

---

* The exception to this rule of irreversibility of a committed cell is the case of experimental fusion (hybridization) of two cells.

Theories of Hemopoiesis[1, 2]

Earlier theories based on study of the cells in smears have been in part confirmed, in part modified by modern experimental studies, although some unsolved problems persist. At the moment, there is general agreement on the following points:

- There exists a totipotential cell which can give rise to all of the blood cells (Fig. 2, I).
- There exists a separate stem cell from which all three cell lines of the bone marrow derive (sometimes referred to as a pluripotential cell) and another cell from which the different lymphoid elements originate (Fig. 2, II and III).
- Finally, committed stem cells exist for the three separate cell lines of the marrow which produce erythrocytes, granulocytes and platelets (Fig. 2, IV–X). Hormones such as erythropoietin and thrombopoietin exert their influence on these committed stem cells.

As indicated in the diagram, monocytes probably originate from the same pluripotential stem cell which gives rise to the three major cell lines of the marrow. Whether the lymphoid stem cell which gives rise to the different types of lymphocytes and plasma cells has separate committed stem cells for each of these lines is still being debated.

Morphology of the Stem Cells[3-7]

Notwithstanding a great deal of effort, the morphology of stem cells is still unknown. While their existence can be inferred from experimental data, we do not know whether the commitment of a cell which restricts its possible progress along different paths is accompanied by morphologic changes. As far as the totipotential or pluripotential stem cell is concerned, it has been possible to fractionate bone marrow in such a way as to increase the percentage of these stem cells by one or two orders of magnitude, judging from their effectiveness in repopulating the marrows of lethally irradiated animals. Certain cells which lack the characteristics of any differentiated cell line and which were increased in these fractions have been postulated to be the pluripotential cells. Because of the remaining uncertainty of the validity of this conclusion, the cells are referred to as "candidate stem cells." A number of other names have been attached to the undifferentiated cells that can be seen in marrows and which may or may not be actual stem cells. A description of these cell types follows.

a) *Hemocytoblast:* It is a large cell approximately 30 μm in diameter with a round to oval nucleus which occupies almost the entire cell. The nucleus is lightly colored and contains several nucleoli. The cytoplasm is a pale blue. It is possible that the cells which answer this description are in fact promyeloblasts or pro-erythroblasts or prolymphoblasts. These cells must not be confused with transformed lymphocytes (see p. 154) which they resemble.
   This cell is called a "myeloblast" by some hematologists.

b) *Lymphoid type stem cell:* The candidate stem cells resemble lymphocytes in the paucity of their organelles, but are of somewhat larger size and have a more finely distributed chromatin (Fig. 3).

Stem Cells of the Peripheral Blood[5, 6]

Pluripotential stem cells are known to be present in the peripheral blood since buffy coat preparations have been used successfully to repopulate the marrow of irradiated animals. Cells from human or peripheral blood can be cultured to give rise to colonies of mature blood cells of all lines. To what extent these cells may or may not resemble the candidate stem cells of the bone marrow is still controversial.

The stem cells of the circulating blood must not be confused with transformed lymphocytes, which can also be found in the peripheral blood as a result of antigenic stimulation (see p. 154).

The frequency of stem cells or transformed lymphocytes is estimated as 0.1 to 0.25% of white blood cells.

Fixed Stem Cells of the Hemopoietic Tissues

Ultimately, the spherical free stem cells must be derived from earlier fixed cells, at least in the embryo. In the past, the reticulum cells of the bone marrow have been thought of as the original stem cells. It now appears that the term reticulum cell (or reticular cell) should be reserved for the cells which provide the network of fibrils in which the hemopoietic elements and fat cells of the marrow are embedded. There is no evidence that these cells have any hemopoietic potential.

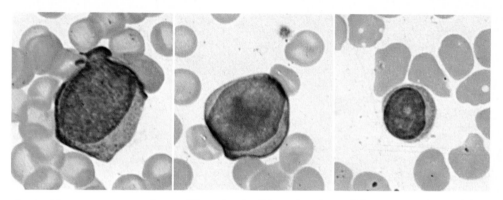

Fig. 1 – *Three appearances of an undifferentiated cell* (possibly stem cells)

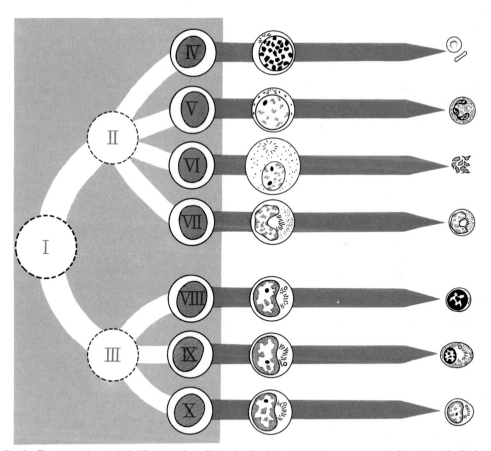

Fig. 2 – *Theoretical model of differentiation of blood cells*. All cells in the grey area at left are hypothetical:
(I) Totipotential stem cell; (II) pluripotential hemopoietic stem cell; (III) lymphoid stem cell: (IV–X) stem cells committed to differentiation into specific lines

9 — Amplification and Maturation

Maturation of the different cell lines follows basically analogous laws of development. The development of the erythrocytic and thrombocytic series illustrates the analogy between two cell lines in which the maturation appears on first sight quite different.

Amplification [1, 2]

In the erythrocytic series, the committed stem cell divides under the influence of the hormone of maturation, erythropoietin, usually four times, which gives an amplification of 16, i.e., 16 red cells appear to originate from a single committed stem cell (see p. 26). Maturation proceeds apace with the successive divisions.

In the thrombopoietic series, the committed stem cell which gives rise to megakaryocytes also divides on the average four times, producing a polyploid cell of 16 and/or 32 N, although 8 N and 64 N cells can also be found. It is only after the completion of these endomitoses that maturation takes place and the cells assume the typical appearance of megakaryocytes (see p. 130).

Disorders of Amplification

The following abnormalities of amplification can be observed in pathologic states:
1) The number of mitoses can be increased or reduced.
2) The sequence of divisions can be asynchronous. (The sequence of divisions as illustrated on p. 20 for the erythroid series implies that the progeny of a single committed stem cell divides synchronously.)
3) Some daughter cells may die, while others continue to divide (ineffective erythropoiesis).
4) Some daughter cells may fail to undergo the same number of divisions as others (skipped division).
5) Nuclear division in the erythroid series may fail to be accompanied by cytoplasmic division (multinuclearity).
6) Cytoplasmic division may occur in the megakaryocytic series resulting in lesser degrees of polyploidy or even diploid nuclei.

The abnormalities which can readily be observed in practice are 3) ineffective erythropoiesis, 5) multinuclearity, and 6). The failure to observe the others is most likely due to our inability to measure the progeny derived from a single cell and does not indicate that they do not exist.

Maturation [3]

Maturation may be thought of as starting with a synthesis of proteins which are specific for a given cell line, such as hemoglobin in the red cell series or the proteins necessary for the assembly of azurophilic granules in the myeloid series.

As the cell matures
- the cytoplasm loses its ribosomes, hence its basophilia disappears and specific morphologic features of the cell appear;
- the size of the cell diminishes with succeeding divisions;
- the nucleus loses its nucleoli and becomes smaller and denser.

In the red cell series, the pyknotic nucleus is eventually extruded while in the megakaryocytic series the nucleus is left behind after the mature cytoplasm has broken up into platelets.

The maturation of the cytoplasm and the nucleus is linked in a characteristic and invariant fashion. A young nucleus always corresponds to a cytoplasm with few specific features and vice versa. This synchronization applies to all cellular organelles.

Disorders of Maturation

1) *Asynchrony* [2, 4]. Theoretically, specific cytoplasmic characteristics may appear too early or too late in relation to nuclear maturation. The commonly observed asynchrony is of the first type, with advanced specific cytoplasmic features (hemoglobin, granules) while the nucleus is relatively immature. This asynchrony most commonly occurs as a result of stimulation of maturation (see p. 88). It can also be due to retardation of nuclear development (see p. 88).

2) *Anarchy*. The maturation is entirely disorganized and the stages of maturation of different organelles have no relation to each other.

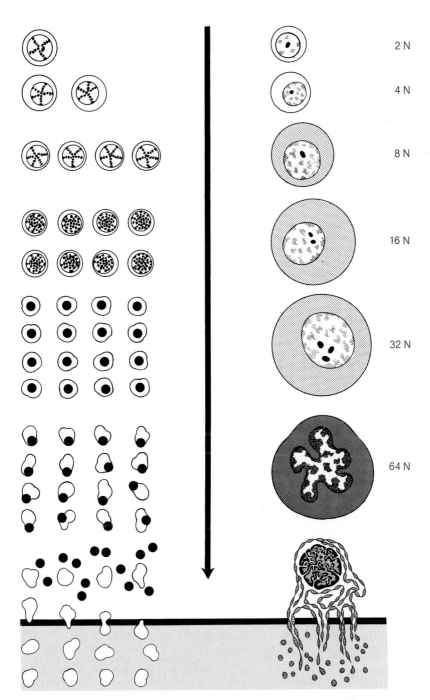

2 N

4 N

8 N

16 N

32 N

64 N

Comparative amplification and maturation of erythroid and megakaryocytic cell lines

As are their conception, their birth, their activity, and their mitosis, so is the time of death of blood cells determined by an inexorable mechanism.

The significance of the death of blood cells has a variety of interpretations. For the organism, it is a means of regulation, since statistically the functioning of a cell will become impaired after a certain period of time. It is also a way to convey food or information to other living cells. The use and recovery of cadavers play an essential part in the economy of the body. As an example, one has only to consider the results of the daily death of 200 billion red cells in a normal adult.

Definition of Cell Death[1]

Except in case of an accident, death is never instantaneous. Like the death of a higher organism, it is a progressive phenomenon. The death of a cell is the miniature image, as it were, of the death of an organism. Like the organs of a living being, the organelles live together but die separately, the death of one organelle leading to the death of others.

Examination of living cells has shown that death is preceded by a "period of agony" and followed by a "period of necrosis".

The three periods of agony, death and necrosis have often been confused because the phenomena were studied only in histologic sections or in smears, that is to say in fixed and stained cadavers. From such preparations it was difficult to reconstruct the events that led to the death of a cell. Two cell cadavers can look alike while the morphologic changes preceding the death may have been quite different. Examination with phase contrast microscopy together with time lapse cinematography has allowed us to study in detail the entire processes of agony, death and necrosis.

The appearances reflecting agony, death and necrosis depend on several factors: the cause of death, the type of cell, the environment and the speed with which the sequence of agony, death and necrosis develops.

Phase Contrast Microscopy[1]

1) *Period of agony*. This period begins when alterations of the cellular organelles become irreversible. It is compatible with the persistence of some functions. The cell, like an organism, battles against death.

As examples we illustrate three types of agony of large lymphocytes on the opposite page. From left to right:

a) Cellular and nuclear edema: the cell and all its organelles swell. Mitochondria become vacuoles with diameters of up to 3 µm. The Golgi and the RER also swell and become small spheres. The cisterna swells to form characteristic vacuoles. The nuclear envelope remains attached at the nuclear pores, resulting in vacuoles of half-moon shapes. The nucleus increases in volume and rounds off. Its chromatin structure becomes effaced. The nucleus may swell beyond the limit of the elasticity of its envelope which breaks. The cytoplasm fragments into multiple droplets.

b) Nuclear pyknosis: the nucleus becomes smaller as a result of expulsion of nuclear sap. One by one, the cytoplasmic organelles become spherical.

c) Karyorrhexis (fragmentation of the nucleus): it begins as an alteration of the surface movements of the cell. When viewed in accelerated time lapse movies, the abnormal nature of cell movements becomes immediately apparent. One may distinguish multiple peripheral blebs giving the entire cell a star-shaped appearance; wheel-like movements of beaded extensions; scarf-like expansions; etc. The nucleus becomes lobulated and individual lobes extend into the cytoplasmic projections. Finally the nucleus fragments and the cell, having lost the offshoots of its cytoplasm, becomes spherical.

2) *Period of death*. Except in case of an accident, the time of death often defies definition. The cessation of respiration is not an adequate indicator of death; the debris of a cell can still consume oxygen.

Customarily, the cytologic criterion of cell death is its staining with certain dyes; the "dead" cell can no longer maintain the equilibrium of its internal milieu against the environment and the dyes can pass through the cell membrane.

3) *Period of necrosis*. Necrosis of the cadaver follows death of the cell. One part of a cell may undergo necrosis, while other parts of the cell still function. General necrosis ensues when a large percentage of membranes have ceased to function.

The different enzymes liberated by the destruction of the lysosomal membranes attack the constituents of the cell and auto-digestion begins which leads to the different appearances so well described by the ancient pathologist: cellular edema, coagulation necrosis, pyknosis of the nucleus, and others.

The period of necrosis may be brief and accompanied by fragmentation, even explosive fragmentation of a cell. The debris may dissolve in the surrounding fluid or may be immediately phagocytosed by histiocytes.

Smears

Of all the dramatic events depicted in time-lapse movies, few signs remain in the smear. Only occasionally is it possible to reconstruct the history from artifacts encountered in the slides.

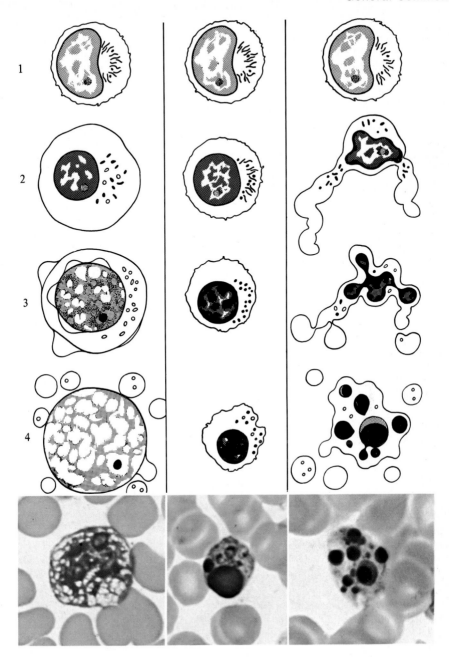

Three sequences of different types of death and necrosis of a lymphocyte. At left, nuclear edema; center: pyknosis; right: karryorhexis

The figure above shows three examples of the agony, the death and the necrosis of a large lymphocyte. The microphotographs indicate the appearance in the smear, the schematics above each microphotograph illustrate the different events that have led to nuclear edema (*left*), pyknosis (*middle*) and karyorrhexis (*right*).

It should be noted that a cell which has fragmented into numerous small pieces may leave only unidentifiable debris on the slide. The examination of sections with the optical or electron microscope and examination after labeling with radioactive markers fare no better. Many deaths occur among cells without leaving cadavers behind.

11 — Sociology of Blood Cells

We have seen that one may approach the study of individual cells, like that of organisms, through anatomy, physiology, and cell pathology. We have been able to study the birth and death of cells. To continue in this anthropomorphic vein, we may attempt to add the notions of demography, ecology, and ethology to the study of cells.

Cell Demography[1]

The different cell types which live together in the bone marrow of an adult constitute a population 100 times larger than the entire human population of the globe. The enormity of this figure has induced some cytologists to apply the general laws of demography to cell populations. The large number of individual cells which renew themselves continuously assures the occurrence of events such as mutation, which are in themselves very rare and improbable.

In theory, each mutation of a single cell should give rise to a subpopulation of abnormal cells. Since nothing like it is observed normally, it follows that the organism has means to dispose of such abnormal cells. It has been suggested that these mutations can explain senility and neoplasia. For the time being, however, we lack essential information on two cardinal points of cytodemography:

1) The mechanisms which maintain an equilibrium between cells being born and cells dying, that is to say, the mechanisms of hemocytostatic equilibrium.
2) The mechanisms of selection or elimination of cells which do not conform to the norm, for instance the still putative process referred to as immunosurveillance.

Cell Ecology

The goal of ecology is the study of a living being in its natural environment. The microenvironment of the cell plays an important part in determining recognition of molecules at the cell surface and ultimately in the interior of the cell. It should be recalled that cells are always in a microenvironment or a microclimate which influences differentiation and maturation. The erythroblastic island is an example of such an ecologic niche (see pp. 38–42).

Cell Ethology[2]

Ethology is the science of the behavior of living beings. It is quite possible to utilize the techniques, findings, and hypotheses of zoologists and sociologists for the study of the life of cells. Conversely, certain behavior of higher organisms might be clarified by the study of the individual cells. Of these newer insights I can give two brief examples:

Homing.[3] It is now well known, although perhaps not appreciated sufficiently, that either cells of a given organ have the faculty to recognize, attract and retain certain circulating cells or the circulating cells recognize a given organ as the most appropriate for their permanent residence. An example of homing familiar to hematologists: the cells of hemopoietic organs may be injected intravenously and localize and proliferate only in hemopoietic organs. The phenomenon of selective recognition plays an important role in the normal organism and in pathologic states: cancer cells appear to have lost this property of recognition and metastasize to different organs.

Necrotaxis.[4] The reaction of living cells to dying or dead cells varies greatly depending on the tissue, the quantity of cells rapidly or slowly transformed into cadavers and the causes of death. A new technique, microirradiation by a laser beam, has allowed us to kill a single cell within a population. As soon as the irradiated cell shows signs of agony, the healthy cells in the neighborhood congregate around the cadaver. In 2 to 10 minutes, the cell is surrounded by phagocytes which divide up the body. One may see cells newly arrived on the stage violently push away the other cells which have formed a rosette around the dead body in order to take their share of the prey. The phenomenon conjures up the picture of a school of sharks who plunge headlong, without paying attention to each other, towards one of the group which has been injured and marked by a trail of blood.

Many features of the social behavior of cells can be impressively demonstrated in time lapse photography. The events remain totally hidden from the hematologist who examines the dispersed cells in smears or the pathologist who looks at sections. They represent but a single fixed moment in the social existence of a cell community.

Chapter 2

Erythrocytic Series

A — Normal Cells

General Comments

Definition

The erythrocytic series consists of a succession of cells which begins with a pronormoblast and ends with the erythrocyte or red blood cell.

Location

In the adult, the erythrocytes are formed in the (red) bone marrow. In the fetus, and under some pathologic conditions, the liver and the spleen become sites of erythropoiesis.

Stages of Maturation and Amplification

One generally distinguishes the pronormoblast*, the basophilic normoblast, the polychromatophilic normoblast**, the reticulocyte***, and the erythrocyte (or mature red cell). Unfortunately, our present criteria do not allow us to identify the first three stages with absolute certainty (see p. 26). Amplification (p. 20) occurs simultaneously with maturation up to the stage of the polychromatophilic normoblast. It appears that normally four mitoses occur between the pronormoblast and the polychromatophilic normoblast (after which there is only maturation and no further amplification). It is convenient to designate the five stages of amplification which are separated by mitoses as E_1 to E_5.

Life Span[1]

Normally, it takes approximately 4 days for a pronormoblast to become a reticulocyte. The individual times in each stage of red cell development are as follows:

Pronormoblast (E_1)	20 h, followed by a mitosis
Basophilic normoblast I (E_2)	20 h, followed by a mitosis
Basophilic normoblast II (E_3)	20 h, followed by a mitosis
Polychromatophilic normoblast I (E_4)	24 h, followed by a mitosis
Polychromatophilic normoblast II (E_5)	30 h, followed by expulsion of the nucleus
Reticulocyte (Rtc)	3 days of maturation
Erythrocyte (E)	about 120 days

The reticulocyte matures in the marrow for approximately 48 h; it is then released into the peripheral blood where it continues its maturation for approximately 24 h.

Death[2]

When erythrocytes reach approximately 120 days of age, they are phagocytized, mostly in the bone marrow and to a lesser extent in the spleen. In contrast, under pathologic conditions, it is the spleen, the liver and other organs which become the principal sites of red cell phagocytosis.

The macrophages phagocytize the intact senescent red cells. It is uncommon for red cells to hemolyze in the circulation or in any organ. When hemolysis does take place, the red cell ghosts are ingested by macrophages.

---

* The term "erythroblast" should be used (and is used in this book) as a generic term to include all normal or abnormal nucleated red blood cells. The term thus includes pronormoblasts, normoblasts, macronormoblasts, megaloblasts, etc.
** The designation "orthochromatic normoblast" has been ommitted. The term indicates that the concentration of hemoglobin has become that of the mature red cell and, therefore, has an identical color. Actually, the reticulocyte always has a lower hemoglobin concentration.
The synonym "acidophilic normoblast" should also be withdrawn from usage.
*** The designation "reticulocyte" is inappropriate because the "reticulum" does not preexist but is precipitated only by action of the supravital dyes (see p. 44). However, the term reticulocyte can probably not be replaced in view of its long and universal usage.

1 – Erythropoiesis

Amplification

Beginning with the pronormoblast, one can generally observe four cell divisions which lead to the production of 16 cells. It is quite possible, however, that this is only an average figure and that the actual amplification may vary even normally between 8 and 32, just as the polyploidy of the megakaryocytic nucleus varies within these limits.

The size of the cells decreases with each division and is, therefore, characteristic of the different stages of maturation. The diameter of the pronormoblast varies from 22 to 28 µm, while that of the polychromatophilic normoblast II is only 9 µm, quite close to the size of the reticulocyte and mature red cell (see p. 28). The cause of this progressive decrease of volume of the cell is not known; it applies not only to the whole cell, but also to the nucleus which condenses without loss of chromatin. The nuclear-cytoplasmic ratio decreases in the process.

Measurements of the volume of the nucleus and spectrophotometric measurement of the content of hemoglobin have not yet led to a better classification of stages of amplification and maturation than the simple observations of the stained smears.

Maturation

Hemoglobin is synthesized by the entire series beginning with the pronormoblast. It is assumed that it is the concentration of hemoglobin which triggers the mitoses. The quantities of hemoglobin of each stage of maturation have been estimated as follows:

E_1	0–14.4 µµg
E_2	7.2–21.6 µµg
E_3	10.8–25.2 µµg
E_4	12.6–27.0 µµg
E_5	13.5–24.5 µµg
Rtc	24.5–29.5 µµg
E	29.5–30.5 µµg

Ineffective Erythropoiesis

It is believed that even in the absence of any pathologic change a small number of cells die in the process of maturation. Ineffective erythropoiesis is certainly very small under physiologic conditions and never reaches 10% as has been claimed. Under pathologic conditions, however, it can attain a very high figure.

Pathologic Erythropoiesis

The normal sequence of progressive changes in nuclear and cell volumes, the progressive change in the color of the cytoplasm as a result of hemoglobin synthesis and eventual loss of ribosomes, all of which normally characterize the different stages of red cell maturation, may be altered. These asynchronies may be characteristic for a particular type of anemia (see p. 90).

Different degrees of ineffective erythropoiesis and changes of amplification may accompany these abnormalities of erythropoiesis.

Ecology of Erythroblasts.
Erythroblastic Islands

In the bone marrow of adult mammals, the erythroblasts appear to be produced in distinct islands around a central histiocyte (see p. 38). The central histiocyte appears to play an important role in the physiology of the cells of the erythrocytic series: it ingests the nuclei extruded by the late normoblasts and phagocytizes some of the red cells that have reached the end of their life span. Possibly it supplies the normoblasts with compounds important for their nutrition or homeostasis. The erythroblastic island appears to represent an "ecologic niche" which maintains the developing red cells in a specific microenvironment which influences their maturation and their release into the circulation.

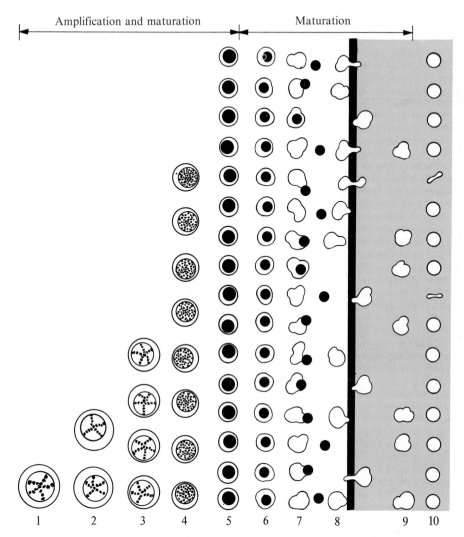

Stages of erythropoiesis. (1) Pronormoblast; (2) basophilic normoblast I; (3) basophilic normoblast II; (4) polychromatophilic normoblast I; (5) polychromatophilic normoblast II; (6–10) maturation without cell division; (7) extrusion of the nucleus; (8) passage into the circulation; (9) reticulocytes; (10) mature erythrocytes

2 – Pronormoblast (E$_1$)

Definition

This is the earliest recognizable cell of the erythrocytic series. The immediately preceeding stage is an "erythropoietin responsive cell" (ERC or E$_0$), the existence of which can be inferred by its response to erythropoietin and from isotopic studies.

Smears

The cell is round or oval with a diameter of 20 to 25 µm. The nucleus occupies 8/10 of the cell. It stains lightly and presents small clumps of chromatin. One or two small nucleoli can be seen, which are, however, frequently obscured by chromocenters. The narrow band of cytoplasm surrounding the nucleus stains a very dark blue. This intense color is very characteristic and due to the abundance of ribosomes. The cytoplasm contains colorless areas: a large one corresponds to the site of the centrosome, small clear areas (due to negative images of mitochondria) sometimes give the cytoplasm a honeycombed appearance.

Living Cells

Mitochondria are readily seen and appear as dark granulations 0.2 to 5 µm in length, usually numbering 20 to 50. They generally surround a clear zone representing the centrosome. E$_1$ does not show ameboid movements. One may, however, see sluggish movements of mitochondria and of the cytoplasm.

Electron Microscopy

The pronormoblast stage is not readily seen. It must be searched for among the large cells. The nuclear structure is uncharacteristic. The centrosome is relatively small, so that the nucleus is not indented. There is an abundance of ribosomes and polyribosomes*. There are a few sacs of endoplasmic reticulum, a few microtubules and microfilaments.

Ferritin. At magnifications of 100,000, ferritin molecules can be seen dispersed in the cytoplasm. In an electron microscope section of a pronormoblast from a normal person, 500 to 1000 ferritin molecules may be counted. The number diminishes in older normoblasts. Since the molecules usually do not coalesce, they cannot be visualized in the light microscope by Perls' reaction (see p. 34). E$_1$ is therefore not a sideroblast.

Rhopheocytosis. Rhopheocytosis is always present. Quite undifferentiated cells may be recognized as belonging to the erythrocytic series by the presence of rhopheocytic vesicles and by the dispersed ferritin molecules.

---

* The number and concentration of ribosomes accounts in stained smears for a blue color which is deeper than that of any other marrow cell, including the plasmocyte (see p. 178).

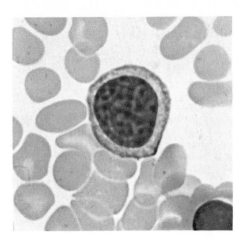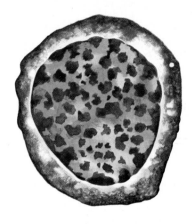

Figs. 1 and 2 — *Pronormoblast*. Note the negative images of the Golgi and the mitochondria

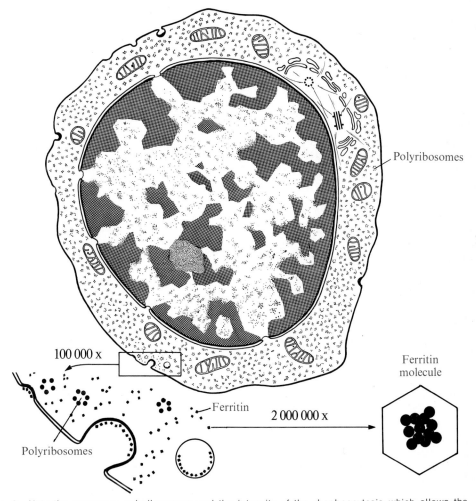

Fig. 3 — Note the numerous polyribosomes and the intensity of the rhopheocytosis which allows the ferritin molecules to penetrate into the cell

3 – Basophilic Normoblasts I and II (E$_2$ and E$_3$)

Definition

These stages are recognized in smears by the progressive loss of the extreme basophilia of the pronormoblast cytoplasm.

Smears

E$_2$ is smaller than E$_1$ and E$_3$ smaller than E$_2$. The size of the nucleus decreases more rapidly than that of the cytoplasm.

The nucleus becomes progressively more characteristic. The chromatin becomes clumped, often in the form of quite regular hexagonal or pentagonal chromatin masses of up to 2 µm in diameter. Their color is often almost black violet against a rose background. They may give the nucleus a cartwheel appearance which is quite characteristic of normoblasts. The number of these chromocenters varies from 15 to 20 but is always above 10, which distinguishes the nuclei of normoblasts from those of plasma cells in which the clumps rarely exceed 9 (see p. 176).

The cytoplasm is a uniform blue with very few colorless areas. The centrosome is seldom recognizable.

Mitoses of Normoblasts (E$_1$ to E$_4$)

Mitoses are frequently seen in normal bone marrow aspirates. They may exhibit some peculiar features:

1) The cytoplasmic basophilia may not be uniform but appear as fine granulations in the colorless background. On rare occasions this phenomenon may also be seen in non-dividing cells. This pecularity is not restricted to the red cell series and may be encountered in other cell lines on rare occasions. It is an artifact identical with the basophilic stippling (see p. 58).
2) On occasion, the nuclear division is not accompanied by cytoplasmic division and one might see basophilic normoblasts with two to eight nuclei in entirely healthy individuals.
3) Occasionally two normoblasts remain linked after division by a fine cytoplasmic bridge which may contain an azurophilic (red) filament. This cytoplasmic bridge contains microtubules that are remnants of the spindle (see Cabot Rings, p. 60).

One may wonder whether the characteristic clumped appearance of the chromatin might be modified during DNA synthesis, allowing one to distinguish that period of nuclear activity. Unfortunately, this does not appear to be the case.

Living Cells

The nuclear structure, provided the cell is not squashed between slide and coverslip, is quite characteristic. The dark grey appearance of the cytoplasm renders the diagnosis of this cell stage easy. A small centrosome and a few mitochondria are present, sometimes concentrated in a small space, sometimes more uniformly distributed through the cytoplasm and surrounding the nucleus (see Ringed Sideroblasts, p. 54).

Electron Microscopy

The nucleus contains large blocks of chromatin (presumably representing inactive chromosomes). The nuclear pores, always located between chromatin clumps, are few in number. The cytoplasm is filled with ribosomes and polyribosomes. Their number diminishes as the hemoglobin accumulates. Special stains permit its characterization and quantification. Approximately 20 to 30 mitochondria are seen per cell; the number diminishes as the cell matures. The centrosome is small and contains two centrioles surrounded by a Golgi complex. Small lysosomes, located near the Golgi complex, may contain ferritin. The microtubules are poorly developed. They can be seen around the centrioles or, in rare circumstances, in the bridge which may unite two normoblasts at the end of their division (see above). Microfilaments are sometimes seen near the nuclear membrane.

Ferritin can be seen at magnifications of 100,000 as dispersed molecules throughout the cytoplasm. In addition, small conglomerates of ferritin molecules may be seen which represent siderosomes, visible in the light microscope after Perls' reaction (see p. 34).

Rhopheocytosis. At adequate magnifications, the cell membrane shows numerous invaginations and the adjacent cytoplasm contains rhopheocytic vesicles.

Normoblasts of all stages are disposed around a central histiocyte (erythroblastic island, see p. 38). They remain in contact with the narrow tongues of cytoplasm extending from the central histiocyte to each of the normoblasts which surround the central cell in one or more concentric rings.

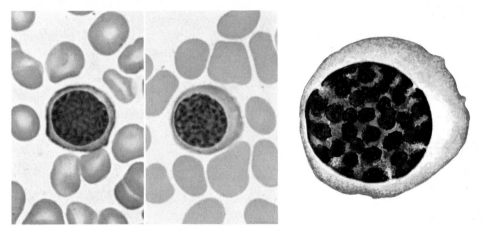

Figs. 1 and 2 – *Basophilic normoblast I and II*. Note the characteristic maturation of nuclear chromatin

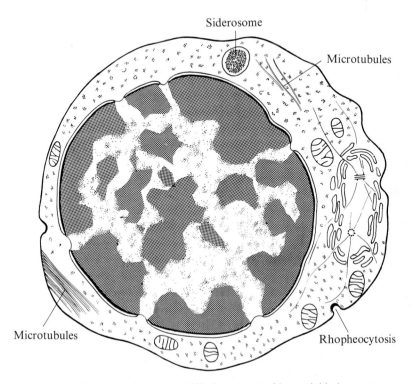

Fig. 3 – The number of ribosomes decreases while the amount of hemoglobin increases

4 – Polychromatophilic Normoblasts I and II (E_4 and E_5)

Definition

These stages of red cell development are defined by the increasing acidophilia of the cytoplasm in Giemsa-stained smears due to the increasing concentration of hemoglobin.

Smears

E_4 is smaller than E_3 and E_5 is smaller than E_4. The volume of the nucleus diminishes more rapidly than that of the cytoplasm and at the end of the E_5 stage the nuclear cytoplasmic ratio is about 1 to 4. The nucleus becomes even darker although it retains its cartwheel appearance due to the clumps of chromatin. However, the number of these clumps diminishes. Eventually the cartwheel gives way to a uniformly dark violet appearance of the nucleus. The nucleus may occasionally have a lobulated appearance. This may progress to a cloverleaf appearance (particularly in marrows with stimulated or stress erythropoiesis).

The cytoplasm in which the hemoglobin concentration gradually increases changes from blue to pink. Occasionally basophilic portions remain. This "punctate basophilia" is due to an artifact in the preparation of the stained smear and is seen particularly during mitoses (see p. 30) and in pathologic states (see p. 58).

After mitosis, E_4 becomes E_5. It has been erroneously called an orthochromatic or acidophilic normoblast. It must be stressed that the cytoplasm of a normoblast never has the exact color of the cytoplasm of a mature red cell. After extrusion of the nucleus, the red cell is still a reticulocyte with some degree of basophilia due to RNA, and with a lower MCHC than a mature red cell (see p. 44).

Living Cells

The cytoplasm of the polychromatophilic normoblasts is dark, due to the increased hemoglobin concentration. The organelles are difficult to see in this dense cytoplasm. However, when the cell is slightly compressed between slide and coverslip, mitochondria appear as white structures against the dark background because the refractive index of the mitochondria is lower than that of the hemoglobinized cytoplasm. Strong compression of the cell results in hemolysis: the cytoplasm becomes transparent, the nucleus can be clearly seen and the mitochondria appear black.

Movements. E_4 and particularly E_5 are engaged in characteristic movements. Round projections appear suddenly in different parts of the cellular circumference and are as rapidly withdrawn. These movements are probably related to the extrusion of the nucleus (see p. 36). Occasionally they are reflected in smears as distortions in the E_5 cytoplasmic contour which appears to consist of multiple half-moons (see Fig. 2, p. 37).

Electron Microscopy[1, 2]

The nucleus becomes smaller and smaller. The chromatin masses condense and become more sharply separated from the nuclear sap which forms the network extending to the nuclear pores. The nucleus disappears. The number of mitochondria and polyribosomes diminishes as the amount of hemoglobin increases. (Hemoglobin can be identified and its amount measured by a specific reaction with diaminobenzidine.) The Golgi complex becomes quite small and may contain lysosomes. Microtubules and microfilaments are few.

Ferritin. The dispersed ferritin molecules have become less numerous and siderosomes (conglomerates of ferritin molecules) have become larger (see p. 34).

Rhopheocytosis is always present and rhopheocytic vesicles are more numerous than in either the preceeding or succeeding stages.

Cell movements can sometimes be recognized as a characteristic irregularity of the cell contour as described under light microscopy, particularly in the last stages of maturation of E_5 (see Fig. 3, p. 37).

In all maturation stages, even E_4 and E_5, the developing cells remain in close touch with cytoplasmic extensions of a centrally located histiocyte (see Erythroblastic Island, p. 38).

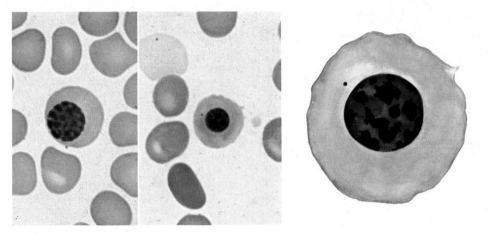

Figs. 1 and 2 — *Polychromatophilic normoblast I and II*. Note the granule, representing a lysosome, a common occurrence at this stage of development

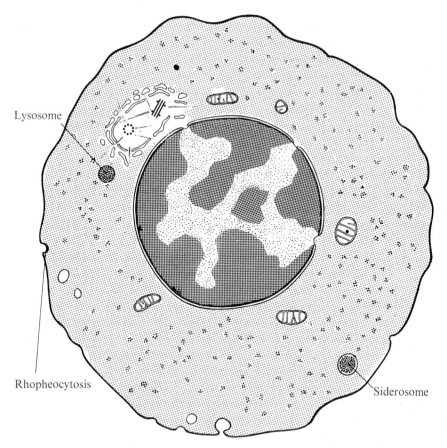

Fig. 3 — The number of polyribosomes is further reduced, and hemoglobin increased

5 – Normal Sideroblasts

Definition[1, 2]

Sideroblasts are red cell precursors in which iron-containing granules can be made visible by Perls' reaction when examined by light microscopy. The granules are called siderosomes, a term which encompasses all iron-containing inclusions (see p. 166).

Smears[3]

Generally, siderosomes cannot be recognized in Giemsa stains. Occasionally, however, small basophilic or azurophilic granules may be seen, which represent Pappenheimer bodies (secondary lysosomes, acid phosphatase positive, see p. 32), which give a positive reaction with Perls': the Giemsa stains the protein matrix, which contains the iron.

Smears Stained for Iron
(Perls' or Prussian Blue Reaction)

After reacting with Perls', smears may be counterstained with Giemsa, but only very lightly in order not to obscure the blue granules, or preferably with safranin, which gives a pale rose color to the nuclei and leaves the cytoplasm yellowish or rose. One to three Prussian blue positive granules may be seen at the limit of the visibility with the light microscope. They rarely exceed 0.3 µm in size.

Normally siderosomes are present in all stages of the developing red cell line after E_1. They are particularly readily seen in stages E_4 and E_5. They do not normally occur in reticulocytes. The percentage of sideroblasts varies between 50% and 70% of all erythroblasts.

In pathologic states, the number of sideroblasts can be diminished (iron deficiency anemia and inflammation, see p. 56) or markedly increased. In such cases, the granules are always more numerous, larger, and may have a quite different origin and significance (see p. 54).

Electron Microscopy[4, 5]

Electron microscopy has allowed a more detailed insight into the composition of siderosomes and has greatly contributed to the separation of normal and pathologic sideroblasts (see p. 54).

In the Normal Individual

a) The siderosomes of sideroblasts constitute aggregates of ferritin molecules.
b) Mature red cells never contain siderosomes. The so-called siderocytes are in fact sidero-reticulocytes (see p. 54).
c) Both erythroblasts and reticulocytes contain dispersed ferritin molecules as well as aggregates (see pp. 28, 32 and 46), but a variable percentage only contains sufficiently large aggregates to be seen as blue granules in the light microscope.

Actually, all normal normoblasts and reticulocytes contain ferritin, when examined with the electron microscope. Consequently, the distinction of sideroblast and non-sideroblastic nucleated red cell is a purely artificial one, as it depends simply on the magnification at which the cell is examined. Nevertheless, examination with the light microscope is very useful, because it permits a rapid diagnosis of pathologic sideroblasts. Hence, the customary designation of sideroblasts (and siderocytes) as red cells which contain iron when viewed with the light microscope needs to be retained. The pathologic features of red cell iron are either an increase or decrease in the numbers of normal sideroblasts or of pathologic sideroblasts in which either siderosomes are larger or, more commonly, iron is found in mitochondria. Mitochondrial iron is never found in normal red cell precursors in man (for details, see pp. 54 and 56).

Origin and Role of Ferritin[4, 6−9]

Isotopic studies have demonstrated that the iron necessary for hemoglobin synthesis is furnished to the developing red cells by means of transferrin. Any excess iron is stored by the cell in the form of ferritin.

Ferritin molecules can also enter the erythroblasts and reticulocytes by rhopheocytosis (see pp. 12 and 40). Although this conclusion is based on electron microscopic evidence which can give only static pictures, a number of cogent reasons support this interpretation. Electron microscopists have carefully studied the process of interiorization of foreign material, when the direction of the movement of the particles was known, i.e. after intravenous injection of colloidal material or extraneous ferritin. The appearances of such interiorization are, therefore, well documented; they are identical with those seen in ferritin rhopheocytosis of erythroblasts. The appearances accompanying expulsion of ferritin, which have also been observed under experimental conditions which determined the direction of the movement, have a slightly different appearance.

The role of ferritin is still controversial. Two hypotheses have been advanced.

First Hypothesis. It serves as a reserve for hemoglobin synthesis when needed.

Second Hypothesis. Ferritin is simply a surplus of iron, which plays no role in hemoglobin synthesis. It is destined to be extruded by the cell or is removed in the normal spleen which has the ability to remove erythrocytic inclusions (pitting function).

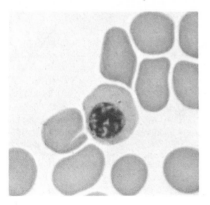

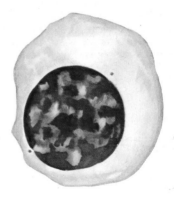

Figs. 1 and 2 – *A sideroblast,* stained with Perls' and counterstained with safranin. In normals, sidero-somes are few in number and their size is at the limit of visibility (cf. Fig. 1, p. 55)

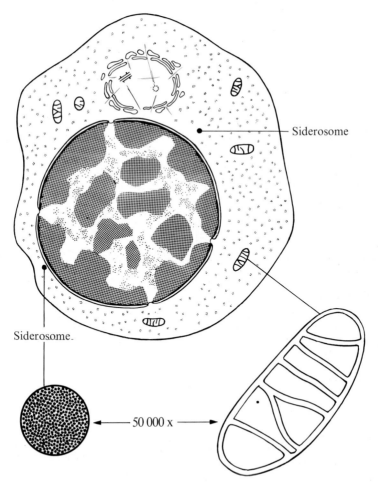

Siderosome

Siderosome

50 000 x

Fig. 3 – Siderosomes consist of masses of ferritin surrounded by a limiting membrane. Mitochondria normally do not contain iron (cf. Fig. 2, p. 55)

6 – Extrusion of the Nucleus

Examination in the Living State[1]

The phenomenon is readily seen in phase contrast microscopy in preparations from patients with acute hemolytic anemia. When blood is examined between slide and coverslip, the nuclear extrusion takes approximately 10 min at room temperature. The cell pushes out and retracts short pseudopodia. One of them contains the nucleus, and this portion of the cell eventually becomes detached from the rest of the cytoplasm.

With time lapse photography with an acceleration of about 100 times, the cell appears to push pseudopodia frenetically in all directions and after a number of apparent convulsions, the nucleus is extruded. This phenomenon has a certain resemblance to the bubbling observed following a mitosis, when the two newly formed daughter cells make efforts to separate.

The question whether nuclear extrusion is a normal event or occurs only in cases of hemolytic anemia or other pathologic conditions has been raised. (Some authors claim that normally the nucleus disappears by dissolution.) It is actually a constant phenomenon. It can be seen in electron microscope pictures of normal bone marrow (see p. 42).

Smears

The different phases of the extrusion of the nucleus can be found in smears of marrow aspirates stained with Giemsa. A comparison with the pictures obtained by phase microscopy clearly indicates that the irregular contour of these cells corresponds to different phases of movement of the cell surface during extrusion of the nucleus (see p. 46). These irregularities have no pathologic significance.

Nuclei without cytoplasm found in smears, particularly in cases of erythroblastosis, are due to an artifact which detaches the nucleus from the rest of the cell during preparation of the smear. (In vivo, free extruded nuclei are not seen because they are immediately phagocytized by histiocytes.)

The extruded nucleus, no matter what its appearance, is always a 2N nucleus when measured isotopically or spectrophotometrically.

Under pathologic conditions one can see fragmentation and even dissolution of the nucleus, a process which generally starts with the appearance of a central vacuole. Only in nuclei undergoing lysis is there a diminution of the DNA below 2N.

Electron Microscopy

In the electron microscope, the extruded nucleus can be seen to be surrounded by a very narrow rim of cytoplasm. Truly naked nuclei do not exist. The electron microscope always shows the nuclear envelope and the narrow border of cytoplasm which may contain a few mitochondria or Golgi vesicles. The rim of hemoglobinized cytoplasm is too narrow to be seen in the light microscope.

The observation is of interest because it shows that the process is more comparable to a separation of two parts of the cytoplasm, one which contains the nucleus, rather than to actual nuclear extrusion. It may also be compared to the liberation of platelets from the megakaryocytes (see p. 136).

Phagocytosis of the Extruded Nucleus. As soon as the nucleus is extruded, it is phagocytized by histiocytes, the histiocytes usually forming the centers of erythroblastic islands (see p. 42).

The small quantity of hemoglobin which surrounds the nucleus is metabolized by the histiocytes and contributes to the formation of the early bilirubin peak (see p. 42).

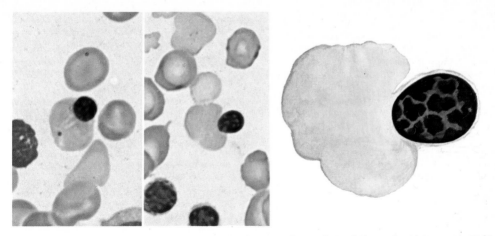

Figs. 1 and 2 — *Expulsion of the nucleus*. Note the irregular outline of the cell which corresponds to the particular movements of the cell at this stage of development

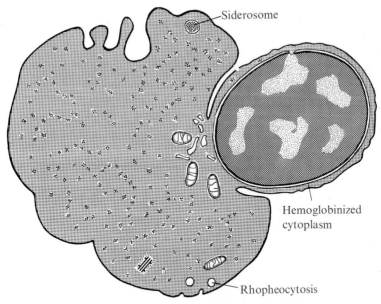

Fig. 3 — The outline of the cell is highly characteristic. The extruded nucleus is always surrounded by a ring of hemoglobinized cytoplasm

7 – The Erythroblastic Island

Definition[1-3]

An anatomic unit consisting of a histiocyte (macrophage), surrounded by one or more concentric rings of maturing erythroblasts*. These islands are always present in the marrow of adult mammals; they do not exist in the embryo and in lower vertebrates. The central histiocyte plays an active role in different stages of the life of the normoblast and of red cells: it phagocytizes the extruded nuclei; it phagocytizes a number of erythrocytes which have reached the end of their life span; and it is possible that it supplies normoblasts with nutrients or other substances during their maturation.

Living Cells[4]

With phase microscopy, erythroblastic islands can be seen when a marrow particle is gently removed from an opened bone and placed between slide and coverslip. Although such preparations are of necessity quite thick, one can see the relationship of the central cell to the concentric rings of normoblasts.

By gentle dissociation of the marrow followed by sedimentation, one can concentrate and isolate the erythroblastic islands. Isolated islands spread out on a slide in such a way that the central histiocyte can be seen in its star-shaped form with erythroblasts attached to the long projections of its cytoplasm. The projections are in continuous move-

ment, now touching one erythroblast, now another. Conversely, each erythroblast is in contact along some part of its surface with projection of the central histiocyte.

Smears[5]

In the preparation of smears, the attachment of erythroblasts to the central histiocyte is ruptured, and the relationships between the cells are destroyed. (Only rarely can an erythroblastic island be found even in the thick portions of the smears where it cannot be adequately studied.) The central histiocyte and its long projections, which are very fragile, break up into several fragments which round off. These bits of cytoplasm are usually overlooked in Giemsa stains (see p. 41, Figs. 1 and 2). However, if these fragments contain iron, they can be recognized for what they are by their diffuse blue color after applying Perls' reaction (see p. 56).

In hemolytic anemias, when erythroblasts are very numerous in the marrow, one may occasionally see an appreciable number of intact erythroblastic islands on smears of marrow aspirates. Occasionally, one may even find them in the buffy coat when entire islands have escaped into the peripheral circulation.

---

* Erythroblast is used as a generic term, see footnote, page 25.

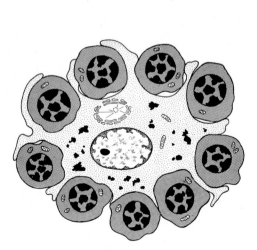

Fig. 1 – *An erythroblastic island*. The central histiocyte is surrounded by a single necklace of normoblasts in the same developmental stage

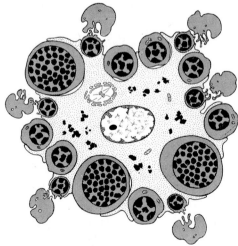

Fig. 2 – *In stimulated erythropoiesis*, the central histiocyte is surrounded by one or more layers of normoblasts in different stages of maturation

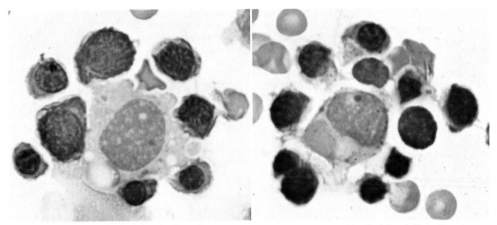

Fig. 1 — *An erythroblastic island.* Note the presence of three generations of erythroblasts. A vacuole at the left lower pole of the central histiocyte probably represents the remnant of a phagocytized red cell

Fig. 2 — *An erythroblastic island.* Note the phagocytosis and digestion of two red cells by the central histiocyte

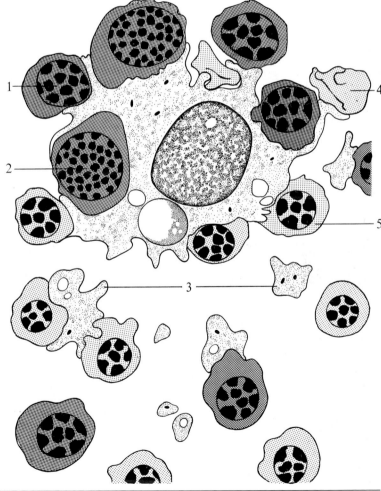

Fig. 3 — Schematic representation of Figure 1, showing the dissociation of the erythroblastic island during preparation of the smear. Fragments of the central histiocyte adhere to individual erythroblasts (see p. 41).
(1) Polychromatophilic normoblast I;
(2) basophilic normoblast; (3) fragments of histiocyte;
(4) reticulocyte;
(5) polychromatophilic normoblast II

39

8 – The Erythroblastic Island (Continued)

Light Microscopy

In sections, erythroblastic islands are commonly seen; the central histiocytes can be recognized only in sections that pass through the center of the island. The fine projections passing between the individual cells of the peripheral crown of erythroblasts cannot be appreciated with light microscopy.

Electron Microscopy

In suitable sections that pass through it, one can see the central histiocyte surrounded by a crown of erythroblasts. The central cell is often star-shaped with the points of the star giving rise to the long projections that embrace even erythroblasts at a considerable distance. It is hardly ever possible to observe an erythroblast without some part of it being touched by a projection from the central histiocyte.

Iron Metabolism

The role of the erythroblastic island in iron metabolism is not yet fully understood. The electron microscope allows one to study the presence of ferritin and hemosiderin as indicators of iron transport or storage.

1) *In the Central Histiocyte.* This cell normally contains ferritin as well as hemosiderin (see p. 166). The iron is derived largely from phagocytized red cells (see p. 42) and to a lesser extent from iron transported to the cell by transferrin (see p. 56).

2) *In Erythroblasts.* Dispersed ferritin molecules and siderosomes are seen in all stages of maturation of the red cell series (see discussion, p. 34).

3) *On the Surface of Erythroblasts.* Ferritin molecules are always present. In pathologic cases these ferritin molecules can become very numerous (see pp. 54 and 56). They are found between the extensions of histiocytic cells and erythroblasts and appear to be most numerous between adjacent erythroblasts.

Three hypotheses have been advanced to account for these appearances and the origin of ferritin. The ferritin may be derived a) from plasma, b) from the central histiocyte which extrudes the ferritin whenever it encounters an erythroblast, or c) it may be synthesized by the erythroblast membrane (see p. 34).

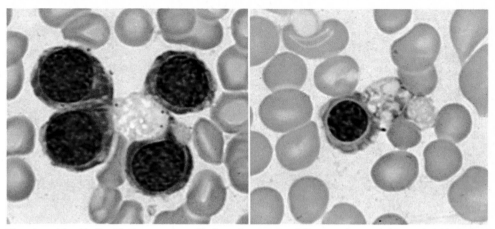

Figs. 1 and 2 — *Erythroblasts,* carrying with them fragments of the cytoplasm of the central histiocyte

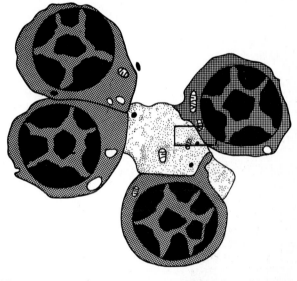

Fig. 3 — *Details of the interface between an erythroblast and the central histiocyte.* The histiocyte (*at left*) contains numerous molecules of ferritin, either dispersed or surrounded by a membrane. In the center, note the free ferritin molecules between the histiocyte and the erythroblast. At right, various stages of rhopheocytosis of an erythroblast. Ferritin molecules adhere to specialized portions of the erythroblast membrane (at 1) which invaginate (at 2 and 3), form vesicles (at 4) which move to the interior of the cell (at 5)

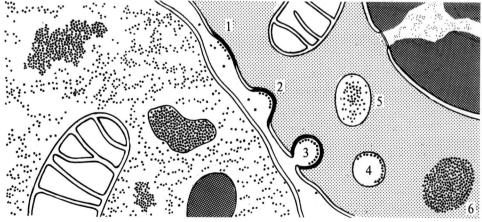

9 – The Erythroblastic Island—Phagocytic Functions

Phagocytosis of Erythrocytes[1-3]

Normally, only red cells which have reached old age at approximately 120 days disappear from the circulation. Simultaneous changes of the cell membrane and within the red cells are responsible for progressive deterioration of cellular functions which make the red cell more and more susceptible to hemolysis and adherence to histiocytes which eventually phagocytize them.

A large number of isotopic investigations have attempted to identify the exact locus of red cell destruction. They have shown that the liver and the spleen play a predominent role in the process. These investigations were undertaken under experimental conditions which do not reflect the normal physiologic processes. Some dealt with the sites of destruction of injected red cells which were either abnormal to begin with or were made abnormal by prior heating, antibodies, or other manipulations. Other measurements were made in patients suffering from hemolytic anemia (to determine whether the spleen was the site of destruction and splenectomy indicated).

Only a few investigations have dealt with the destruction of normal red cells under physiologic conditions. They indicate that the bone marrow is the site of destruction of red cells that have reached the end of their life span by normal wear and tear.

In smears, it is possible to see phagocytized red cells or fragments of red cells which still contain hemoglobin (see p. 39, Fig. 2). Subsequently, phagocytized erythrocytes loose their color and eventually only vacuoles remain which are difficult to identify as being derived from phagocytized red cells (see p. 39, Fig. 1). With the electron microscope, one can easily follow the successive stages of ingestion. They are fully described on p. 164.

Phagocytosis of the Extruded Nuclei

The extruded nucleus, or to be precise, the part of the cell which contains the nucleus after separation from what is now a reticulocyte, is immediately phagocytized by the central histiocyte of the erythroblastic island, which always contains normoblast nuclei in different stages of digestion.

The Early Bilirubin Peak[4,5]. Hemoglobin containing cytoplasm surrounding the phagocytized nucleus probably contributes to the "early bilirubin peak". When precursors of hemoglobin are tagged with ^{14}C or ^{13}N, one subsequently finds a constant level of the isotope used in the circulation passing for the duration of the life span of red cells. However, approximately 11% of the isotope can be recovered as bilirubin or fecal stercobilin within the first 3 days after administration of the tracer. Several hypotheses have been advanced to explain the "early peak":

1) Destruction of erythrocytes just before or immediately after their release into the circulation (ineffective erythropoiesis)
2) Metabolism of myoglobin, catalases, peroxidases, or cytochromes which are labeled along with red cell hemoglobin
3) Synthesis of heme not used for hemoglobin formation
4) Direct synthesis of bilirubin in the liver without passing through the porphyrin-hemoglobin cycle
5) Excess synthesis of heme without immediate hemoglobin formation now appears to play the largest part under normal physiologic conditions.

The metabolism of hemoglobin phagocytized along with the extruded nuclei of erythroblasts makes a minor contribution to the early peak. Hemoglobin may also be present in the nucleus, but if so in very low concentrations.

Extrusion of the Nucleus Under Pathologic Conditions[6,7]

After stimulation of erythropoiesis or in pathologic states, the normoblasts may loose contact with the central histiocyte of the island before the process of nuclear extrusion is completed. Enucleation may then take place while the nucleated red cell attempts to pass through the wall of the sinusoid into the circulation. The nucleus which represents the most rigid portion of the normoblast may have difficulty in passing through the sinusoidal wall and may be separated from the rest of the cell in the process. Those normoblasts which enter the circulation intact, subsequently loose their nuclei when attempting to pass through the extrasinusoidal channels of the spleen with its network of fibrils (pitting function).

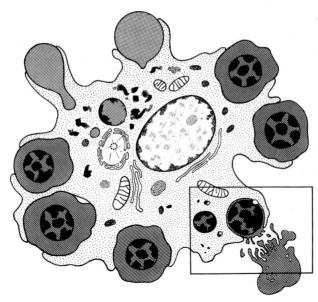

Fig. 1 — *An erythroblastic island. At upper left,* stages of phagocytosis and digestion of aged red cells. *At lower right,* stages of digestion of an extruded erythroblast nucleus

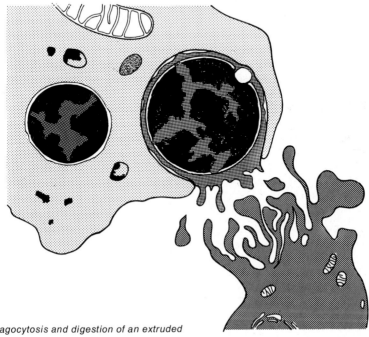

Fig. 2 — *Phagocytosis and digestion of an extruded nucleus*

Definition[1]

The reticulocyte* is that stage of maturation of the red cell which begins with the extrusion of the normoblast nucleus and ends when the reticulocyte has lost its organelles (mitochondria and ribosomes) at which time the erythrocyte assumes its disc shape and becomes a mature red cell.

Reticulocytes are identified by so-called supravital dyes. Actually, the dyes kill the cells and then stain them.

Vital Staining[2]

The effect of basic dyes on living cells is a function of their concentration. In weak concentration, they color vacuoles without interfering seriously with cellular functions, including the movements of the cell. When brilliant cresyl blue is used, the vacuoles stain metachromatically, a ruby red. One can also use neutral red, old or new methylene blue, or fluorescent dyes for the purpose.

Supravital Staining – Smears

At higher concentrations, the basic dyes kill the cells and a network appears, called "substantia reticulofilamentosa". This reticulum has given the reticulocyte its name, although it can be observed in any cell rich in ribosomes (e.g. normoblasts). After supravital staining and precipitation of the ribosomes, one can prepare smears and stain them with Giemsa. During fixation, the ribosome network is decolorized by the alcohol. It is subsequently stained by the methylene blue of the Giemsa.

Electron Microscopy[1]

Young reticulocytes can be identified readily by their bizarre outlines which reflect their continuous movement (see p. 46).

The reticulocytes contain portions of the Golgi complex, mitochondria and a variable number of mono- and polyribosomes. Ferritin molecules are present singly, in small clusters and in siderosomes. Rhopheocytic vesicles are seen near the surface. They are the more numerous the younger the cell.

After supravital staining, the ribosomes can be seen to have agglutinated into a network which varies with the age of the reticulocytes. Mitochondria are swollen and degenerated as a result of the toxicity of the dye. (Compare fig. 6, p. 45 and fig. 4, p. 59.)

Disappearance of Organelles
Autophagosomes[3, 4]

As the reticulocyte matures, ribosomes and organelles disappear progressively. The cristae of mitochondria become effaced and the number of mitochondria diminishes. Remnants of the Golgi complex, the centrioles and the aggregates of ferritin similarly disappear.

Autophagic vacuoles (autophagosomes or secondary lysosomes) can be found occasionally in normals, more frequently in pathologic conditions. They contain ribosomes, mitochondria and ferritin.

These autophagosomes can be extruded from the red cell in the marrow or they can conglomerate into a single aggregate and become separated from the rest of the red cell as it passes through the narrow meshes of the splenic network of reticular fibers. This has been referred to as the "pitting function" of the spleen.

Passage Into the Circulation

Reticulocytes enter the bloodstream by diapedesis. The reticulocyte advances a pseudopod and passes through a small opening in the wall of a capillary or a sinusoid. While the process can be seen only rarely in electron micrographs from normal subjects, it can be readily observed whenever erythropoiesis is stimulated, e.g. in hemolytic anemias. One may wonder whether the peculiar movements of mature normoblasts and reticulocytes associated with the extrusion of the nucleus may not aid diapedesis. In fact, however, the movement of the reticulocyte disappears rapidly after it becomes separated from the erythroblastic island. Moreover, although the movements are rapid and impressive, they do not impart to the reticulocyte any capability of locomotion.

---

* Advancing knowledge should improve our nomenclature (see p. XIII). However, a new nomenclature carries the danger of creating confusion by the simultaneous existence of a new and better name while the old, inappropriate yet time-honored name persists. I will let the reader weigh the advantages and inconveniences of the proposed changes in the nomenclature of the reticulocyte. *Reticulocyte:* the name may lead to confusion with the reticular cells of the histologists. The name *proerythrocyte* proposed some years ago has not found favor with hematologists.

Reticulum of reticulocytes: this term is confusing in view of the existence of an endoplasmic reticulum in many other cells though not in the reticulocyte. There is no need to use this term and it should be abandoned.

Substantia reticulofilamentosa: as noted, this is not a substance but an artifact. I propose to refer to it as precipitated network of ribosomes.

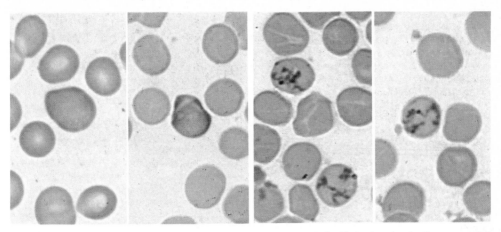

Figs. 1 and 2 — *Reticulocytes I* in a Giemsa stain. Note basophilia

Figs. 3 and 4 — *Reticulocytes I* after supravital staining, smear counterstained with Giemsa

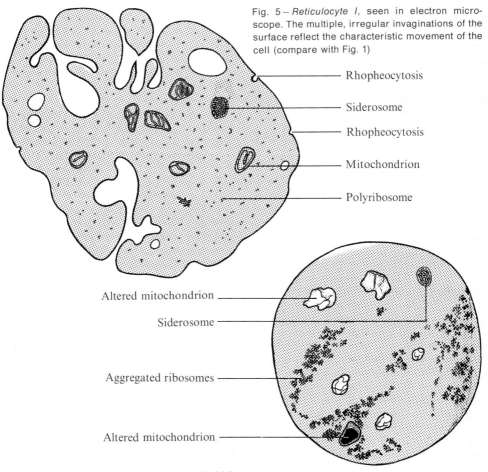

Fig. 5 — *Reticulocyte I*, seen in electron microscope. The multiple, irregular invaginations of the surface reflect the characteristic movement of the cell (compare with Fig. 1)

Rhopheocytosis

Siderosome

Rhopheocytosis

Mitochondrion

Polyribosome

Altered mitochondrion

Siderosome

Aggregated ribosomes

Altered mitochondrion

Fig. 6 — The supravital dye has killed the cell which has become spherical (cf. Fig. 4)

45

11 – Maturation of Reticulocytes

A classification of reticulocytes according to the magnitude of the precipitated network of ribosomes has been proposed. A preferable classification would appear to be the distinction between reticulocytes that exhibit the peculiar movement characteristic of them and those that do not. The former type (type I) is normally restricted to the bone marrow and the latter (type II) to the peripheral blood. Stimulation of erythropoiesis which in turn leads to early release of reticulocytes results in the passage of the first type of reticulocyte into the peripheral blood (shift reticulocytes)[1].

Living Cells[2, 3]

Type I has a lobulated outline and may resemble 3, 4 or 5 leafed clover, due to contraction of one part of the cytoplasm with corresponding expansion in adjacent parts of the cell as may already be seen in polychromatophilic normoblasts (see pp. 32 and 36). These movements diminish as the cell matures. The movements can be best followed in time lapse movies (Fig. 3) and can be readily inferred from scanning electron microscope pictures (Fig. 4).

When reticulocytes are kept for prolonged periods between slide and coverslip, 2 to 10 vacuoles appear and often congregate in the center of the cell. When the reticulocyte dies, movements cease, the irregularity of its cell surface disappears and the cell becomes round.

Type II has a cup-shaped form, and careful examination shows that it has 1 or 2 granules (remnants of organelles) and sometimes a refringent granule (granule of Isaacs).

Hemolysis. When a reticulocyte hemolyzes, due to a hemolytic agent or due to pressure between slide and coverslip, 5 to 10 black granules appear in its stroma which undergo Brownian movements. They represent mitochondria and sometimes remnants of other organelles.

Smears

The experienced eye can recognize reticulocytes in Giemsa stains without supravital staining on a normal bone marrow smear. Reticulocytes of type I are readily recognizable by their irregular outlines as well as by their polychromasia. It should be noted, however, that these irregularities of the contour can only be seen in a thick portion of a properly prepared smear. In its thin portion all red cells will be rounded off (Fig. 2).

Reticulocytes of type II are not easily recognized in smears. At first, they are more or less polychromatophilic. In the more mature forms, however, the few ribosomes present cannot impart sufficient polychromasia to the red cells to be appreciated in Giemsa smears, although the ribosomes are still precipitable into one or several small granules.

In the Soret band, type II reticulocytes can still be recognized, particularly in some areas, by the negative images of mitochondria or the presence of a "granule of Isaacs" (see Fig. 2, p. 46).

Sidero-reticulocytes[4]

As already noted (p. 34), a small number of reticulocytes may contain 1 or 2 siderosomes. The removal of siderosomes from circulating reticulocytes in the spleen has been demonstrated by transfusing blood from donors with increased numbers of sidero-reticulocytes into normal recipients. Under these circumstances, siderosomes rapidly disappeared from the red cells which continued to circulate as could be ascertained by isotopic tagging. In splenectomized recipients the siderocytes persisted.

Achromocytes

When smears are examined in dark field illumination or in phase contrast, one can recognize reticulocytes with a markedly decreased hemoglobin content which might normally be overlooked. These cells are called achromocytes. They probably correspond to cells which were hemolyzed during preparation of the smear, since they are more numerous in thin portions of a smear and particularly at the feather edge.

Shift Cells and Macroreticulocytes[1, 3, 5, 6]

In the presence of an intense response to an acute anemia or large injections of erythropoietin, reticulocytes may reach up to two times their normal size with a corresponding increase in their hemoglobin content. This is a normal response of the erythropoietic system to severe stress and these large reticulocytes are referred to as "stress" reticulocytes. Whether the doubling of the size is due to one less mitotic division than normal in the process of amplification or some other phenomenon is not yet clear. In contrast, even under moderate erythropoietic stress, some of the reticulocytes in the bone marrow pool are shifted to the circulating pool to compensate the anemia and these reticulocytes are referred to as "shift" reticulocytes. Shift reticulocytes can be either of normal size or large depending on whether the cells being shifted from the marrow are of normal volume or macroreticulocytes.

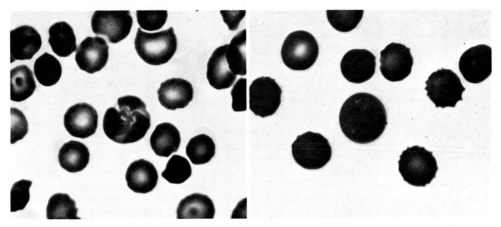

Fig. 1 — *Best portion of the smear*. The characteristic shape of a reticulocyte I is seen (Soret band)

Fig. 2 — *End of the smear*. The reticulocyte I is rounded off. *Fine white specks* represent the negative images of mitochondria

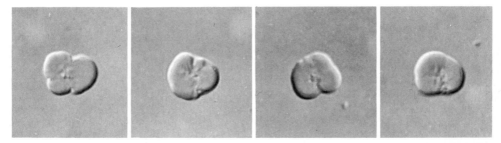

Fig. 3 — *Movements of reticulocytes I*. Photographs taken at 20-s intervals (Nomarski interference microscope)

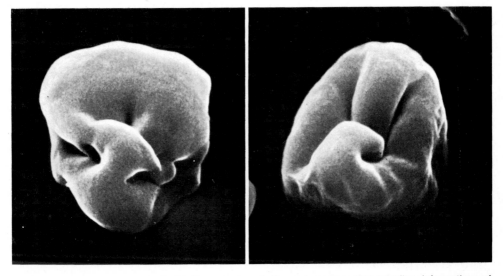

Fig. 4 — *Reticulocytes I* examined with the scanning electron microscope. The peculiar deformations of the surface can be clearly seen

12 – Normal Erythrocytes: Discocytes

Definition[1-3]

In the circulation mature red cells have the shape of a biconcave disc, hence the name discocyte. Their diameter is approximately 8 µm. Their thickness is 2.5 µm at the periphery and 1 µm in the center. The surface is 160 µm² and the volume 90 µm³. The weight is 30×10^{-12}g. Each of these values varies by 5–10% in normals.

Living Cells

Erythrocytes have a yellowish color when viewed in the light microscope. With phase microscopy they appear dark grey, with a clear center and a surrounding halo. In the interference microscope, the halo is absent. A false impression of relief is produced (see p. 226).

Rouleau Formation. In the circulation, in a test tube or in a thick preparation between slide and coverslip, red cells form rouleaux (Fig. 2). The length of the rouleaux determines the speed of erythrocyte sedimentation (ESR) used clinically. Rouleau formation is altered in pathologic conditions:
1) Rouleaux are long and less readily dispersed when there is a marked increase in fibrinogen (e.g. in infections) or globulins (e.g. in infections or multiple myeloma). Rouleaux are seen most readily in thick portions of the smear (Fig. 1).
2) Rouleau formation is decreased because of shape changes in red cells, so that they cannot adhere to each other in the usual fashion of a stack of flat plates or poker chips. This is the case, for example, in the presence of spherocytes due to hereditary spherocytosis, or to antibodies (Erythroblastosis fetalis).

Agglutination. In the presence of red cell antibody, red cells form clusters rather than rouleaux. The agglutinated red cells are difficult to disperse, particularly in the presence of high antibody titers. Agglutinated red cells seen in the phase microscope can sometimes be partially separated by a light tap on the coverslip. In that case, red cells assume a fusiform shape and remain connected to neighboring cells by thin filaments which are extensions of the tapered ends of each cell (Fig. 4). These filaments are very elastic. When they break off, they retract. The resulting myelin figures consist of parts of the red cell membrane. They contain a minute amount of hemoglobin.

In Smears, the fusiform shapes of red cells can be seen in thick portions (Fig. 3). They are indicative of irregular antibodies or cold agglutinins.

Deformability. Normal red cells are easily deformable and return as easily to their biconcave shape. This deformability is physiologically very important; it allows a red cell with a diameter of 8 µm to pass through 3 µm capillaries. Under circumstances which render the red cells less deformable, they may fail to pass through capillaries, block the circulation and occasionally precipitate thrombosis or hemolytic events. Studies of red cell deformability should become a routine part of any thorough hematologic examination (see p. 102).

Smears

Red cells are pink in well-stained Giemsa smears. Improper staining procedure may color them blue, yellow, greenish, etc. In the center of a well-made smear, the red cells do not overlap and each red cell has an area of central pallor which fades gradually toward the more deeply stained periphery. In badly made smears, the area of central pallor appears as an entirely colorless center separated by a sharp and occasionally irregular line from a uniformly staining peripheral ring of hemoglobin (torocyte). The appearance is artifactual. At the sides and the feather edge of the smear red cells have been flattened and appear round without an area of central pallor. Red cells may also appear in smears as echinocytes, target cells or stomatocytes even though their appearance in vivo is normal. Conversely echinocytes or acanthocytes may appear as flat, round cells at the sides or the feather edge of a smear. Recognition of these artifacts is important to avoid diagnostic errors (see pp. 58 to 100).

Examination in the Soret Band. Examination in the near ultraviolet (at 415 nm) brings out minor differences in the hemoglobin concentration in different parts of the cell and reveals a small area of central pallor in cells that may otherwise be mistaken for spherocytes. Thus illumination in the Soret band is particularly well suited for black and white photography of red cells (see pp. 58 to 100 and p. 228).

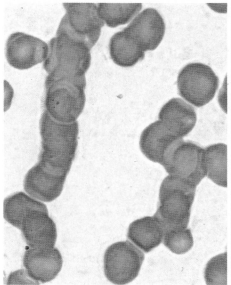

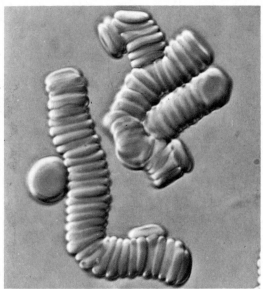

Fig. 1 — *Rouleau formation in a smear.* This appearance is seen in thick portions of smears or even in thin portions when the plasma viscosity is markedly elevated (as in hypergammaglobulinemia or hyperfibrinogenemia)

Fig. 2 — *Rouleaux seen in the living state* (interference microscope)

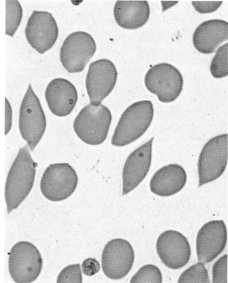

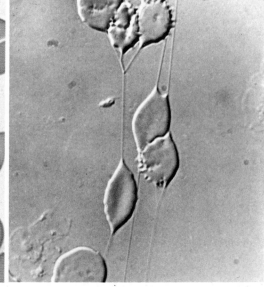

Fig. 3 — *Fusiform red cells seen at the thin end of some smears.* They are due to agglutination of red cells and are seen in the presence of autoantibodies or increased plasma viscosity

Fig. 4 — *Red cells agglutinated by antiserum.* Seen in the living state (interference microscopy). When the agglutinated cells are dispersed, the cells remain attached to each other by filaments

Definition[1-5]

Under the influence of intrinsic or extrinsic factors, the discocytes can assume the appearance of echinocytes (crenated cells) or stomatocytes (cup-shaped cells). These transformations are normally reversible.

The recognition of this reversible equilibrium state between discocytes on the one hand and echinocytes or stomatocytes on the other has clarified a number of artifactual appearances in smears and allowed to distinguish them from similar changes which occur under experimental or pathologic conditions in vivo (see pp. 52 and 64 to 100).

Echinocytes and Sphero-echinocytes

They are subdivided into five classes: echinocytes I, II, III and sphero-echinocytes I and II (Fig. 1). Echinocytogenic agents include: fatty acids, lysolecithin, anionic compounds, elevated pH (see figure below).

Stomatocytes and Sphero-stomatocytes

They are divided into five classes: stomatocytes I, II, III and sphero-stomatocytes I and II (Fig. 3).

Stomatocytogenic agents include cationic drugs and low pH (see diagram below).

Influence of the Tonicity of the Medium

The echinocytic and stomatocytic shapes of red cells are not related to the tonicity of the medium. Moderately hypertonic solutions reduce the water content of all forms of red cells, but do not alter their original shape: discocytes become flattened, echinocytes become contracted and stomatocytes increase the depth of their concavity.

Moderately hypotonic solutions swell all forms, but do not alter their original shape, with one exception: biconcave red cells become uniconcave and resemble stomatocytes.

Abnormal Red Cells

Echinocytic and stomatocytic transformations can be induced not only in normal discocytes, but in all sorts of abnormal red cells. These changes are superimposed on the initial pathologic shape and are also entirely reversible (see for example pp. 66 and 74).

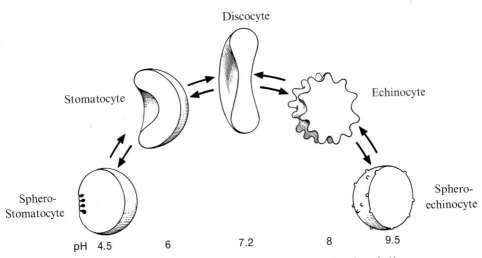

Diagram of the discocyte-echinocyte-stomatocyte transformation as a function of pH

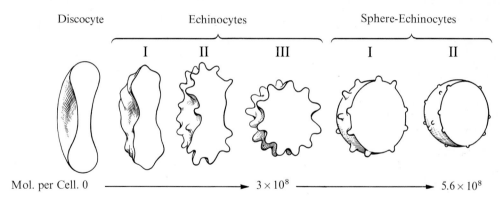

Mol. per Cell. 0 ——————————→ 3×10^8 ——————————→ 5.6×10^8

Fig. 1 – *Schematic representation of discocyte, echinocyte and sphero-echinocyte transformation* (by anionic phenothiazine derivative)

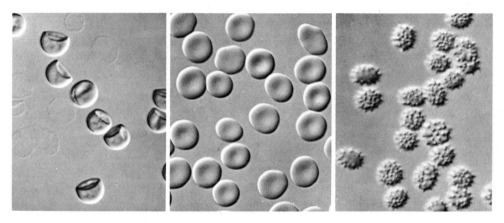

Fig. 2 – *Stomatocytes (left), discocytes (center) and echinocytes (right)* in the living state (interference microscope)

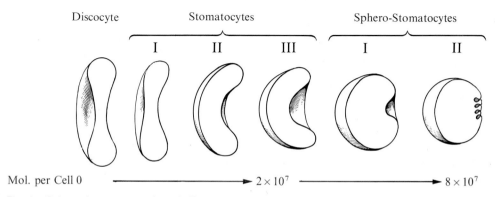

Mol. per Cell 0 ——————————→ 2×10^7 ——————————→ 8×10^7

Fig. 3 – *Schematic representation of discocyte, stomatocyte and sphero-stomatocyte transformation* (by a cationic phenothiazine derivative)

51

14 — Echinocytes

Smears[1]

Artifacts. In a large number of poorly made smears, but also in a number of well made ones, there are areas in which some or all erythrocytes are transformed into echinocytes. This is not due to dehydration of the cells in the drop of blood during preparation of the smear as is commonly thought; it is probably a "glass effect" (see below).

Echinocytosis. Occurrence of echinocytes in vivo is rare. It has been reported in uremia, pyruvate kinase deficiency and in newborns with liver disease. Echinocytes may also occur due to a high level of fatty acids in the blood of heparinized patients. Under these circumstances, echinocytes are observed after blood is allowed to stand for 15 min in the test tube, and more rarely in vivo.

Frequently, echinocytes will only be present in the center of the smear. At the feather edge increased surface tension will flatten the cells and round off their surface.

Living Cells[1, 2]

Artifact. The Glass Effect. When the red cells are washed in physiologic saline and examined between a glass slide and coverslip, they are always transformed into echinocytes. If they remain between slide and coverslip for more than a few minutes, they become sphero-echinocytes and eventually hemolyze. Between plastic slide and coverslip, the same cells remain discocytes. The so-called glass effect is actually due to the diffusion of basic substances from the glass, with resulting elevation of pH of the saline (near pH 9). The buffering action of plasma on unwashed red cells delays the development of the same phenomenon which then requires several hours rather than minutes and usually occurs only in part of the preparation.

One can demonstrate the glass effect by inserting, with the aid of a micromanipulator, a fine glass rod between a plastic slide and coverslip into a suspension of washed red cells in saline. As the glass rod is made to approach an individual red cell to within 1 or 2 µm, the discocyte transforms progressively into an echinocyte I, II, III and a sphero-echinocyte I. The echinocyte reverses to a discocyte as the rod is removed. The echinocyte transformation and reversal can be repeated any number of times by repetition of the experiment. The spicules appear always in the same location both in normal and abnormal cells.

Echinocytosis. To diagnose the presence of echinocytes in the circulation, one must examine a drop of fresh blood which has not been in contact with any glass surface. The drop of blood must be transferred from the test tube to a plastic slide and coverslip, or enclosed between 2 plastic coverslips and then attached to a slide with a drop of oil. Alternatively, one can use a very thick drop of blood between ordinary glass slide and coverslip. In this case, the pH is not much altered.

Acanthocytes must be distinguished from echinocytes (see p. 64).

Echinocytes in Stored Blood[3, 4]

After a variable time, depending on the temperature (3 weeks at 4°C, 24 h at 37°C), a higher and higher number of echinocytes appears in stored blood. This phenomenon is due a) to a progressive depletion of ATP and b) to formation of lysolecithin in the plasma. These echinocytes are reversible. They disappear within a few minutes after transfusion into a normal recipient and, in vitro, after replenishing the cellular ATP and/or after washing in fresh plasma.

Action of Stored Plasma[4]

Plasma kept at 37°C for 24 h or at 4°C for several days becomes echinocytogenic. Before the effect of storage on plasma was appreciated, the plasma of certain patients was reported to crenate red cells and anti-A and anti-Rh antibodies were reported to modify the red cell shape. Probably stored plasma was responsible for these erroneous interpretations.

One can prevent plasma from becoming echinocytogenic by heating it to 56°C for 30 min before storing it. This is believed due to the destruction of lecithin-cholesterol acyl transferase, which produces lysolecithin in normal stored plasma.

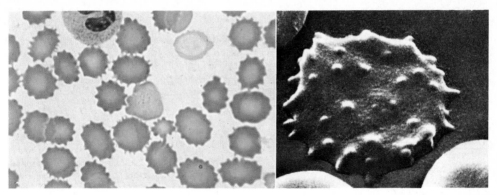

Fig. 1 – *Smear of normal blood,* in which the red cells have assumed an echinocytic form, probably due to the pH of the glass. Note a reticulocyte I in the center

Fig. 2 – *An echinocyte in a smear,* examined with the scanning electron microscope

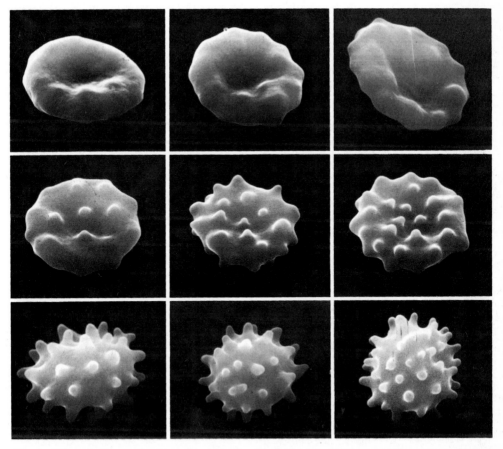

Fig. 3 – Different shapes of echinocytes I (*top*), echinocytes II (*middle*) and echinocytes III (*bottom*)

1 – Pathologic Sideroblasts and Siderocytes (Sidero-mitochondria)

Iron overload in hemochromatosis or hemosiderosis is associated with a slight increase in the number of sideroblasts and the size of the siderosomes. In contrast, some disorders of hemoglobin synthesis lead not only to a remarkable increase in the number of siderosomes (up to 20 per cell) but also to modification of their size and shape. Electron microscopy reveals that these siderosomes are, for the most part, iron-containing mitochondria.[1-3]

Smears

Such pathologic erythroblasts are usually hypochromic. No other anomalies are seen with Giemsa stains, except some Pappenheimer bodies (see p. 34) and occasionally inclusions which stain with PAS (see below).

Smears Stained for Iron (Perls' Reaction)[4]

The blue-stained siderosomes are more numerous, larger than normal, and of irregular form, most commonly oval. These pathologic siderosomes are sometimes distributed in a circle around the nucleus (ringed sideroblasts).

Electron Microscopy[1-3, 5]

The iron found in pathologic sideroblasts assumes several forms:
1) Sidero-mitochondria
2) Hemosiderin granules of various shapes (see p. 166)
3) Ferritin molecules dispersed in the cytoplasm or present in aggregates (see p. 34).

The sidero-mitochondria are characteristic of pathologic erythroblasts. They are never present in any other cell in the organism, even if the iron overload is very great. The iron contained in the mitochondria is rarely ferritin. Most frequently it is formed by very fine particles, which are tightly packed into micelles.

Disorders of globin synthesis manifest themselves by the presence of multilobed inclusions, which correspond to the PAS positive inclusions seen with the light microscope. Their size varies from 0.1 to 3.0 or 4.0 μm. They may contain hemosiderin (see p. 166). They may represent autophagosomes containing mitochondria, ribosomes, degenerated membranes, myelin figures, sometimes mixed with iron micelles, or ferritin.

Reinterpretation of Iron-Stained and Giemsa-Stained Smears

The pathologic sideroblasts can be recognized by light microscopy. The irregular appearance of siderosomes is due to the inclusion of hemosiderin, their oval aspect to the presence of iron

in mitochondria. The ringed sideroblasts are due to the presence of iron in virtually all mitochondria of erythroblasts, which form a crown around the nucleus. PAS positive inclusions correspond to pathologic precipitates of globin or to autophagosomes.

Sidero-reticulocytes

A large number of siderocytes are usually present along with pathologic sideroblasts. The electron microscope reveals that these are, for the most part, sidero-reticulocytes containing sidero-mitochondria and all the other iron-containing inclusions described in sideroblasts.

Sidero-erythrocytes[5]

Occasionally sidero-mitochondria and dispersed ferritin are present in erythrocytes, which have all signs of maturity, absence of ribosomes and a disc shape. These erythrocytes are always hypochromic. Frequently they contain vacuoles or inclusions, which resemble Heinz bodies.

Occurrence

Disease states in which pathologic sideroblasts occur are those associated with severe disturbance of hemoglobin synthesis. The cells are generally hypochromic, even though they contain an excess of iron. Dyserythropoietic anemias, sideroachrestic anemias, thalassemia, particularly after splenectomy, and refractory anemias, particularly those preceding the development of leukemia, are the conditions most frequently accompanied by the appearance of pathologic sideroblasts and siderocytes.

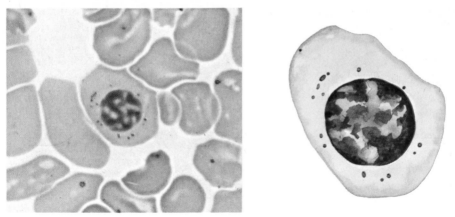

Fig. 1 – *A pathologic sideroblast*. Perls' reaction shows the iron stained blue. The counterstain is safranin. In this case, the iron granules are numerous and large, occasionally oval (cf. Fig. on p. 35)

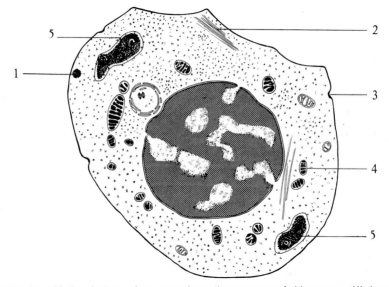

Fig. 2 – *Examination with the electron microscope* shows the presence of siderosomes (1), iron containing mitochondria (4) and pathologic inclusions (5). The polyribosomes are dispersed into monoribosomes. The microtubules (2) have a normal appearance. Rhopheocytosis (3) is increased

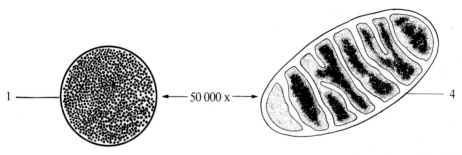

Fig. 3 – *Siderosome filled with ferritin*   Fig. 4 – *Mitochondrion filled with micelles of iron*

2 – Iron in Inflammation

In "inflammatory reactions" such as bacterial infection, sterile abscesses, tuberculosis, acute rheumatic fever, chronic rheumatoid arthritis, or generalized carcinomatosis, iron metabolism is commonly affected. The serum iron is low, but the tissue iron is elevated. The changes in iron metabolism are reflected in the appearance of cells in light and electron microscopy.

Smears Stained for Iron (Perls' Reaction)[1, 2]

1) *The number of sideroblasts is low:* 5% to 10% compared with a norm of 70%. Frequently, sideroblasts are entirely absent.

2) *The iron content of histiocytes is increased.* As we have described earlier (see p. 38), one may occasionally see an entire erythroblastic island, when one carefully examines the thicker portion of some aspirates. The central histiocytes of the erythroblastic islands are colored a light blue in all their fine projections when stained with Perls' and, in addition, contain numerous large inclusions of hemosiderin. More usually, the islands are dispersed in the process of preparing the smears and the central histiocyte, a fragile cell, is broken up into several fragments. They can round off or become oval. They are readily identifiable by their light blue color when stained for iron (Fig. 1).

3) *The normoblasts may have some blue cytoplasmic fragments attached to them.* Careful examination reveals that many normoblasts carry portions of the cytoplasm of the central histiocytic cells attached to their periphery. Normally they are easily overlooked (see p. 40), but when they do contain iron and are stained with Perls', the fragments of histiocyte can be seen readily. Sometimes they are reduced to a small blue sphere at the edge of the cell and may be mistakenly thought to be intracellular. They must not be confused with siderosomes.

Electron Microscopy[1]

Increased Rhopheocytosis. The entire surface of normoblasts is covered by ferritin molecules and rhopheocytic vesicles are more numerous than normal. Rhopheocytosis is very readily seen in reticulocytes.

Increased Normoblastic Ferritin. When examined under 100,000 magnification, a large number of dispersed ferritin molecules can be seen. They are also seen in the reticulocytes (but not in the mature erythrocytes, in contrast to sideroblastic anemia, see p. 54).

Absence of Siderosomes in Normoblasts. Conglomeration of ferritin molecules are never seen in inflammation.

Increase of Histiocytic Iron. The central histiocyte contains not only numerous ferritin molecules but also many clumps of hemosiderin.

Reinterpretation of Smears

1) The absence of sideroblasts is misleading. Many normoblasts contain a large number of ferritin molecules, but they always remain dispersed in the cytoplasm and do not aggregate in compact masses. Consequently, the Perls' reaction does not visualize the iron in the light microscope. Clearly, the presence and the number of sideroblasts and siderocytes are of no value in a quantitative evaluation of non-hemic iron in normoblasts.

To explain the aggregates of ferritin which are large enough to be made visible in the light microscope by the Perls' reaction, the following hypothesis may be advanced. When the molecules of ferritin are rapidly utilized, they remain in a dispersed stage until they disappear. In contrast, when the ferritin molecules persist for a considerable time in the cell, they are subject to the cytoplasmic movements and aggregate, as happens with any foreign substances. Thus the existence of larger aggregates of ferritin molecules indicates that the number of such molecules entering the cell is greatly in excess of the number that disappears.

2) The small blue sphere, which one can see close to the surface or at the edges of some normoblasts, represent a portion of the cytoplasm of the central histiocyte, which is fragmented during the preparation of the smear. Sky blue color in the cytoplasmic fragments is due to the increased numbers of dispersed ferritin molecules.

Comparison with Iron Deficiency Anemia[3]

In smears one sees neither sideroblasts nor histiocytic iron. With the electron microscope, ferritin is not seen in either normoblasts or histiocytes. Rhopheocytosis persists, but the vesicles produced contain no ferritin. Possibly apoferritin or substances utilized by the normoblasts are carried into the cell in this fashion. Ferritin is also absent from histiocytes which phagocytize senescent red cells. Probably their iron is immediately reutilized and is not stored.

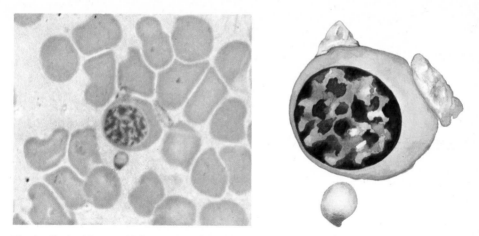

Fig. 1 – *Erythroblast* to which a portion of the central histiocyte has remained attached. The cytoplasmic fragment contains iron (stained with Perls')

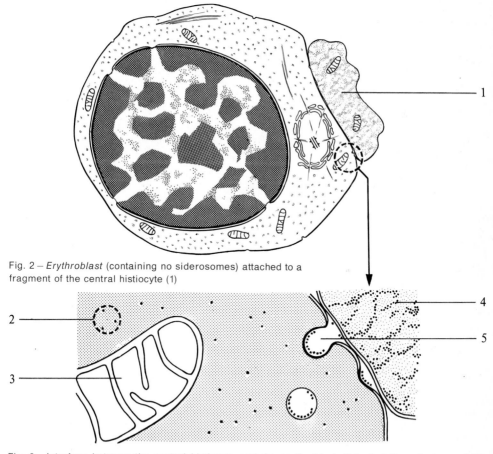

Fig. 2 – *Erythroblast* (containing no siderosomes) attached to a fragment of the central histiocyte (1)

Fig. 3 – *Interface between the central histiocyte and the erythroblast.* Note that the cytoplasm of the histiocyte is filled with ferritin (4) which can also be seen in the intercellular space from where it enters the erythroblast by rhopheocytosis (5). All of the ferritin is in dispersed form (2) which accounts for the absence of sideroblasts on examination with the light microscope. Mitochondria (3) do not contain iron in this case

3 — Reticulocyte and Erythrocyte Inclusions

Type of inclusion	Appearance in Giemsa	Present in	Living cells	Reinterpretation	Observed
Substantia reticulofilamentosa	Invisible except after supravital staining	Erythroblasts Reticulocytes	Invisible except after supravital staining	Precipitation of ribosomes by the dye	Normally
Howell-Jolly bodies	Purple or violet, spherical granules	Reticulocytes Erythroblasts Erythrocytes	Faintly visible except after hemolysis	Nuclear fragments (aberrant chromosomes)	Anemias
Azurophilic granules	Purple dust	Reticulocytes Erythroblasts	Invisible except after hemolysis	Remnants of nucleus (after karyorrhexis)	Severe anemias (sometimes artifacts)
Cabot rings	Bright red rings or threads	Reticulocytes Erythroblasts	Invisible except after hemolysis	Remnants of fused micro-tubules	Severe anemias
Siderosomes	Invisible except after Perls' reaction	Erythroblasts Reticulocytes Erythrocytes	Visible but not identifiable	1) Ferritin containing	Normally
Pappenheimer bodies	Blue or violet granules (identifiable only after Perls')	Erythroblasts Reticulocytes Erythrocytes	Visible but not identifiable	2) Mitochondria containing iron micelles 3) Hemosiderin granules	Sideroblastic anemias Thalassemias
Heinz bodies	Invisible except after Nile blue or supravital dyes	Erythroblasts Reticulocytes Erythrocytes	Visible	Denatured hemoglobin	Enzymopathies Toxins affecting hemoglobin

Basophilic Stippling

Smears. These are round or irregularly shaped granules of variable number and size which are stained blue by Giemsa.

Electron Microscopy.[1] In a well-made preparation one does not find any structure which corresponds to the punctate basophilia seen in smears. Nor do such granulations exist in the living cell. They are formed during the drying of the smear. Their formation depends on the exact manner in which the cell is dried. Hence, the variability of results with different preparations. One can in fact obtain a punctate basophilia by drying a smear rich in (normal) reticulocytes slowly. By using special technics it has been possible to produce basophilic stippling in red cells in a test tube and to prepare them for electron microscopy. They can be seen to represent aggregates of ribosomes and polyribosomes, which are sometimes associated with mitochondria and siderosomes.

Occurrence.[2,3] Basophilic stippling is very unusual in the normal adult, but common in the fetus and infant. It is frequent in lead intoxication, in a number of other anemias associated with disordered hemoglobin synthesis and in cases of a rare deficiency of pyrimidine 5'-nucleotidase.

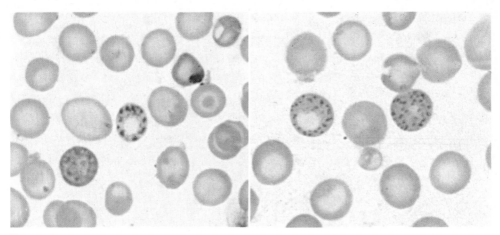

Figs. 1 and 2 – *Basophilic stippling in reticulocytes stained with Giemsa*

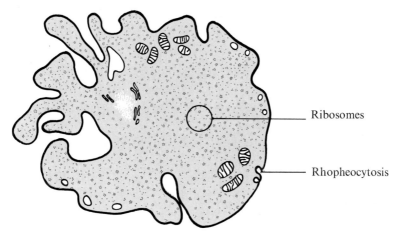

Fig. 3 – *Normal reticulocyte,* as seen with the electron microscope

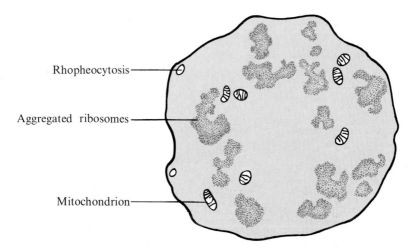

Fig. 4 – *Basophilic stippling* as seen with the electron microscope after slow drying

Howell-Jolly Bodies

Smears

These are small, round bodies, approximately 1 µm in diameter. They give a positive Feulgen reaction, i.e. they contain DNA. They are usually located eccentrically in reticulocytes and erythrocytes (or normoblasts).

Electron Microscopy

In the electron microscope they may be seen to be bordered by a membrane of the same composition as the nuclear envelope.

Formation[1, 2]

Howell-Jolly bodies are generally derived from chromosomes which have become separated from the spindle in the process of abnormal divisions. On occasion similar small nuclear fragments are produced by pathologic fragmentation of a nucleus (karyorrhexis).

Occurrence

Howell-Jolly bodies are very rare in normals. They are common after removal of the spleen, whether the spleen was normal or not, in congenital absence of the spleen or splenic atrophy or functional asplenia, in some hemolytic anemias, and in megaloblastic anemia. They may occur in very severe anemias.

In megaloblastic anemias and rarely in other severe anemias, one may occasionally see multiple Howell-Jolly bodies in normoblasts.

Cabot Rings

Smears

These are rings, figures-of-eight, incomplete rings or similar configurations of a reddish-violet fine filament in smears stained with Giemsa. Occasionally, several may exist in a single cell.

Formation[3, 4]

Some authors have interpreted Cabot rings as remnants of the nuclear membrane and believed them to represent evidence that the nucleus disappears by dissolution rather than by extrusion. This hypothesis must be abandoned, because Cabot rings can be seen in normoblasts with intact nuclei.

There is a more likely explanation for the actual origin of Cabot rings: fibres of the spindle radiating from the centrosome can be seen on occasion to have the same coloration as the Cabot rings with Giemsa (see p. 3, Fig. 2). These fibres can remain visible at the end of the telophase as a fine red line between the nuclei of two daughter cells that have just completed their mitosis. When this filament remains within the cell it may then assume the ring-like configuration as a result of intracytoplasmic streaming.

Occurrence[5]

Cabot rings may, therefore, be interpreted as indicating an anomalous mitosis and somewhat abnormal stainability of the microtubules which form the spindle. One may see them in severe anemias, and they are frequent in dyserythropoiesis.

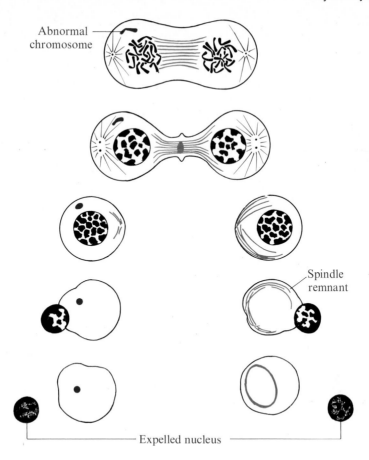

Abnormal chromosome

Spindle remnant

Expelled nucleus

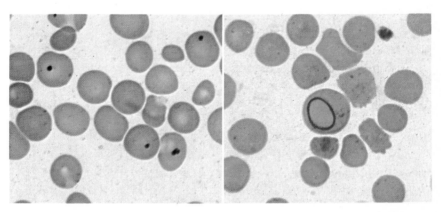

Fig. 1 – *Formation of a Howell-Jolly body* due to condensation of an aberrant chromosome left behind during division

Fig. 2 – *Formation of a Cabot ring* due to abnormal persistence of microtubules of the mitotic spindle

5 – General Remarks on Pathologic Erythrocytes

Reinterpretation of Poikilocytes

The term poikilocyte indicates a variable shape of red blood cells in smears, which may include round, oval, pear-shaped, crenated, and many other shapes. One must make certain that what appears as poikilocytes are not artifacts due to the preparation of smears. (The examination under the phase microscope carried out with appropriate precautions can usually determine, whether the deformations pre-existed or developed during the preparation of the smear (see pages 64 to 100).

Little by little we have been able to classify an increasing number of poikilocytes by their mode of origin and behavior. Only 65 years ago sickle cells were described by their discoverer as "freakish erythrocytes". Today we can not only distinguish sickle cells but, among spiculated cells, differentiate acanthocytes, echinocytes, and keratocytes. A catalogue of a variety of peculiar shapes is gradually being replaced by a specific nomenclature, which classifies these shapes by origin and behavior (see below).

Shape and Pathophysiology of Erythrocytes

Examination by the scanning electron microscope (SEM) has renewed the interest of hematologists in the shape of pathologic erythrocytes. It has become clear that in many instances the pathologic shapes can be explained by recent discoveries on the structure and changes in the biochemistry of the membrane or the interior of the red blood cell. Moreover, it has been possible to establish correlation between the shape of erythrocytes and their deformability, which plays an important role in the physiology of the micro-circulation: alterations in the rheologic properties of red blood cells can explain a number of hemolytic anemias and thrombotic disorders.

The clinician seeing a bizarre-shaped red blood cell in a smear can now translate that appearance into the true shape of that cell in three dimensions. In many cases he can also infer the mechanism that produced the bizarre shape; he can anticipate the behavior of the cell as it traverses arteries, capillaries and veins (see p. 102).

Reinterpretation of Anisocytosis

Diameter. Historically, the term anisocytosis indicates that on examination of the smear red cells have markedly differing diameters (diameters of cells vary more than normally).

The causes of this increased variability may be quite different:

1) *Red cells with normal hemoglobin content.* The diameter may be small because cells are spherocytic (as in many hemolytic anemias) or larger than normal because they are leptocytic (in obstructive jaundice).

2) *Red cells with a larger than normal content of hemoglobin.* Megalocytes or macrocytes (see p. 88) have always a greater diameter than normal, except when they become crenated or spherocytic.

3) *Red cells containing less hemoglobin than normal.* Their diameter is usually smaller than normal. Schizocytes (see p. 92) and codocytes (see p. 68) belong to this group, but their diameter is difficult to determine.

4) *Reticulocytes* are always slightly larger than mature red blood cells. When they become markedly larger, they are macroreticulocytes (see p. 88).

Volume. Because the measurement of diameters is cumbersome and readily affected by artifacts, it has been replaced by volume measurements which can be readily made by automated instruments. Mean values (MCV) may, however, be misleading as a given MCV may represent a mixture of different populations.

Multiple Populations of Erythrocytes

They are often difficult to distinguish in smears if the populations overlap. A second population can be diagnosed safely only when no intermediate forms exist, (e.g. in some hypersideremic anemias with a small number of hypochromic, macrocytic cells admixed to a large population of normal appearance). Physical methods of differential agglutination, differential hemolysis, filtration through Millipore, gravimetric methods, etc., can be used to isolate single populations. Microspectrophotometric methods have been used successfully in studying multiple populations in smears (e.g. G6PD deficiency in homozygotes) or elution of hemoglobin A (to identify fetal cells in the circulation of the mother or to distinguish different types of persistence of fetal hemoglobin).

Multiple Diseases[1–5]

In rare cases multiple diseases coexist and may manifest themselves in the same cell, leading to very unusual cells or in different cells, leading to a double population. The cause may be the inheritance of two congenital abnormalities or an acquired disease developing in a congenital red cell abnormality. Examples are:

Involvement of the same cell. Hereditary spherocytosis combined with thalassemia or C hemoglobinopathy.

Double population. Spherocytes and sickle cells in the same slide in acquired spherocytosis in sickle cell anemia.

Compensation of the shape change. Normal appearance of cells in hereditary spherocytosis complicated by a retention jaundice (see p. 98).

Note on Nomenclature

We have introduced a number of Greek names for poikilocytes for two reasons. They can be internationally used without having to find corresponding words for such descriptive names as "burr cells" or "spur cells", for which there may be no equivalent expression in another language. More importantly, the proposed nomenclature lends itself readily to describe forms in which multiple anomalies coexist. For instance, a "spur cell" undergoing echinocytic transformation is described in a single word: echinoacanthocyte, rather than by an entire sentence describing its origin.

Ideally one may wish to have a nomenclature which classifies the cells by their mode of origin; however, multiple pathophysiologic changes may give rise to the same form. A stomatocyte, which is characteristic of one type of hereditary hemolytic anemia is identical in shape to a red cell which has been subjected to an appropriate pH change.

While the proposed nomenclature, as every classification, must remain provisional until more is known about shape changes, it has the advantage of being flexible. For instance, if biochemical changes that lead to the same appearance of the cells can be distinguished, the two identically shaped cells can be designated by the addition of separate prefixes.

A number of prefixes are already in general use:

*Macro*cytes or *micro*cytes are red cells of increased or decreased volume. A *sphero*cyte is a cell with a lower than normal surface to volume ratio; it is a cell that is thicker than normal rather than a true sphere, but the name spherocyte is retained for historical reasons (see p. 96). It is the opposite of a *lepto*cyte, a cell that is thinner than normal due to an increased surface to volume ratio.

Specific Names

Name	From the Greek	Seen in
Acanthocyte	Spicule	Abetalipoproteinemia, cirrhosis and rarely other liver diseases
Codocyte	Hat	Thalassemia, retention icterus
Dacryocyte	Tear	Thalassemia, hemolytic anemia, anemia with Heinz bodies (myelofibrosis)
Drepanocyte	Sickle	Sickle cell anemia (trait after deoxygenation)
Keratocyte	Horn	Intravascular coagulation, fibrin deposits in blood vessels or vascular prosthesis
Knizocyte	Dimple	Hemolytic anemia, hereditary spherocytosis
Megalocyte	Large	Megaloblastic anemia
Schizocyte	Cut	Same as keratocyte
Stomatocyte	Mouth	Hereditary or acquired hemolytic anemia

Spiculated Red Blood Cells[1]

In a variety of diseases, spiculated red blood cells have been described and given a variety of names. Although some of them are still unidentified, careful observation allows one to classify the majority of these spiculated cells as one of the following:

Names	Comments
Echinocyte (p. 52)	Red cells with 10 to 30 spicules distributed regularly over the surface, mostly seen in vitro, may occur in vivo (see below)
Echinocytic forms (pp. 66 to 100)	All pathologic red cell shapes, even already spiculated red cells, such as acanthocytes or keratocytes, can still undergo secondary transformation into their echinocytic counterparts
Acanthocyte (pp. 64, 66)	Spheroidal cells with 2 to 20 spicules distributed irregularly over the surface
Dacryocyte (p. 74)	A single spicule which deforms the erythrocyte (so as to give it the shape of a chinese spoon or a tear drop)
Drepanocyte (pp. 76 to 80)	Discocyte with 1 to 10 rigid spicules
Keratocyte (p. 84)	Discocyte with 1 to 6 spicules
Schizocyte (pp. 92, 94)	Fragments of discocytes with 2 to 4 angular or pointed projections (created by the fragmentation of the whole cell)

Acanthocytes

Definition

Acanthocytes are red blood cells which have lost their discoid shape and which have 3 to 12 spicules of uneven length irregularly distributed over the red cell surface. They are therefore quite different from echinocytes. Nevertheless, they can resemble them in certain circumstances, and the separation of individual cells from echinocytes may be difficult if both types of cells coexist in a smear. They can be most readily distinguished by utilizing the facts that acanthocytes develop secondary spicules in the presence of echinocytogenic substances, and that they cannot be reversed to discocytes by normal plasma.

Smears

Acanthocytes appear smaller than discoid cells, because they have assumed spheroidal shape. Characteristically, the spicules are distributed irregularly over the surface, and the individual spicules differ from each other. Most of them have club-like rather than pointed tips.

At the edge of the smear, acanthocytes become distorted under the influence of surface tension and the spicules frequently disappear. At times, some of the spicules are forcibly detached from the body of the cell at the feather edge. The detached spicules round off and appear as very small granules or bodies of pink color surrounding the red cells. They are best seen in the Soret band (see p. 228 and Fig. 3).

Echinocytes[2—6]

We have described (p. 52) the precautions necessary to distinguish echinocytes already present in the circulation of a patient from echinocytes appearing during the preparation of a smear. The rare cases of in vivo echinocyte formation are:
1) Uremia: they disappear rapidly after dialysis.
2) Certain neonatal hepatitides: they must not be confused with acanthocytes.
3) After injection of heparin, due to fatty acids produced by lipoprotein lipase, when the normally present clearing factor is decreased.
4) Certain congenital anemias, due to a lowering of intraerythrocytic potassium (in a small percentage).
5) Pyruvate-kinase deficiency, attributed to ATP depletion (in a very small percentage).

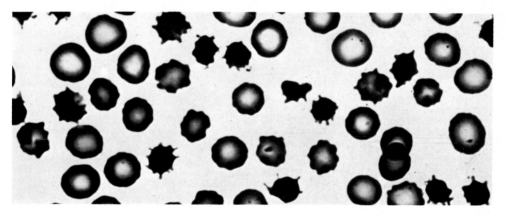

Fig. 1 — *Good portion of the smear*. The discocytes have normal central pallor and acanthocytes have clearly visible spicules

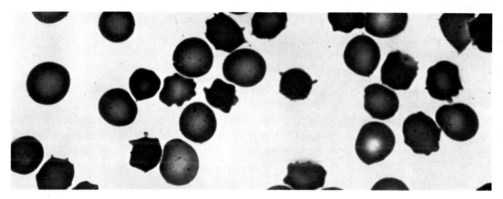

Fig. 2 — *End of smear*. The discocytes no longer have the normal central pallor and acanthocytes appear to have lost some of their spicules

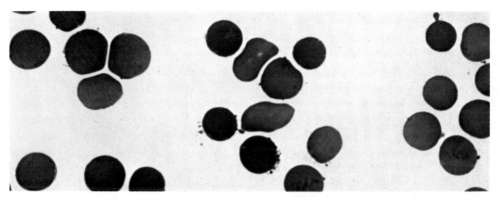

Fig. 3 — *Extreme feather edge*. The discocytes are flattened and distorted. The acanthocytes have lost their spicules and are rounded off. Some of the spicules have become detached and surround the individual red cells

7 — Acanthocytes (Continued)

Phase and SEM

The appearance is characteristically that of spheroidal* cells with spicules of differing length irregularly spaced over the surface. The number of spicules varies generally from 3 to 12.

Mechanism of Formation

Nothing is known about the subject. Normoblasts and reticulocytes appear normal in these patients. The incubation (up to 12 h) of normal erythrocytes in the plasma of patients with abetalipoproteinemia does not result in the formation of acanthocytes. In acquired acanthocytosis and cirrhosis, acanthocytes appear after 5 days in the blood of the patients following exchange transfusion with normal blood.

Physical Characteristics[11]

The acanthocytes do not appear to differ from normal red cells by their volume, surface, or the type, content and concentration of hemoglobin, osmotic fragility, or life span. The acanthocytes in liver disease differ from those in congenital abetalipoproteinemia; they have a different behavior in some tests (i.e. azide hemolysis).

Echino-acanthocytes and Stomato-acanthocytes[8, 9]

Echinocytogenic agents produce secondary spicules on the primary spicules of the cells. The acanthocytes become echino-acanthocytes and eventually sphero-echino-acanthocytes. Stomatocytogenic agents produce a concavity in acanthocytes, resulting in stomato-acanthocytes. As the concavity deepens, the spicules disappear and the red cell eventually becomes a sphero-stomato-acanthocyte. The echino- and stomato-acanthocytes can be reversed to acanthocytes by washing in fresh plasma.

Occurrence[7 – 11]

The presence of a large number of acanthocytes is a valuable diagnostic sign. In congenital acanthocytosis of abetalipoproteinemia, 50 to 100% of the circulating red blood cells have a typical appearance. In disorders of lipid metabolism, alcoholic cirrhosis and in rare cases of neonatal or other acquired hepatitis, one can find 10 to 50% and, rarely, higher percentages of acanthocytes.

Following splenectomy a small number of acanthocytes, usually between 1 and 5%, appear in the peripheral blood. The mechanism of their formation is, once again, unknown.

---

* A special form of acanthocyte is called *astrocyte*. It has been reported in cases of hemangioma of the liver. They are flat rather than spheroidal cells.

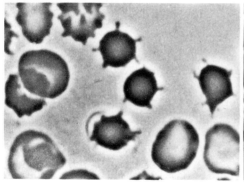

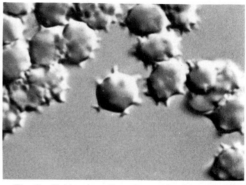

Fig. 1 — *Living acanthocytes* (phase contrast microscope)

Fig. 2 — *Echino-acanthocytes and echinocytes* (interference microscope). Note the two-pronged spicules, due to secondary spicule formation

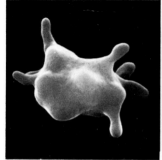
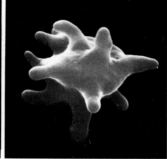
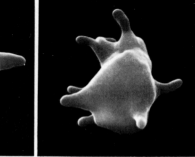

Fig. 3 — *Acanthocytes* (SEM)

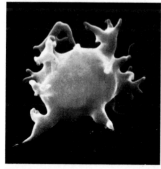
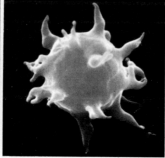
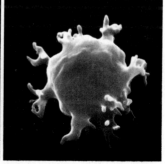

Fig. 4 — *Echino-acanthocytes* (SEM)

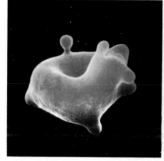

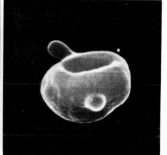

Fig. 5 — *Stomato-acanthocytes* (SEM)

8 – Codocytes and Target Cells

Definition[1]

These are bell-shaped red cells, which differ from cup-shaped stomatocytes by the thinness of their walls. Codocytes may be classified as codocytes I and II depending on the depth of their concavity.

Smears

When codocytes I are spread on a slide, they assume the appearance of target cells*, also called mexican-hat cells. Codocytes II assume the appearance of a Greek helmet (see p. 70). Echinocyte formation may be induced artifactually during preparation of the smears, probably due to the glass effect (see p. 52). One may thus find echino-codocytes (see Fig. 2).

It is, as always, essential to examine well-spread portions of the smear in which individual cells are seen sufficiently separated from each other to reveal their target shape. At the feather end of the smear, where all cells are flattened, target cells are no longer found (see Fig. 3). Nor can they be seen in thick portions where cells overlap and their individual shape is obscured.

Mechanism of Formation

Codocytes appear when the red cell surface is increased disproportionately to its volume. The excess membrane may be due to a diminution of the hemoglobin or an increase in the surface membrane or both.

A disproportion in the ratio of sodium to potassium, which leads to a reduction in cell water and hence in cell volume, may have similar results.

Physical Characteristics

The codocytes, in view of their increased surface-to-volume ratio, are more resistant to hypertonic solutions than normal red cells. Hemolysis may not occur until concentration of saline reaches values of 1.5 to 2.0%. They may be considered cells with an envelope too large for their hemoglobin content.

Hence, the hemoglobin concentration is lower than normal. The absolute hemoglobin content, the type of hemoglobin and the life span vary with different pathologic conditions (thalassemia, retention icterus, etc.).

Echino- and Stomato-codocytes

Superimposition of echinocytic and stomatocytic features on codocytes will result in a variety of bizarre shapes which may be difficult to recognize for what they are.

Occurrence[2]

Microcodocytes occur in thalassemia, hemoglobinopathies (C, S) and iron-deficiency anemia. Co-docytes with a normal volume, but increased surface membrane, are frequently seen in obstructive jaundice where they represent 75% of all red cells. They have been reported in the rare congenital deficiency of lecithin-cholesterol acyl transferase (LCAT), an observation which needs confirmation. Their increased surface is believed to be the result of relative increase in unesterified cholesterol and hence an excess of membrane.

The existence of diverse forms of codocytes, their transformation by artifacts, and the existence of multiple populations in some pathologic conditions make a diagnosis on the identification of codocytes sometimes difficult and their differentiation of only limited clinical diagnostic utility.

---

* A target cell appearance may be seen with phase microscopy in living red cells under exceptional circumstances. In particular when a knizocyte transforms into a discocyte it passes through a target cell stage. The same appearance may be produced by an echinocyte I which remains disc-shaped and has a single spicule in the center of its concavity.

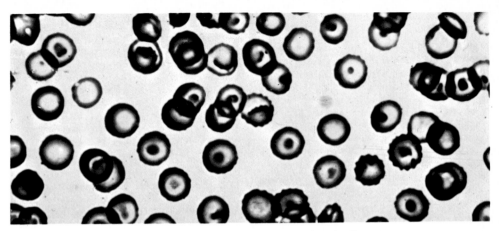

Fig. 1 – *Good portion of smear*. Almost all codocytes appear as target cells

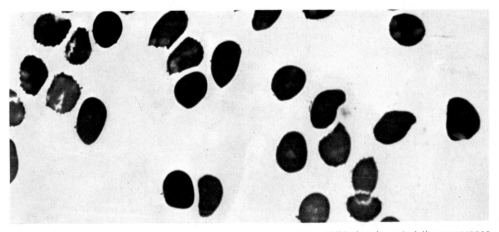

Fig. 2 – *Thin portion of the smear*. Several codocytes are transformed into codo-echinocytes, due to the pH of the glass

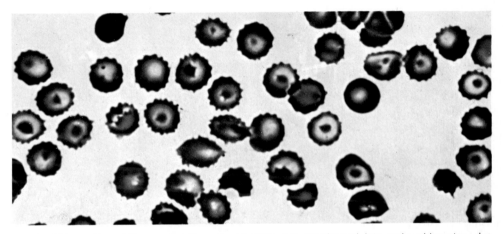

Fig. 3 – *Extreme feather edge*. The flattened codocytes no longer exhibit the characteristic appearance of target cells

9 – Codocytes and Target Cells (Continued)

Phase Contrast and Scanning Electron Microscopy

The cells can readily be seen to be thin walled and uniconcave. The concavity may be as shallow as in a soup plate or as deep as in a beaver hat. While these shapes are stable, they are deformable. These deformations must not be confused with the erroneously named parachute shapes seen in the microcirculation in vivo and demonstrable by high-speed microcinematography. Such red cells resume their normal biconcave shape as they pass from capillaries into larger venules. Hemolysis does not alter the shape of codocytes and the ghosts appear bell-shaped.

Smears Examined with SEM

Examination with the SEM readily shows the artifactual deformation of the codocytes. In the different portions of the smears the artifacts differ:
In the middle (Figs. 1 and 2) the cells are spread on their side (A) or with their concavity inverted and projecting in Mexican hat shape above the center (B).
At the feather edge (Fig. 3) they are completely flattened and they appear larger.

Note that the codocytes of thalassemia frequently contain vacuoles, autophagosomes or inclusions. At the end of the smear, the autophagosomes may rupture and the inclusions may be extruded resulting in further artifacts of apparent holes in the erythrocyte membrane (C).

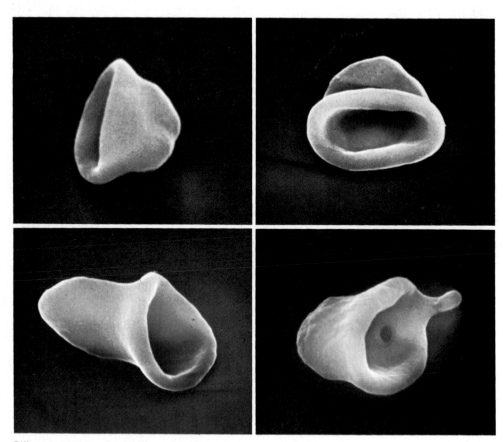

Different appearances of codocytes II. When spread in a smear, the codocytes assume the appearance of a Greek helmet (p. 71, Fig. 1A) or a target (Fig. 2B). Note a hole in the inner surface of the cell on the lower right: it is probably an artifact (cf. p. 71, Fig. 3C)

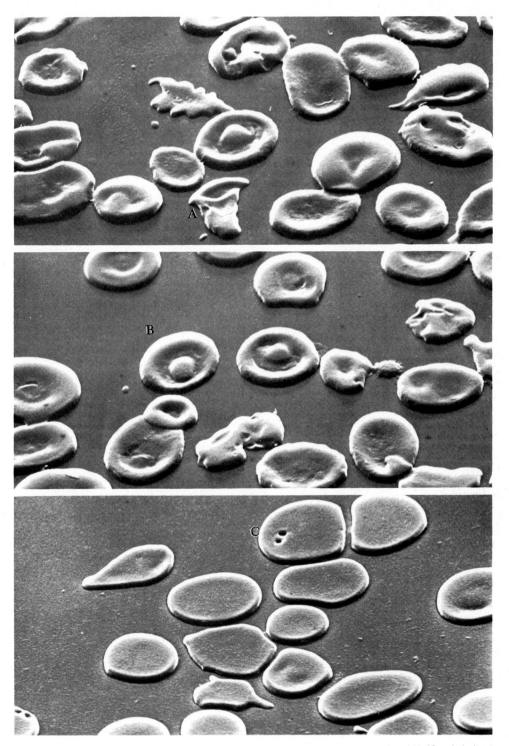

Codocytes in smear examined with the SEM. A (*at top*) a codocyte spread on its side (Greek helmet cell); B (*middle panel*) target cell; C (*bottom panel*) thin portion of the smear, the surface tension has flattened the codocytes. One of them contained vacuoles which have burst through its surface

71

10 – Diagnosis of Uniconcave Red Cells

It is possible to distinguish different types of codocytes and stomatocytes* in smears. This is only
possible, however, in a good portion of the smear
as artifacts complicate the picture in the thick or
thin fringes of the smear (see figure, p. 71).

All shapes may be of normal, decreased (micro)
or increased (macro) volume. All may undergo
echinocyte or stomatocyte transformation and all
combinations with other pathologic shapes are
possible, e.g. drepanocytes or elliptocytes (see
pp. 80, 82).

Name	True shape	Appearance in smears	Comments
Codo-cyte I	Soup plate	Target cell	The wall of the codocyte is always thin
Codo-cyte II	Beaver hat	Greek helmet cell**	
Stomato-cyte I	Thick-walled cup	Elongated central pallor (mouth)	The wall of the stoma-tocyte is always thick
Sphero-stomato-cyte	Sphere with central depres-sion	Reduced central pallor	

---

* The reinterpretation of volume and shape of abnormal
cells in smears may appear as unwieldy refinements appropriate for research and useless for the clinic. I do not
share this view. At any rate, one cannot form an authoritative opinion of the significance of the various shapes
except after a study of their origin and behavior. Such
studies presuppose separation of the various shapes
which is greatly facilitated by an appropriate nomenclature (see pp. 62, 64). Only on the basis of such knowledge
will we eventually be able to decide which of these forms,
if any, deserve further investigation or may be of interest
for the clinic. The present classification of the various
shapes serves this purpose and is not an assertion of
their importance.
** Not to be confused with fragmented red cells (schizocytes, see p. 92) which resemble the side view of a tropi-
cal helmet and to which the term of helmet cell is commonly applied.

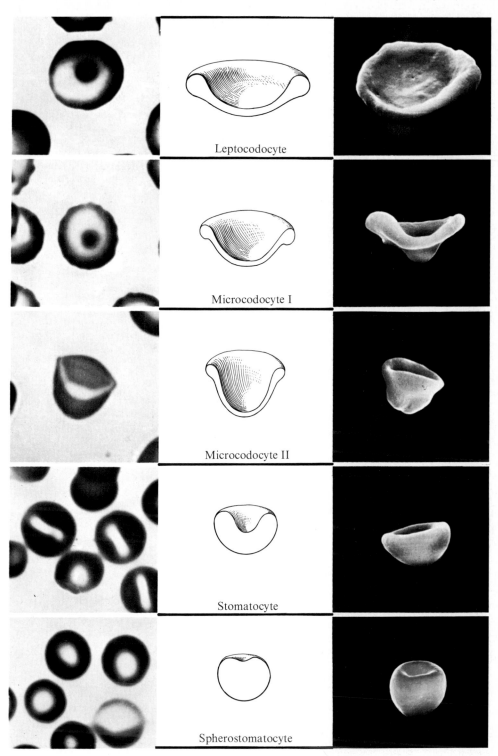

Leptocodocyte

Microcodocyte I

Microcodocyte II

Stomatocyte

Spherostomatocyte

Different forms of uniconcave erythrocytes. On the left: the appearance of cells in smears (Soret band). *On the right:* appearance of corresponding cells in the scanning electron microscope

11 – Dacryocytes

Definition

The name dacryocyte is a Greek translation of the commonly used English name "teardrop cell." They are, in fact, discocytes drawn out into a spicule.

Smears

Dacryocytes are pear-shaped, with a more or less extended tail which sometimes ends in a swelling. They can be of normal, reduced or increased size.

Living Cell and SEM

The cells can be seen to have retained their biconcave shape. Only the extended portion has lost its central depression. Some have been described as racquet shapes or chinese spoon shapes, but these are merely different names for the same basic change.

The preexisting teardrop shapes must not be confused with the artifactual production of pear shapes by adhesion of the red cell at one point to the glass or to another red cell. When such cells are detached from each other or the glass, they return to their discoid shape, in contrast to the dacryocytes, which retain their teardrop shape even after hemolysis.

Mechanism of Formation[1]

The normal discocyte is very deformable and can easily traverse the fenestrations which separate the splenic cords from the sinuses (see pp. 48 and 102). However, when red cells contain rigid inclusions, a portion of the red cells cannot be deformed normally and may be unable to pass through that filter. The portions of the red cells containing the inclusions may be separated from the rest of the red cell and left behind. The phenomenon has been described as the pitting function of the spleen. In these circumstances, a large number of Heinz bodies may accumulate in the splenic cords. In thalassemia, in which precipitation of alpha chains leads to the formation of inclusions, this phenomenon can be observed regularly. It remains to be explained why the shape of a teardrop, which is assumed by the red cell as it is pulled away from the part of its body containing the Heinz body or inclusion, becomes permanent. The most likely explanation is that stretching of the red cell membrane beyond a certain point will result in loss of deformability and loss of ability to return to its original shape. Experimentally, however, red cells which attach to the fibrin strands and which pull loose do not readily form teardrops. Possibly the time during which cells were kept in a teardrop shape in the experiments performed to date was too short to cause permanent deformation.

Echino- and Stomato-dacryocytes

The superimposition of spicules or transformation to a uniconcave shape occurs in dacryocytes as in other pathologic shapes. Echinocytic spicules may form on the pointed portion of the dacryocyte as well as on the rest of the cell.

Occurrence

Dacryocytes are seen primarily in myelofibrosis or myeloid metaplasia, cancers metastasizing to the bone marrow, tuberculosis, after Heinz body formation due to the ingestion of certain drugs, and in thalassemia.

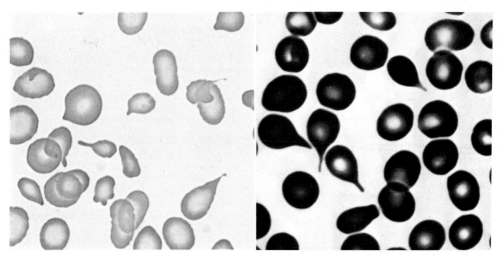

Figs. 1 and 2 — *Dacryocytes* (tear drop cells) *in smears. On the left,* stained with Giemsa; *on the right,* as viewed in the Soret band

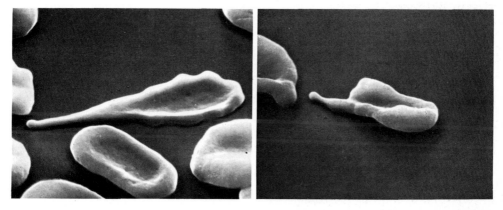

Figs. 3 and 4 — *Dacryocytes,* seen in the scanning electron microscope

Figs. 5 and 6 — *Echino-dacryocytes* (blood kept at room temperature for a few hours)

Drepanocytes

Definition

Drepanocytes (or sickle cells) are discocytes deformed by rod-like polymers of abnormal hemoglobin (HbS).

Smears

In homozygotes one can see sickle-shaped erythrocytes in small numbers. In heterozygotes smears appear normal, except if the smear is made under unusual conditions of lack of oxygen (see p. 77).

The drepanocytes seen in smears only rarely have the typical appearance of "sickles." They are for the most part boat-shaped (see Fig. 1). Occasionally, the ends are slightly curved. They probably correspond to "irreversible" drepanocytes. The classical appearance of sickles is only seen between slide and coverslip or in smears in cells submitted to low oxygen tension.

Electron Microscopy[1]

Hemoglobin S can be seen to have polymerized into filaments approximately 18 nm in diameter, which can attain a length of 5 to 15 µm. These filaments are usually arranged in parallel or slightly twisted bundles, oriented in the long axis of the cells. Studies with very high resolution show the filaments to consist of helicoidal arrangements of hemoglobin molecules.

This appearance is quite characteristic and differs from crystallization of hemoglobin in erythrocytes (see below).

Crystallization and Polymerization of Intra-erythrocytic Hemoglobin

Normal hemoglobin can crystallize or polymerize under special conditions, such as sojourn of blood between slide and coverslip for 24 h in man or for a few hours in the rat. This rearrangement of hemoglobin molecules depends on a variety of factors and results either in intracellular crystals or a fusiform appearance.

Intracellular Crystals[2, 3]

All hemoglobins can crystallize in vitro. The crystals have a characteristic form, cubical in normal man, star-shaped in the rat.

Certain abnormal hemoglobins can form crystals in vivo. In HbC disease, one may note tetragonal crystals in smears made immediately from fresh blood. They occur in about 2% of cells, which can go up to 10% after splenectomy. Inclusions in many other hemoglobinopathies do not correspond to crystals of hemoglobin, but to a special form of Heinz bodies.

It should be noted that in hemoglobinopathies (particularly one involving hemoglobin C) erythrocytes may become spherical with many folds and invaginations, resembling reticulocytes; however, the folds are permanent and do not change shape.

Intracellular Polymerization[2-4]

Only hemoglobin S* polymerizes into rods and results in true drepanocytes. However, all hemoglobins can, under certain experimental circumstances, form tubules which organize themselves into bundles and deform the cell by elongating it. Occasionally, they can give the discocyte a fusiform appearance (pseudo-sickles). They must not be confused with true sickles or with artifactual, mechanical elongation of erythrocytes by heat, freezing, or tensile forces during preparation of a smear.

---

* There are, however, rare exceptions to this rule. Recently an unstable hemoglobin has been reported to polymerize into rods quite comparable to hemoglobin S, forming true sickles.[5]

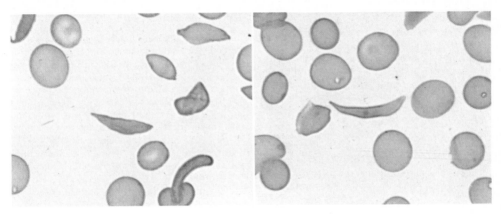

Figs. 1 and 2 — *Irreversible drepanocytes* (sickle cells). Blood taken from a cubital vein

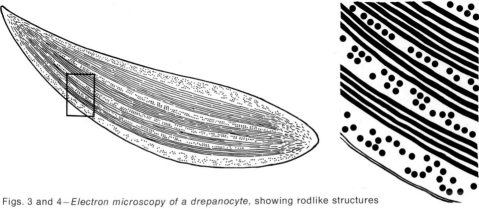

Figs. 3 and 4 — *Electron microscopy of a drepanocyte*, showing rodlike structures of polymerized hemoglobin S

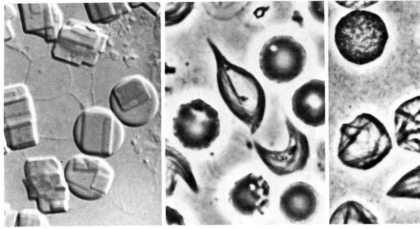

Fig. 5 — *Crystals of hemoglobin A* seen between slide and coverslip in normal blood after 24 h (interference microscopy)

Figs. 6 and 7 — *Pseudodrepanocytes which* appear after 24 h in normal blood kept between slide and coverslip, due to polymerization of hemoglobin A (phase contrast)

Phase Microscopy and SEM

In the blood of carriers of the sickle cell trait 100% of red cells transform into sickle cells when the oxygen tension is lowered. To perform the sickle cell test, one may 1) put a tourniquet for 10 min around the finger before taking blood; 2) seal a drop of blood between slide and coverslip for 24 h before examination, or 3) add Na-dithionate to a drop of blood to reduce oxygen tension before sealing it and examining it. Only the last one is routinely used.

The cells take on the appearance of sickles, sometimes with spicules at either end. Sometimes holly-leaf forms result (Figs. 1 and 5). These forms correspond to echinocytes I, which have been deoxygenated: the polymerization of hemoglobin into rigid rods follows the outline of the cell as may be seen when sickle cells transform into echinocytes and stomatocytes (see p. 80). Between slide and coverslip echinocytes and holly-leaf forms develop readily (see p. 52).

The phenomenon is reversible: as soon as the oxygen tension increases, the red cells resume their normal biconcave shape.

Mechanism of Sickle Cell Formation

In the absence of Na-dithionate it may take a considerable time before one may see a particular red cell begin to sickle. However, once the sickling process starts, it is complete within 1 to 2 s. First, one sees slight shape changes under the influence of rearrangement of the hemoglobin within the cell. The cell loses the vibratory movement (flicker) characteristic of the normal discocyte; then the spicules appear and deform the outline of the discocyte and the entire cell becomes rigid.

When hemoglobin rigidifies in the spicules, they stretch the cell membrane like fingers projecting on opening a closed fist. The thinned portion of the cell membrane can then no longer be seen with the optical microscope and some authors believed the stroma to be frayed. The scanning microscope demonstrates clearly that the cell membrane is merely stretched.

The observations suggest that initially hemoglobin forms monofilaments which coil into helices and form rods. The rods form bundles along the cell wall. As these bundles grow in length, they deform the cell. Because the initial shape of the cell determines the direction of the rigid rods, different shapes result from deoxygenation: discocytes become sickles, echinocytes I become holly leaf shapes.

In homozygote infants, one finds only 1 to 30% of the cells to sickle, and it is only at the age of 4 months that the percentage of sickle cells reaches the normal adult figure of 90%. This has been ascribed to the protective effect of fetal hemoglobin in the neonate.

Irreversible Drepanocytes[1, 2]

In sickle cell disease, a small number of cells, up to 10% in some patients, fail to resume their discocyte form on reoxygenation. They either remain typical sickle cells or, more commonly, assume a fusiform shape. In the phase microscope, they may be seen to be semi-rigid and to exhibit the flicker phenomenon only in a portion of the cell. These are the cells which are recognizable in smears made without special precautions.

Cross transfusions have shown that these cells are rapidly eliminated from the circulation. The mechanism of their formation is probably complex and due to the interaction of a number of factors including the percentage of hemoglobin F, metabolic changes such as a loss of ATP, alteration in the permeability of the membrane due to accumulation of calcium, and the number of preceding reversible transformations of discocytes to drepanocytes, during which the cell may loose some of its membrane.

Physical Characteristics

The average red cell volume of sickle cell is normal, although it should be noted that only average volumes have been reported. During sickling, the surface area and the mechanical fragility are greatly increased.

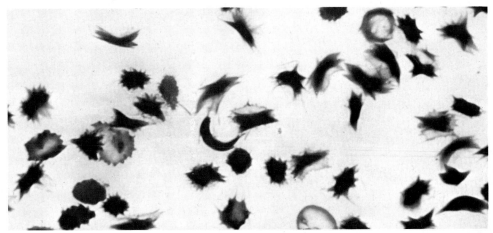

Fig. 1 – *Drepanocytes obtained by deoxygenation in vitro* (smear, Soret band)

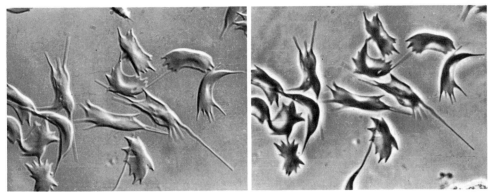

Figs. 2 and 3 – *Drepanocytes obtained in vitro.* Note presence of sickle and holly leaf forms. Living cells viewed with the interference microscope (*left*) and with phase contrast (*right*)

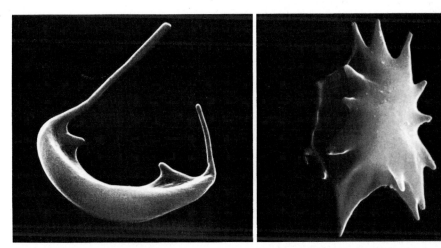

Fig. 4 – *A drepanocyte obtained by deoxygenation of a discocyte*

Fig. 5 – *A drepanocyte obtained by deoxygenation of an echinocyte*

79

In the presence of oxygen, the behavior of HbS containing red cells does not differ from that of normal red cells; they transform into echinocytes and stomatocytes under the same conditions as discocytes.

When a red cell has transformed into an echinocyte and the oxygen pressure is reduced, whether it contains hemoglobin SA or SS, the hemoglobin S polymerizes in the spicules and elongates and deforms them (drepano-echinocyte)[1].

The appearance differs with the type of echinocyte. The spicules end in truncated cones. With the electron microscope, one can demonstrate that the ends of their spicules contain 5 to 50 rod-like hemoglobin bundles.

When oxygen is removed from HbS containing stomatocytes, the hemoglobin filaments again follow the contour of the cells, become rigid and deform the cell. The general cup shape of the cell is maintained[2].

Consequences of Sickling in the Circulation[3]

When sickle cells form in the circulation, their rigidity leads to their entanglement and to blockage of capillaries, which may result in thrombosis. The fine long spicules are fragile, break off the main body of the cell and hemolysis and severe anemia result.

Occurrence[4]

The formation of sickle cells within the circulation is pathognomonic of sickle cell anemia, the homozygous form of the trait. In the heterozygote, or carrier of the trait, formation of the sickle cell in the circulation is rare and restricted to severe hypoxic states.

While 100% of the cells sickle in the absence of oxygen in SS or SA, a variable number sickle in SC disease or S thalassemia and all other heterozygotes.

Drepanocytes which do not contain HbS can be found in very rare circumstances.

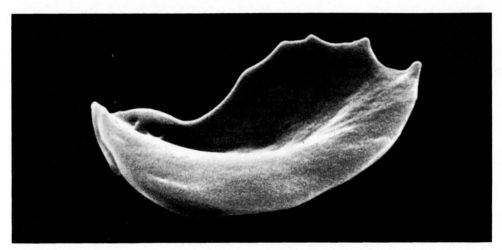

Fig. 1 – *Drepanocyte obtained by deoxygenation of a discocyte*

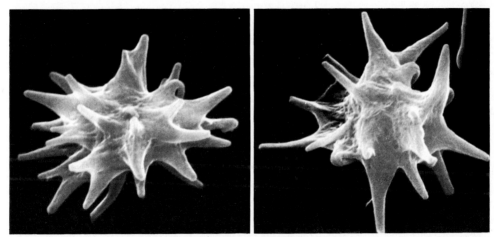

Figs. 2 and 3 – *Drepanocytes obtained by deoxygenation of echinocytes III*

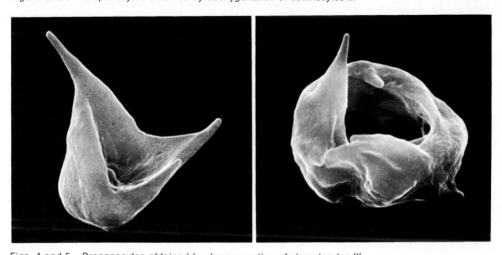

Figs. 4 and 5 – *Drepanocytes obtained by deoxygenation of stomatocytes III*

15 — Elliptocytes

Definition

Elliptocytes are oval cells which are classified into elliptocytes I, II, III and IV, depending on the ratio of the short to the long axis of the ellipse. Elliptocytes I deviate only slightly from the spherical form, while elliptocytes IV are almost rod-like.

Smears

The two ends of the ellipse may stain more darkly due to a central biconcavity which is comparatively small and round in relation to the elongated cell (see Fig. 3). Elliptocytes IV may be 15 µm in length and only 3 µm in width.

Phase Microscopy and SEM

Normal elliptocytes are biconcave. Human elliptocytes resemble the oval cells of camels only superficially. Human elliptocytes have the same osmotic fragility as normal cells. In contrast, camel red cells have an increased resistance to hemolysis, so that camels can loose one-third of their weight in the desert and subsequently drink up to 130 liters of water in 10 min without hemolysis. In man, the absorption of 6 liters suffices to provoke hemolysis.

Mechanism of Formation

It is unknown. The normoblasts and reticulocytes (Fig. 2) have a normal shape in elliptocytosis.

Echino- and Stomato-Elliptocytes

Echinocytogenic and stomatocytogenic compounds lead to the expected many possible forms such as echino I—elliptocytes I to IV, sphero-echino I—elliptocytes I to IV, etc.

Physical Characteristics

Except in case of hereditary elliptocytosis (see below) or superimposed erythrocytic anomaly, elliptocytes seem to behave like normal red cells. However, the studies should be repeated after separation of homogeneous populations of different types of elliptocytes.

Occurrence

Less than 1.0% of red cells are oval in normal man. In anemias of almost any type, the number of oval red cells can increase up to 10%. They are particularly common in thalassemia, iron deficiency, megaloblastic anemia, and anemia associated with leukemia. In these instances their elliptical shape is simply added to whatever other abnormalities are characteristic of that anemia. *Hereditary elliptocytosis* is characterized by the presence of a high percentage of elliptocytes, usually of type III. One can distinguish three clinical types: 1) elliptocytosis without any abnormalities, 2) elliptocytosis with compensated hemolytic anemia, and 3) hemolytic anemia. In that case, 90% of the erythrocytes are elliptocytes, some rod-shaped. Their osmotic fragility is increased and so is their autohemolysis at 48 h, a defect which is corrected by the addition of glucose and ATP. Commonly some spherocytes, stomatospherocytes, and ellipto-stomatocytes are present.

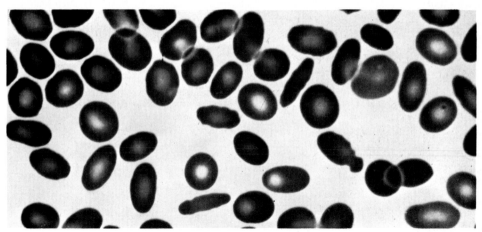

Fig. 1 – *Congenital elliptocytosis* (smear viewed in the Soret band)

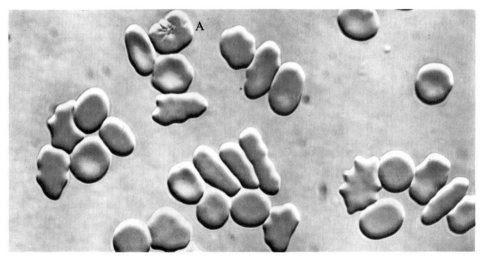

Fig. 2 – *Congenital elliptocytosis* (living cells in the interference microscope). A reticulocyte I is seen at A. Some cells begin to assume echinocytic shapes

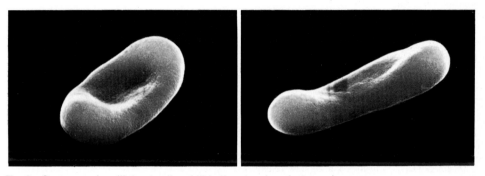

Fig. 3 – *Same sample:* elliptocytes II and III in the scanning electron microscope

16 – Keratocytes

Definition

Keratocytes are red cells with one or several notches. The border of the notches forms projections whick look like horns, hence the name of the cell. Keratocytes have a hemoglobin content which is normal or only slightly lower than normal. This will be apparent from their mechanism of formation.

Smears

Keratocytes as a rule still have an area of central pallor. The spicules or horns are more or less spread out. Occasionally only one spicule remains and the cell may then be mistaken for a dacryocyte. Occasionally, when four spicules result from two notches the cell may be mistaken for an acanthocyte.

Mechanism of Formation[1-4]

The fragmentation of red cells used to be attributed to collision with damaged portions of the vessel walls or with foreign material of a prosthesis, the collisions becoming more damaging if the circulation is turbulent, particularly at the edges of a prosthesis torn loose from its attachments. More recent evidence indicates that fibrin deposition is a necessary intervening stage and keratocytes (like schizocytes, see p. 92) result most commonly from the impalement of red cells on fibrin strands. When a red cell impinges upon a strand of fibrin, two saddlebags form on either side of the fibrin strand which attempts to bisect the red cell (see figure, p. 92). The opposing interior walls of that portion of the red cell adhere to each other along the fibrin strand. When the pull of the bloodstream frees the red cell to reenter the circulation, the fused portion of interior surfaces is colorless and may look like a pseudo-vacuole (Fig. 3). This stage may be referred to as a pre-keratocyte. Within minutes the pseudo-vacuole ruptures, leaving a notch with its bordering spicules or horns, the keratocyte.

Alternatively when the red cells are forced against the fibrin strand, one of the saddlebags ruptures and the portion of the red cell that is hemoglobin free due to the adhesion of the walls in this area appears as a pseudo-vacuole.

Pseudo-vacuoles can similarly be produced by a microbeam of a laser in a portion of the red cell. It thus appears that whenever parts of the red cell membrane come in apposition, they may fuse whatever their original mechanism of injury and create the appearance of a pseudo-vacuole.

Living Cells and SEM

Keratocytes have the appearance of normal discocytes except for a rupture of the annulus at one point of the cell (Figs. 6 to 9). The formation of the two horns is unexpected as it is known that the biconcavity and the peripheral annulus of the red cell conform to no preformed structure and can be formed by any part of the membrane. Apparently, the fusion of the two membranes has caused the rigidity of this zone which resulted in this permanent appearance. When the annulus ruptures at two points, four horns result and the cells may resemble acanthocytes or other spiculed cells with which they must not be confused.

Physical Characteristics

It appears that keratocytes and schizocytes are more fragile than normal cells as they do not survive in the circulation for more than a few hours once they have formed. They usually evolve into sphero-keratocytes or sphero-schizocytes and hemolyze rapidly.

Occurrence

Keratocytes may be found under the same conditions as other schizocytes, i.e. in all syndromes involving intravascular coagulation, microangiopathic hemolytic anemia, some cases of glomerulo-nephritis, during rejection of renal transplants, cavernous hemangiomas, hemolytic anemias due to anomalies of cardiac valves, whether infiltrated by calcium or the result of malfunction of valvular prostheses.

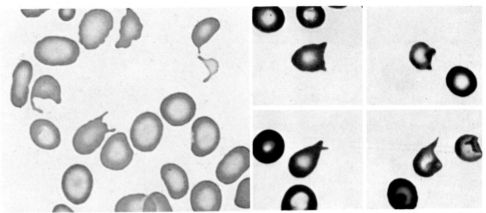

Figs. 1 and 2 – *Different types of keratocytes in smears. Left,* with Giemsa. *Right,* in the Soret band

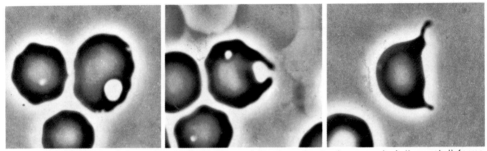

Figs. 3–5 – *Stages in the formation of a keratocyte* (phase contrast microscope). A "vacuole" forms in the interior of the erythrocyte, it opens and its walls separate

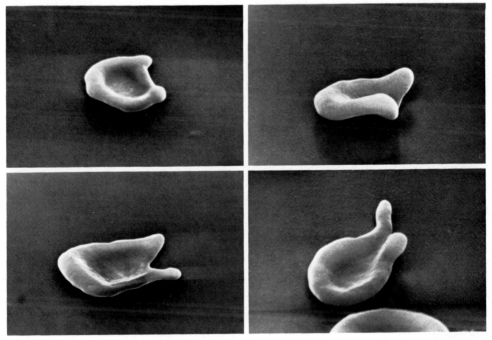

Figs. 6–9 – *Keratocytes in the SEM* (cf. Fig. 2)

17 — Knizocytes

Definition[1]

Knizocytes are erythrocytes with more than two concavities; they are generally triconcave cells.

Phase Microscopy and SEM

It is only in the living state and in the SEM that knizocytes can be properly appreciated (Figs. 2 and 3). They are permanent forms which must not be confused with similar shapes produced by shear stresses (see below).

Smears

Knizocytes are difficult to recognize in smears stained with Giemsa, because they are usually very pale. They can be best seen in the Soret band. Even then, they can only be seen in the thick portion of the smear. The diameter of these cells is small, with a band of hemoglobin across its center. This appearance is due to the red cell's having come to rest on its flat surface. If it comes to rest on one of the two more pronounced concavities, two areas of pallor are seen on the same side of the band of hemoglobin (Fig. 1, extreme right).

Mechanism of Formation[2]

The reason why knizocytes form in the circulation or in smears is still unknown. Transient knizocytes can be produced experimentally by first allowing a red cell to attach itself to a glass slide at multiple points and then attempting to move the red cell to-and-fro by hydraulic force in a specially constructed thin chamber. If the red cell remains attached, only the hemoglobin is moved backward and forward. This gives the appearance of a wave of hemoglobin, which results in the shape of a knizocyte when the crest of the wave reaches the midline of the red cell. One can also observe knizocytes in the ektacytometer (see p. 104) at low shear forces before elongation of the red cell takes place.

These observations are of interest, because they have to be taken into account when one attempts to produce a model of the structure of the red cell and its discoid biconcave shape.

Occurrence

One finds knizocytes in all instances in which spherocytes are formed (see p. 96).

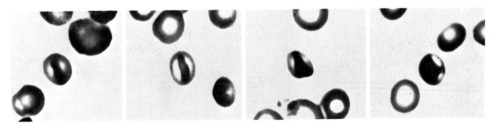

Fig. 1 — *Different aspects of knizocytes in smears* (Soret band)

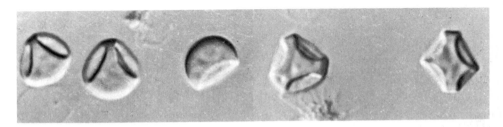

Fig. 2 — *Knizocytes* in the living state (interference microscope)

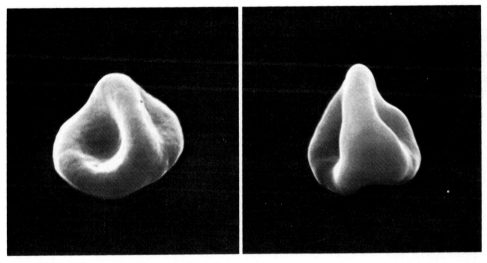

Fig. 3 — *Knizocytes* in the scanning electron microscope

18 — Macrocytes and Megalocytes

Macrocytes

Definition

Macrocytes are erythrocytes of increased volume (greater than 100 µm³). One must not confuse macrocytes with reticulocytes, in which their increased size is a result of their immaturity (see p. 46).

Smears

The diameter of macrocytes can be smaller, larger, or equal to that of normal red cells. This is so, because the diameter of the red cell in smears depends on the degree to which it is spread rather than on its true volume. Thus cells described as "thin macrocytes" are in fact leptocytes (see p. 90). In general, however, they are larger than normal erythrocytes and stain darker. Macrospherocytes have an almost normal diameter, but a very small area of central pallor (Fig. 4).

Physical Characteristics

The hemoglobin content is always increased and can attain twice normal figures. Their other parameters are not well known.

Mechanism of Formation[1-4]

Most macrocytes are due to stress erythropoiesis. It can be shown experimentally that hemolysis and acute blood loss result in an increased production of erythropoietin, which in turn acts on developing red cells at two levels: 1) red cells are released at an earlier stage of their maturation than normal, and 2) hemoglobin synthesis in normoblasts is accelerated, so that macronormoblasts and macroreticulocytes develop, which eventually lead to the formation of mature macrocytes.
Macrocytes can also result from alteration in the permeability of the red cell membrane, so that sodium accumulation leads to "hydrocytosis."

Occurrence[5,6]

Macrocytes are typically seen in hemolytic anemias, during recovery from acute hemorrhages, and in hyperthyroidism (see p. 46).
One of the most typical macrocytosis in man is seen in erythroblastosis fetalis. Careful examination of the blood smears is necessary to avoid errors. One may observe red cells with normal diameter or even less than normal diameter, as well as large polychromatophilic red cells. These represent macro-reticulocytes. The others are macro-spherocytes, macrocytes which have become spherical under the influence of anti-Rh antibody, resulting in a normal or even smaller than normal diameter. They can be recognized for what they are by the decreased central pallor (Figs. 2 and 4). Macrostomatocytes may also be found.

Megalocytes

Definition

This is a large red cell, often oval in shape with an increased hemoglobin content. Since only their oval shape, if present, differentiates the megalocytes from macrocytes, one can in fact only speak of megalocytes if the bone marrow aspiration shows the presence of megaloblasts. Megaloblasts are characterized by asynchrony of maturation of nucleus and cytoplasm. Hemoglobinization proceeds in the cytoplasm, while the nucleus remains relatively immature. Thus, megaloblasts typically have a very advanced hemoglobinization similar to that of a normal reticulocyte when the nucleus is still immature as in a pronormoblast.

Smears

In pernicious anemia, oval red cells which measure up to 12 µm in their greater diameter are usually present. Their central pallor may be less distinct than normal, anisocytosis is always very marked, and the mean corpuscular volume (MCV) is always increased.

Mechanism of Formation

The common denominator may be inhibition of DNA synthesis and mitosis during hemoglobin synthesis. How this leads exactly to the appearance of megaloblasts is unknown.

Occurrence

Megalocytes are found in all megaloblastoses, whether due to B 12 and folic acid deficiency or associated with leukemia or refractory anemia.

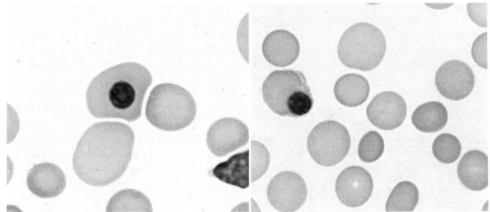

Fig. 1 — *Megalocytes* (pernicious anemia)

Fig. 2 — *Macrocytes* (hemolytic disease of the new-born—erythroblastosis fetalis)

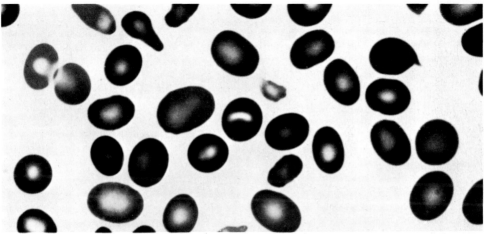

Fig. 3 — *Megalocytes* viewed in the Soret band

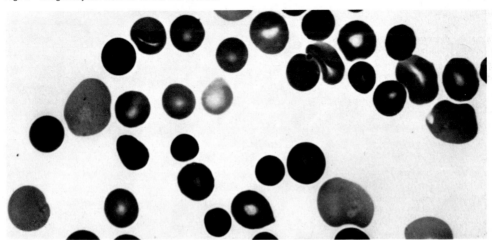

Fig. 4 — *Macrocytes* in the Soret band. The reticulocytes are greatly increased in size. Many macrocytes have formed macrostomatocytes, due to the action of anti-Rh antibodies. Note that they have markedly reduced central pallor

19 — Microcytes and Leptocytes

Microcytes

Definition

These are red cells with a reduced red cell volume (less than 80 μm³).

Smears

Microcytes in smears can have reduced, normal or increased diameters. The area of the central pallor is usually increased if hypochromia coexists. Occasionally, hemoglobin is only present in a thin rim at the periphery. However, the transition between central pallor and the peripheral annulus is not sharp as in torocytes (see pp. 48 and 224) which must not be mistaken for evidence of hypochromia.

Mechanism of Formation[1]

These red cells are usually due to iron deficiency, which prevents the red cell from having its normal content of hemoglobin. The microcytosis can be of metabolic origin. For instance, in pyruvate-kinase deficiency, the loss of ATP results in loss of potassium and cell water and formation of microcytes with increased viscosity (dessicocytes). This special sort of microcyte has a normal hemoglobin content.

Leptocytes

Definition

Erythrocytes that are thinner than normal; they can be macro- or micro-leptocyte or have a normal volume.

Smears

Leptocytes are paler than normal cells and their center is colorless. They must not be confused with torocytes (see p. 224). Frequently these cells, whether macro-, micro- or normoleptocytes, are cup-shaped, with little depth of the cup. When the cup increases in depth, they become codocytes I, which result in the formation of target cells in some parts of the smear (see p. 72).

Mechanism of Formation

Leptocytes can be due to lack of hemoglobin as in iron deficiency and thalassemia or reduced amount of water (dessicocytes). They can be due to increase of the red cell membrane as in retention icterus. Both mechanisms lead to the same end result of a disproportionately large surface compared to the volume of the cell, but the mean hemoglobin concentration is decreased in iron deficiency and increased in dessicocytes.

Occurrence

Microcytes (with hypochromia) are found in all sorts of iron deficiency anemias. Slight degrees of microcytosis (with or without hypochromia) may occur in inflammation and other disorders and can usually only be appreciated by measuring the MCV or, even better, cell distribution in modern electronic instruments. Leptocytes are common in biliary obstruction, certain cirrhoses, and some steatorrheas.

Microleptocytes occur in iron deficiency anemia, thalassemia, hemoglobinopathies and in hypersideremic (sideroblastic) anemias, usually as a minor population. The MCV in such sideroblastic anemias is often elevated even in the absence of reticulocytosis or megaloblastosis.

Macroleptocytes are rare. One finds them in thalassemia and in dyserythropoiesis.

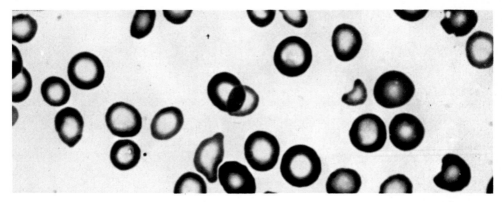

Fig. 1 — *Microleptocytes* (in iron deficiency anemia). The cells are smaller than normal and have a disproportionately reduced amount of hemoglobin. Consequently, the cells are usually thin and in smears have a larger than normal area of central pallor

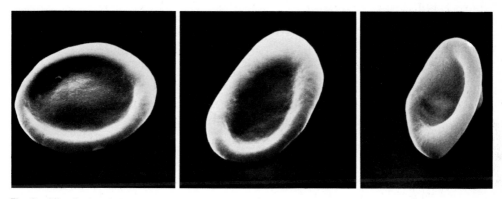

Fig. 2 — *Microleptocytes* as seen in the scanning electron microscope

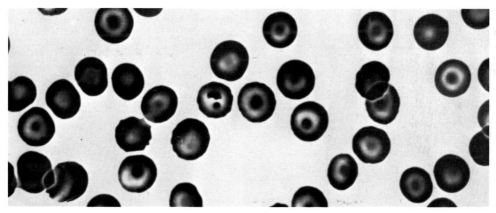

Fig. 3 — *Leptocytes* (from a patient with obstructive jaundice). The red cells have a normal volume, but increased surface. In smears, target cells are formed

20 — Schizocytes

Definition

Fragments of red cells due to a localized membrane damage.

Smears

Depending on the location of the damage which the red cells have suffered one can distinguish quarter, half or three-quarter discocytes*. More bizarre forms also exist, which may be triangular or comma-shaped.

Schizocytes very easily assume echinocytic forms during the preparation of a smear (Figs. 1 and 2).

At the feather edge of the smear, schizocytes may be round and give only the appearance of anisocytosis.

Mechanism of Formation[1–4]

It is identical to that of keratocytes (see p. 84). Erythrocytes are stretched when arrested in their progress through the circulation by fibrin strands on the altered wall of vessels or damaged protheses. When a red cell becomes attached to a fibrin strand, saddlebags may form on either side under the shear of the blood stream. Sometimes the cell is cut along the line of adhesion to the fibrin strand; frequently the cell is not actually cut, but by some irregular movement of the blood stream or formation of eddies, the red cell is freed from its attachment. The line of attachment to the fibrin strand is now marked by a fusion of the two opposing internal surfaces of the red cell membrane. It takes the appearance of an elongated "pseudo-vacuole" (see p. 84) and after a while (or after passage through the spleen) the red cell splits into two parts. Damage to the membrane along the line of fusion produces a localized decrease in deformability. The remaining flexible membrane will arrange itself so as to accomodate the stiffened patch in a minimum membrane bending energy configuration (which implies a dimple and an annulus).

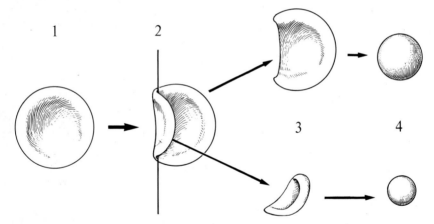

Production of schizocytes due to bisection of discocyte (1), by a fibrin thread (2). The schizocytes (3) become sphero-schizocytes (4), which hemolyze rapidly. Another mechanism of formation is described on p. 94

Survival of schizocytes in the circulation is short when damage is severe; they transform into spheroschizocytes and then hemolyze. If the cell has suffered only minimum damage, it may survive several days.

Spheroschizocytes (true microspherocytes, see pp. 94 to 96) can also result from two very different mechanisms: fragmentation of red cells due to burns, or incomplete phagocytosis (see p. 94).

---

* So-called helmet cells, not to be confused with "Greek helmet cells" (see p. 72).

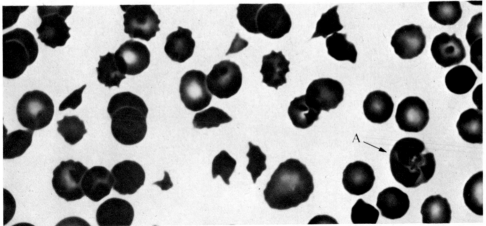

Fig. 1 – *Good portion of the smear*. Schizocytes in a patient with metastasis to the marrow. Many have transformed into schizoechinocytes. A reticulocyte is seen at A

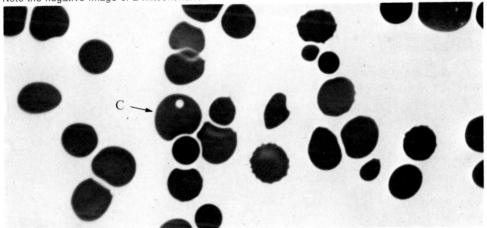

Fig. 2 – *Thin portion of the same smear*. A reticulocyte at B has become flattened and rounded off. Note the negative image of a mitochondrion

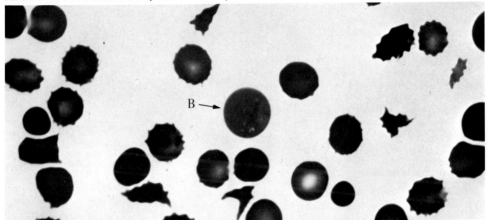

Fig. 3 – *Extreme feather edge*. All the schizocytes are rounded off. At *C*, a Howell-Jolly body in a reticulocyte. Since it contains no hemoglobin, it appears white in the Soret band

Phase Microscopy and SEM[4]

The red cell fragments are still deformable. Before hemolysis they become spherocytic, but the ghost resumes the shape of the original schizocyte. Between glass slide and coverslip, they rapidly become echinocytic, then spherocytic.

In the scanning electron microscope, one can readily distinguish the different types of echinocytes. The maintenance of the relative position of the annulus and the central concavity in the red cell fragment (Fig. 4) should not be taken as evidence for the existence of a cytoskeleton. It can be shown experimentally that the stretched part of the red cell fragment will have a fixed relationship with the concave and convex portions of the red cell (as in keratocytes, see p. 84).

Spheroschizocytes can be formed by a totally different process, namely phagocytosis. Not infrequently a phagocyte cuts a red cell in two in the process of phagocytosis[5]. The portion of the red cell which remains in the circulation rapidly becomes spherical (Figs. 1 and 2), and remains so after it becomes detached from the phagocyte.

Spheroschizocytes can also be formed by yet a different mechanism, as a result of the action of heat. The cells first become spherical; then, by a budding process, they are divided into small spherical hemoglobin-containing spherules, soft and labile. These fragments are deformed by the slightest current flowing between slide and coverslip.

Physical Characteristics

Of necessity the schizocytes are always microcytes, since they represent a fragment of a red cell. In principle the relation of surface to volume of the two fragments is normal at the moment the cell is cut and before the fragments sphere. Other parameters such as hemoglobin content and concentration depend on associated diseases and the size of the fragments. Their fragility is always increased, regardless of the mechanism (osmotic, mechanical, heat).

Echino- and Stomatoschizocytes

The forms that echinocytogenic or stomatocytogenic substances may produce are those expected (Fig. 6).

Occurrence

Schizocytes occur in disseminated intravascular coagulation, microangiopathic anemias, thrombotic thrombocytopenic purpura, valvular lesions, primarily calcifications or valve prostheses. In all of these instances, fibrin strands are present in the circulation and are responsible for the production of the schizocytes. They also have been seen in march hemoglobinuria, in which case it is uncertain whether the fragmentation is purely mechanical or fibrin strands are also formed in the vessels. Schizocytes can be observed in uremic syndromes. Their mechanism of formation is unknown. In severe burns only spheroschizocytes are observed, as a result of heat.

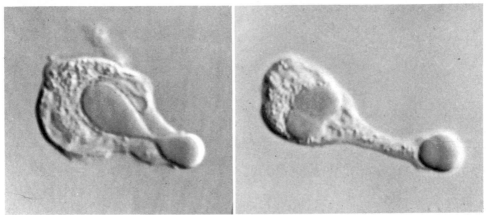

Figs. 1 and 2 – *Bisection of red cell in the process of phagocytosis.* The portion of the red cell outside the neutrophil can be separated and become a sphero-schizocyte

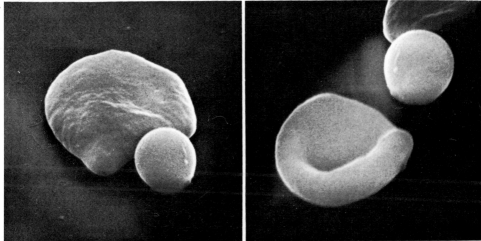

Figs. 3 and 4 – *Disco-schizocytes and sphero-schizocytes* (SEM)

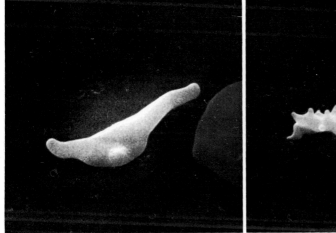

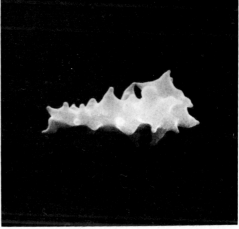

Fig. 5 – *Schizocyte* on which an echinocytic spicule begins to form

Fig. 6 – *Echino-schizocyte.* Obtained by aging of the preparation

Definition

Notwithstanding the time-honored use of the word, most spherocytes are not truly spherical cells. Usually they merely have a much increased thickness, so that the central concavity is much reduced and may be readily overlooked. They should be called pachycytes in contrast to leptocytes.

The proportion of thickness to diameter of the red cell is fixed with a mean ratio of 3.4. When this ratio is less than 2.4, a spherocytosis exists; if it is above 4.2, leptocytosis is present.

Truly spherical forms occur only at the prelytic stage, no matter what their origin. Echino- or stomatocytic factors in sufficient strength, antibodies or hypertonicity all lead to the same end result.

Different Types of Spherocytes. I have retained the word spherocyte in this book because the term is sanctioned by long usage. One needs to be aware, however, that the term spherocyte is applied both to truly round cells and to cells that only appear to be round on examination with the light microscope.

Apart from the echino- and stomatospherocyte resulting from reversible transformation of normal red cells (see p. 50), one can distinguish:

1) Spherocytes (pachycytes) of hereditary spherocytosis (p. 98)
2) True macrospherocytes in some hemolytic anemias due to combined action of antibodies and stress erythropoiesis (see p. 88), or those resulting from increase of intracellular water
3) True microspherocytes which are quite rare except for schizospherocytes (see pp. 92 and 94).

Smears

If the spherocyte has a normal volume, the diameter is of necessity smaller than that of discocytes. Because the volume and hemoglobin content of spherocytes are normal, they appear in smears as microcytic and hyperchromic. The mean corpuscular hemoglobin concentration (MCHC) is above 36, a hallmark of the disease. Since spherocytes have a normal volume (except for schizospherocytes), they should not be called microspherocytes.

The recognition of spherocytosis in smears is more difficult than would appear from textbooks. One must guard carefully against misinterpretation because of the different appearances of red cells in different portions of normal smears (see p. 98). The maintenance of a markedly reduced central pallor is more readily appreciated in the Soret band than in the Giemsa stained smear.

Depending on the ratio of diameter to thickness, spherocytes can be classified provisionally as spherocytes I, II and III (see figure on the opposite page). The spherocyte I is a discocyte with increased thickness, the spherocyte II corresponds to stomatocytic forms (see pp. 98 and 100) and the rare spherocytes III are cells with a very small umbilical depression. The central pallor is minimal and can only be seen in the Soret band (Fig. 3).

Occurrence

The presence of spherocytes always indicates a hemolytic process. Hereditary spherocytosis is the most frequent cause of spherocytes in smears (see p. 98) but they are frequently present in other hemolytic anemias, particularly acquired hemolytic anemias due to autoimmune or isoimmune antibodies. Macrospherocytes are seen in hemolytic anemias with stress reticulocytosis and in the syndrome of hydrocytosis.

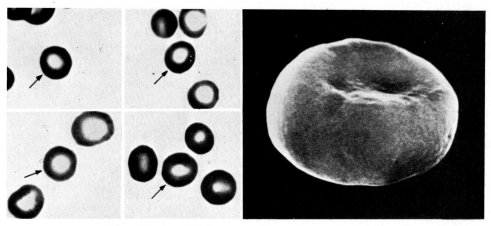

Fig. 1 – *Spherocytes I* (Soret band and SEM)

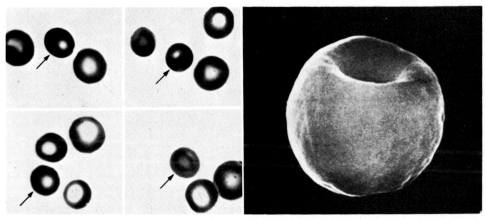

Fig. 2 – *Spherocytes II*

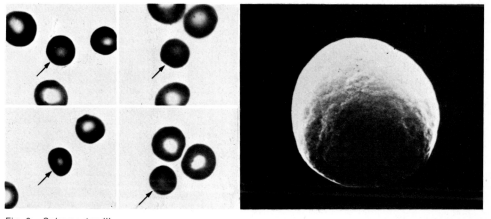

Fig. 3 – *Spherocytes III*

Smears[1]

Usually the inexperienced hematologist who expects to see spherocytes is disappointed. The erythrocytes appear and in fact sometimes are normal. In most cases, however, careful examination of a portion of the smear where the cells are optimally spread allows the identification of at least some spherocytes (see opposite page). All forms of spherocytes as described on p. 96 may be seen, as well as some knizocytes and reticulocytes.

After splenectomy, spherocytes persist unchanged. In addition, one observes acanthocytes and Howell-Jolly bodies.

Mechanism of Formation

The cause of spherocyte formation in hereditary spherocytosis is not known. It appears to be a primary defect of the red cell membrane, in contrast to hemolytic anemia due to antibodies in which it is acquired. The following have been noted: an increase in the activity of ATPase, an alteration in the lipid and protein content of the membrane, reduced resistance to entry of sodium, and other abnormalities. However, none of them can be considered specific or suffices to explain the formation of spherocytes.

Mechanism of Destruction[2]

The spherocytes are sequestered and destroyed in the spleen. It is assumed that the diminution of the deformability of the erythrocytes, which in this case is due to their shape (see p. 102), prevents them from passing through the network of fibrils that surrounds the sinuses and separates them from the splenic cords through which normal red cells must pass. It is possible that this mechanical action is aided by a metabolic change of the spherocyte, which hastens the destruction.

Physical Characteristics

The fragility of the spherocyte is increased. In the majority of cases the hemolysis begins at a concentration of NaCl of 0.6%, occasionally even at 0.8%. The hemolysis is total at 0.4% or 0.3% of NaCl.

The mean corpuscular volume of spherocytes is usually normal, often slightly below normal, and occasionally greater (note that the *diameter* is always decreased). The mean corpuscular hemoglobin concentration is always increased (36–38 g/100 ml). Autohemolysis is accelerated and partly corrected by the addition of glucose. The shortened survival of spherocytes depends on the presence of a spleen. Survival is normal in splenectomized individuals.

Echino- and Stomatospherocytes

Echinocytogenic agents do not reverse spherocytes in hereditary spherocytosis to discocytes (although they do so reverse stomatocytes induced by chlorpromazine). The cells of hereditary spherocytosis become echinocytes without passing through a discocyte stage.

Stomatocytogenic agents tend to increase the concavity of the spherocyte and make it more cup-shaped. If spherocytes are exposed to stomatocytogenic agents for a prolonged period, additional concavities may appear.

Combinations of Spherocytes and Leptocytes[3, 4]

Icterus due to obstruction of the bile duct results in interesting changes in the form and survival of red cells. It compensates in some fashion for the fact that in spherocytosis the amount of membrane is decreased in proportion to its volume. In a normal individual the increase of lipids in the membrane caused by obstruction of the bile duct leads to the formation of leptocytes. In hereditary spherocytosis, it tends to normalize the appearance of the cells. Osmotic fragility decreases, their deformability and life span increase.

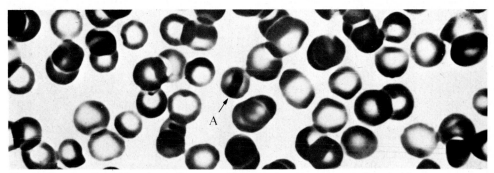

Fig. 1 — *Good portion of the smear*. Several spherostomatocytes and one knizocyte (*A*)

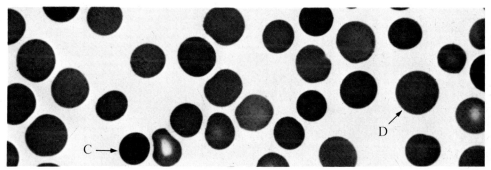

Fig. 2 — *Good portion of the smear*. Note that even in the small spherocytes, some central pallor (*B*) is still visible if viewed in the Soret band

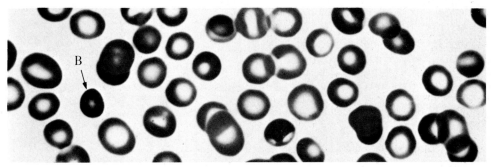

Fig. 3 — *Feather edge*. Note the difference in densities between spherostomatocyte III (*C*) of small diameter and the other erythrocytes (*D*)

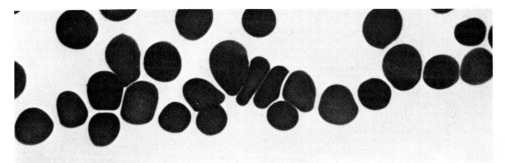

Fig. 4 — *Extreme feather edge*. The cells are deformed in the process of maximum spreading and their original shape can no longer be recognized

24 – Stomatocytes

Definition[1-4]

These are uniconcave red cells of the shape of a very thick cup. There are different types of stomatocytes which must not be confused with each other. Apart from the stomatocytes due to the reversible discocyte-stomatocyte transformation (see p. 50), there exist:

1) The stomatocytes of hereditary spherocytosis (p. 98)
2) Stomatocytes of hereditary stomatocytosis
3) Stomatocytes associated with Rh Null cells
4) Stomatocytes produced by antibodies (p. 88)
5) Stomatocytes produced by increased water content of the cell (hydrocytosis).

Smears

Stomatocytes are small red cells with an oval, triangular or slitlike, more or less narrow central pallor, as discussed on page 72. The specific appearance is due to the cup-shape of the cell, its wall thickness and the preparation of the smear. Stomatocytes are usually seen only in some parts of the smear.

Phase Microscopy and SEM

The appearance always conforms to that of very thickwalled cups.

Occurrence

One encounters some stomatocytes in all cases of spherocytosis, but the typical stomatocyte appearance is only characteristic of that particular type of hemolytic anemia called hereditary stomatocytosis, which is distinct from hereditary spherocytosis. Stomatocytes have also been described in some cases of alcoholic cirrhoses and as a result of congenital or acquired decreased effectiveness of the sodium pump.

A large number of stomatocytes are seen in the anemia associated with Rh Null. It is believed that there exists in this blood group an anomaly of lipoproteins.

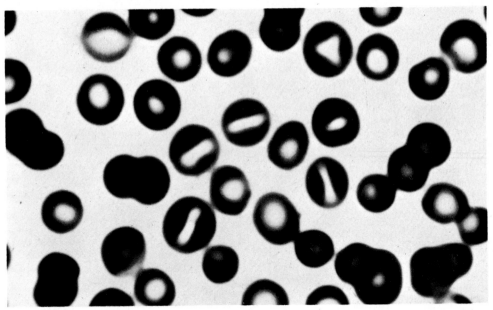

Fig. 1 — *Hereditary stomatocytosis* (Soret band). Good portion of the smear

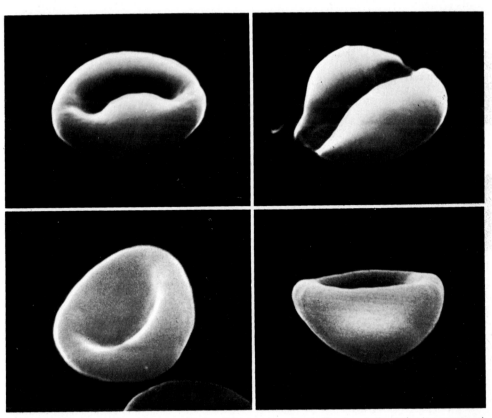

Fig. 2 — *Different stomatocytes seen with the SEM* (same patient). The two cells at the top represent fixation artifacts which resemble the appearance of stomatocytes in smears

Mean Cell Diameter

Price-Jones Curve[1]

By means of an occular micrometer, one can prepare a distribution curve of the diameters of individual red cells in a smear, called the Price-Jones curve. Although it was the subject of many investigations in the past, it has recently been criticized and little used. Red cells in smears are deformed, hence their diameter is altered and differs in different portions of the same smear. Many anemias contain poikilocytes, which are difficult to classify by diameter. The procedure is time-consuming and impractical for daily clinical use.

The Price-Jones curve is too cumbersome to be clinically useful except in cases where it deviates quite clearly to the left or right. Hence, in order to judge clinical cases rapidly, it might be well to replace it by the diffractometric method, which is simple and has been known for a long time: the first article on the subject dates back to 1813.

Diffractometry[2, 3]

When a spot of light is viewed through a thin film of red blood cells on a slide, iridescent rings are seen. The rings are due to the diffraction of the light by the red cells; their diameters are inversely proportional to the diameters of the red cells. With a little practice one can, by this simple test, determine whether the red cell diameter is normal, markedly increased as in pernicious anemia, or markedly decreased as in hereditary spherocytosis. The only precaution necessary is to view the light spot through a portion of the smear where the red cells are close to each other without overlapping.

The diffraction rings of white light have the sequence of colors of the spectrum starting with violet and proceeding through blue, green, yellow, orange and red. The last ring is followed by a green ring, which belongs to the second order diffraction spectrum (Fig. 1). By using a monochromatic light source, the reading can be made easier (Fig. 2).

While the Price-Jones curve measurement requires hours for 500 individual cells, the diffractometric method examines thousands of cells in seconds. In general, the diameter of the light beam covers 10,000 to 50,000 blood cells.

A number of instruments (called halometer, hemodiffractometer) have been constructed for these measurements. Nevertheless, notwithstanding its simplicty, the method has fallen into disuse, because the diameter of the cells in smears does not accurately represent their volume, and modern electronic instruments are now available that allow the measurement of volume distribution with greater accuracy.

The diffractometry can occasionally be useful, but I mention it here primarily because of the present development of a new method of diffractometry, which allows one to measure red cell deformability.

Red Cell Deformability

The Physiologic Role of Deformability[4-6]

Red cells must circulate for 120 days through blood vessels, very fine capillaries in hematopoietic organs, and the splenic filter. This requires a great deformability as well as a considerable resistance to external mechanical forces. The discocyte with its diameter of 8 µm passes through capillaries which have diameters of 3–10 µm. It must also pass through openings in the basal membrane, which separates the cords from the sinuses of the bone marrow, and the spleen with openings of 0.5–5.0 µm in diamter.

The factors which determine the deformability of red cells have been studied in detail and can be classified in three categories:

1) *A form factor.* The relationship of the surface to volume of the cell. A rigid sphere, a spherocyte, although smaller than a discocyte, is undeformable and cannot pass through capillaries with a smaller diameter than its own. The normal discocyte, in contrast, is comparable to a collapsed sphere and has an "excess of surface" which allows it to be deformed readily.

2) *Internal factors.* These depend primarily on the viscosity of the hemoglobin. For example, cells containing hemoglobin S become rigid or less deformable when the oxygen tension drops.

3) *Rigidity of the membrane.* This may be due to functional alterations or anatomical lesions, for example multiple small Heinz bodies attached to the interior of the red cell membrane.

Measurement of Deformability by Diffractometry[5-8]

A number of techniques have been developed to evaluate deformability of red cells. In the ektacytometer (from the Greek ekta, elongation), the red cells are exposed to shear stresses in a cylindrical viscosimeter. A laser beam provides a diffraction image which is circular when the red cells are undisturbed. As the viscosimeter is rotating, the red cells become elongated and the diffractometric images are deformed. These images can be photographed or directly analyzed by an appropriate electronic system. When the shear stresses cease, the diffraction image returns to the normal circular form in less than a second.

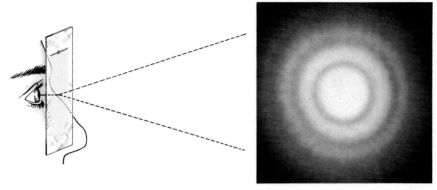

Fig. 1 — *Diffraction rings* appear when a point source of white light is viewed through a blood smear

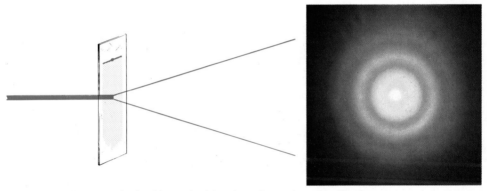

Fig. 2 — *Diffraction rings* obtained by projecting a laser beam through a blood smear

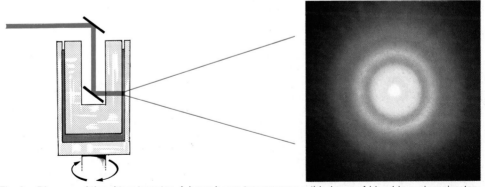

Fig. 3 — *Diagram of the ektacytometer.* A laser beam transverses a thin layer of blood in a viscosimeter. At rest, the diffraction rings are circular. When one of the cylinders of the viscosimeter rotates, the cells become deformed and a more or less elliptical image results (see p. 104)

We illustrate here two examples of the utility of this new technique which should be useful in the clinic.

1) During the discocyte-echinocyte and discocyte-stomatocyte transformations induced in vitro, changes of deformability are uniquely due to the form factor. As the cells become more and more spherical, they become less and less deformable. While the normal discocyte elongates from a diameter of 8.0 µm to a maximum length of 17.0 µm by a shear stress of 700 dynes/cm², the spheroechinocyte only elongates from a diameter of 5.5 to 8.0 µm, and the spherostomatocyte even less, from 5.5 to 7.0 µm, under the same shear stress.

2) In hereditary spherocytosis, the ektacytometer shows the presence of multiple subpopulations according to their deformability. The decrease of deformability parallels the increase of osmotic fragility.

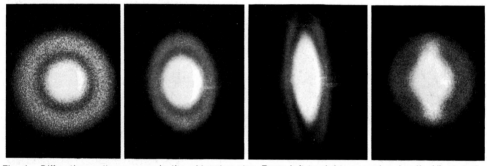

Fig. 1 — *Diffraction patterns seen in the ektacytometer.* From left to right, normal red cells (*1*) at rest, (*2*) at 10 dynes/cm², (*3*) at 70 dynes/cm². At the *extreme right,* an image indicating the presence of a double population of red cells in a pathologic blood sample (at 70 dynes/cm²)

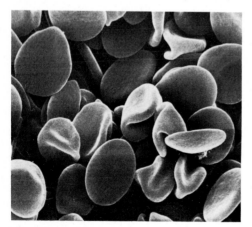

Fig. 2 — Appearance of red cells at low shear forces (SEM)

Fig. 3 — Appearance of red cells at high shear forces (SEM)

Chapter 3

Granulocytic Series

General Comments

The three types of granulocytes (neutrophils, eosinophils and basophils) have a parallel development. The stages of maturation will be described in detail in the neutrophilic series only, since they are quite similar for all three.

Definition

The granulocytic series starts with the myeloblast and develops through the promyelocyte, myelocyte, and metamyelocyte stages to the mature granulocyte. The mature granulocyte has a polylobed nucleus which has earned it the name of "polymorphonuclear" leukocyte.

Location

In the normal adult granulocytes are produced in the bone marrow alongside cells of the erythrocytic and thrombocytic series and become intermingled with them. During their life, granulocytes pass through three compartments: the bone marrow, where they mature, the blood, where they are in transit, and the tissues, where they commonly fulfill their function.

Maturation Stages, Amplification and Life Span[1, 2]

One generally recognizes seven stages of granulocyte maturation. Because different schools differ slightly in the use of the same terms, it is most convenient to adopt the symbols M_1 to M_7 to designate the successive stages of maturation.

Amplification takes place concurrently with maturation in the first stages (M_1 to M_4), after which there is maturation only and no further division.

Senescence and Death of Granulocytes

Granulocytes disappear randomly from the peripheral blood with a halftime of $7^1/_2$ h, in contrast to red cells which normally leave the circulation as a function of age. The exponential disappearance of granulocytes is so rapid that it does not appear to be modified by an aging process.

The largest numbers of granulocytes seem to be lost from the blood through the gastrointestinal tract. However, granulocytes also pass from the blood vessels into the tissues, attracted by bacterial and other chemotactic substances, and die there quite rapidly. The death can occur by fragmentation or the cells may be phagocytized and rapidly digested by macrophages, and the cadaver may not be identifiable (see p. 22). Only a small number of granulocytes die in the blood. Death and necrosis can sometimes be inferred from pyknocytic nuclei in which the lobes of the nucleus have separated, rounded off and lost their intrinsic structure. The cytoplasm is usually dark and the granules are absent.

On rare occasions, one can see granulocytes phagocytized by macrophages in the bone marrow. Their number may be increased in severe infections.

Stage of maturation	Life span
Myeloblast (M_1)	1 day
Promyelocyte (M_2)	2 days
Myelocytes (M_3 and M_4)	4 days
Metamyelocyte (M_5)	2 days
"Band form" (M_6)	2 days
Polymorphonuclear (PMN: M_7)	5 days (of which 8 h only are spent in the blood vessels)

1 – Granulocytopoiesis

Amplification[1-3]

It is generally thought that four sequential divisions take place in the granulocytic series so that 16 granulocytes are produced from 1 myeloblast. However, it seems quite likely that amplification varies with the needs of the organism.

Amplification may be quite disorderly in pathologic states. In cultures, the number of divisions can increase to 12 or more.

Maturation

Since maturation is a continuous process, the successive stages are difficult to separate. In general, however, seven separate stages can be recognized in smears stained with Giemsa. This classification is based on the nature and the number of granules and the characteristics of the nucleus. All of this actually applies only to the normal physiology: in stress or pathologic states, the synchronism of maturation of the organelles is altered and the classification becomes imprecise (see p. 208).

Stage of maturation	Granules (visible in smears)	Nucleus
(Myeloblast) M_1	Azurophilic (few)	Large, round
(Promyelocyte) M_2	Azurophilic	Large, oval, indented by centrosome
(Myelocytes) M_3 and M_4	Azurophilic and neutrophilic	Medium or small, oval
(Metamyelocyte) M_5	Neutrophilic	Kidney shaped
(Band form) M_6	Neutrophilic	Sausage shaped
(Polymorpho-nuclear)PMN, M_7	Neutrophilic	Polylobed

It is important to remember that stained smears do not allow recognition or precise characterization of all types of granules. For instance, we now know from electron microscopic studies that the azurophilic granules do not disappear with maturation: they become less numerous and lose their tinctorial affinities, at least in part. It is, however, possible to recognize some azurophilic granules in normal neutrophil PMN. The preceding table is, therefore, inexact, because it accounts only for granules seen in Giemsa-stained smears (see pp. 112 and 122).

Passage Into the Circulation[4]

PMN pass from the bone marrow into the circulation by active diapedesis. Some investigators maintain that the PMN leave the bone marrow just like erythrocytes when they become sufficiently deformable, and are forced into the circulation by the higher pressure existing in the marrow. This hypothesis is unproven*.

Regulation of Granulocytopoiesis

A number of factors have been isolated which stimulate granulocytopoiesis of in vitro cultures. "Granulopoietin" has not been characterized as definitively as erythropoietin. Some authors believe that two regulators exist, one controlling production, the other release of cells from the bone marrow.

Ecology of PMN

There exist several pools of mature PMN which are of great importance for the study of PMN physiology.

The Bone Marrow Pool consists of PMN which have fully matured but do not pass immediately into the circulation. They furnish a reserve that can be discharged into the circulation in case of need.

The Marginal and Circulating Pool. In the vessels, there are circulating cells and cells which adhere to the vessel walls, referred to as the marginating pool. Normally, approximately one-half of the PMN are in the marginating pool at any one time. Consequently, leukocytosis can be due to the mobilization of the marginating pool or due to the sudden exit of mature cells from the bone marrow pool which is 10 to 20 times larger than the peripheral pool.

The Tissue Pool. The PMN which leave the blood to migrate into the tissues or the body cavities are generally believed not to be capable of returning to the circulation. It also appears that they are destroyed or die quite rapidly once they leave the circulation. This is in contrast to the recirculation of lymphocytes (see p. 145).

---

* It is essential that one distinguishes *passive* deformation, very marked in mature erythrocytes and limited in leukocytes, even mature ones, and the *active* deformation associated with cytoplasmic contraction which may lead to cell locomotion. It is marked in mature leukocytes.

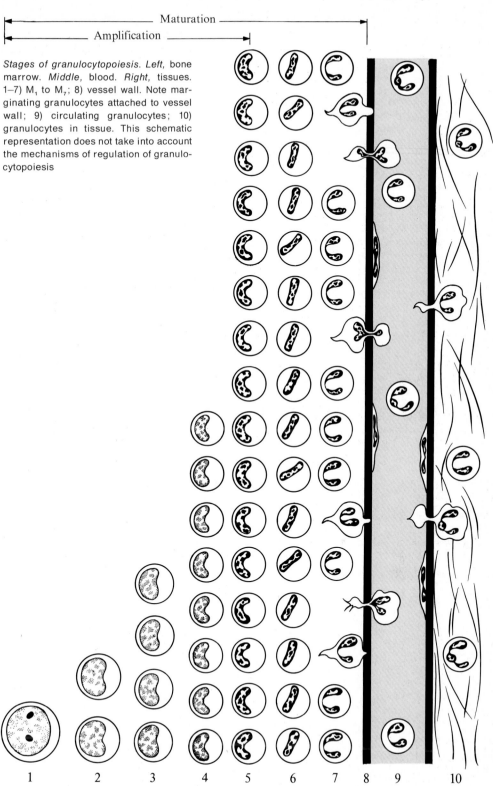

Maturation

Amplification

Stages of granulocytopoiesis. Left, bone marrow. *Middle,* blood. *Right,* tissues. 1–7) M_1 to M_7; 8) vessel wall. Note marginating granulocytes attached to vessel wall; 9) circulating granulocytes; 10) granulocytes in tissue. This schematic representation does not take into account the mechanisms of regulation of granulocytopoiesis

1   2   3   4   5   6   7   8   9   10

2 – M₁ (Myeloblast)

Definition

The myeloblast is the first cell of the granulocytic series.

Nomenclature

Some authors define a myeloblast as the undifferentiated stem cell of the bone marrow series which we have called a hemocytoblast (see p. 18). Others define it (as we have done) as the first stage of the granulocytic series (M₁), but they describe it as a cell without any granules.

In order to avoid confusion, we will reserve the name myeloblast for cells which already have a few azurophilic granules and, therefore, are identifiable as belonging to the granulocytic series. The limitation of this definition must be recognized. With the electron microscope, it is sometimes possible to identify an earlier stage as belonging to the granulocytic series by a positive peroxidase reaction of the perinuclear cisterna, though azurophilic granules are not yet present.

Smears

The myeloblast is a cell of 15 to 20 µm in diameter. The nucleus is large, round or oval. It is sometimes indented by the centrosome. It has a very fine chromatin structure, with 2 to 5 nucleoli, which are round and pale blue. The cytoplasm is clear blue, more heavily colored at its border and contains at least some azurophilic granules. They are small and rare in very young cells. They measure up to 0.5 µm and are more numerous in older cells (Fig. 1).

Cytochemistry

The azurophilic granules are peroxidase positive. Occasionally, the peroxidase reaction identifies a myeloblast although the azurophilic granules are not yet formed (see below: Electron Microscopy).

Living Cells[1]

The myeloblast has a clear nucleus with two or three dark nucleoli. The cytoplasm is fairly dark blue (due to the number of ribosomes). The many mitochondria are not readily distinguished from the azurophilic granules. The myeloblast is capable of very slow locomotion. In leukemia, comparable cells occasionally emit veils and may be phagocytic (see p. 208).

Electron Microscopy

The cytoplasm of the myeloblast is characterized by the presence of numerous ribosomes and polyribosomes, free or associated with sacs of endoplasmic reticulum (RER). It is the large number of ribosomes which is responsible for the blue color of the cytoplasm in Giemsa stains.

The nucleus has a clear appearance and contains large nucleoli which frequently have a reticulated appearance. The centrosome is large. Some small granules can be seen in the interior or attached to the cisternae of the Golgi complex. The primary, azurophilic granules are present in variable numbers, 0 to 10 per section. With the usual EM staining, they appear grey, homogeneous or slightly floccular. Their formation is described on page 110.

The peroxidase reaction is positive in the perinuclear cisterna, the RER sacs, the Golgi complex and its granulations. This positivity is sufficient to identify a myeloblast, even if it is limited to the perinuclear cisterna. No other bone marrow cell gives this reaction (see p. 190), except the promonocyte. One may see bundles of fibrils, generally in the neighborhood of the nucleus. They are quite rare.

108

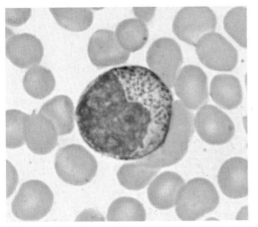
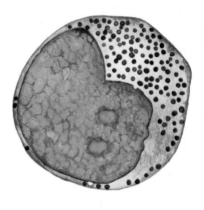

Figs. 1 and 2 – *A myeloblast (M₁)*. The cell depicted contains a large number of azurophilic granules. This stage begins theoretically as the first azurophilic granulation appears (see text)

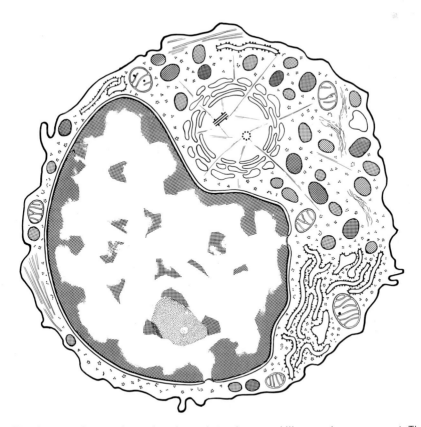

Fig. 3 – The electron microscopic section shows that only azurophilic granules are present. The many ribosomes (some of them attached to the ER) explain the basophilia in this stage of maturation

3 — M₂ (Promyelocyte)

Smears

Promyelocytes are 15 to 20 μm in diameter, have an oval nucleus, sometimes with a slight, concave depression. The chromatin begins to condense in some parts of the nucleus without forming compact blocks. Occasionally, small lighter zones are seen, which represent nucleoli hidden behind the perinucleolar chromatin.

The cytoplasm is very slightly basophilic. Usually, a large centrosome is discernible as a lighter juxtanuclear zone, often adjacent to a slight depression of the nucleus. Granules are not present in that zone, but outline its periphery. The azurophilic granules of the promyelocyte vary in size, form (rodlike or comma-shaped) and color (wine red to bright red). They contain peroxidase, lysosomal acid hydrolases, antibacterial cationic proteins, lysozyme, and acid mucosubstance.

Living Cells

The numerous granulations often hide the nucleus if the preparation is not very thin. The neutrophilic granules are difficult to distinguish from the azurophilic granules and the mitochondria. The juxtanuclear clear zone is usually well outlined and free of granulations which arrange themselves around it in a radial fashion. The nucleus shapes itself around the centrosome which appears to have a firmer consistency. Often the centrosome is separated from the nucleus by a layer of granules.

Electron Microscopy[1-3]

The nucleus has a fine structure and contains one or two nucleoli, usually surrounded by denser chromatin. The amount of RER is decreased. The Golgi complex is well developed and its lamellae actively manufacture the azurophilic and neutrophilic granules. In the center of the large Golgi complex, the satellites of the centrioles give rise to numerous microtubules. Microfibrils are scarce. The granules are either round or ellipsoid and frequently contain a crystal.

Cytochemical Reactions

These are described on p. 112.

Granulogenesis[1-3]

The formation of granules is best demonstrated in electron micrographs of marrow cells that have been reacted for peroxidase. The promyelocyte stage of maturation is characterized by the accumulation of a large population of peroxidase-positive granules which vary in contour: most are spherical, approximately 500 nm in diameter, but ellipsoid, crystalline forms are also present. Peroxidase can be seen throughout the entire secretory apparatus, the sacs of the RER, the Golgi and its vesicles, and all developing granules, indicating the pathway by which the enzyme is synthesized and packaged into granules.

In contrast, the entire secretory apparatus is peroxidase negative in later stages of granulocyte development, beginning with the myelocyte, when production of peroxidase-positive granules stops and peroxidase-negative granules are first produced*.

The peroxidase-positive granules are called "azurophilic" granules because of their staining characteristics in promyelocytes or alternatively "primary" granules because of the time of their appearance. The granules developing in the myelocyte stage are neutrophilic, eosinophilic or basophilic in the Giemsa or Wright stains**.

---

* The notion that a sharp division exists between promyelocytes and myelocytes in the type of granules produced is convenient, but it is unlikely that such sharp division reflects the biologic events. A few neutrophilic granules may well be produced in the promyelocyte stage (see p. 111, Fig. 3) without being visible in the light microscope. Another example of differences in nomenclature which probably produce more heat than light.

** In this book, I have used the term azurophilic for the peroxidase positive primary granules which are produced in myeloblasts and promyelocytes. The term specific granule refers to the granules produced in the myelocyte stage which may be neutrophilic, eosinophilic or basophilic. Some overlap in the use of the terms neutrophilic and specific granules is thus unavoidable.

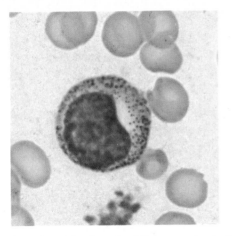

Figs. 1 and 2 — *A neutrophilic promyelocyte (M₂)*. Note the clear space occupied by the centrosome which indents the nucleus and displaces the granules

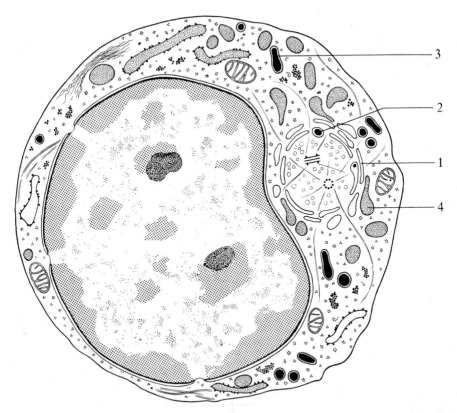

Fig. 3 — Note the beginning of the formation of neutrophilic (1, 2, 3), between azurophilic (4) granules

111

4 – M₃, M₄ and M₅ (Myelocytes and Metamyelocyte)

Myelocytes (M₃ and M₄)

Smears

The term myelocyte is used for the two successive generations of M₃ and M₄. The cytoplasm and the nucleus of M₄ are more mature than those of M₃, and the size of M₃ is from 16 to 20 µm in diameter and that of M₄ from 10 to 16 µm.

Myelocytes have a round or oval nucleus, occasionally slightly indented. It is smaller and darker than that of the promyelocyte and has sharply defined chromatin masses. The nucleolus can no longer be seen.

As maturation progresses, the cytoplasm loses its basophilia and becomes slightly acidophilic. Specific granulations appear, first around the centrosome. It is not uncommon to find myelocytes in which the central zone is already acidophilic and contains neutrophilic granules, while the peripheral zone appears still basophilic and has mainly azurophilic granules.

Electron Microscopy[1]

The M₃ myelocyte stage is characterized by the production and accumulation of a second population of granules which are peroxidase negative. During the myelocyte stage, the only peroxidase positive elements present are the azurophilic granules, formed earlier. All organelles of the secretory apparatus lack the enzyme. However, a second population of granules, the "specific", or "neutrophilic" granules which are peroxidase negative are now formed by the Golgi complex. They vary in size and shape but typically appear as spheres (approximately 200 nm) or rods (130 × 1000 nm). In the human, these granules lack lysosomal acid hydrolases. Their contents remain largely unknown, but they have been demonstrated to contain lactoferrin and lysozyme. Further work is necessary to identify the localization of alkaline phosphatase, although in the rabbit this enzyme is present in granules that correspond to the specific granules in man.

Cytochemistry[1−3]

Cytochemical reactions change from that in the promyelocyte when specific granules appear in M₃ and increase in M₄ myelocyte cells.

Azurophilic granules are no longer produced, but they persist and are reduced by half with each division. They remain peroxidase positive, though they usually lose their affinity for the azure component of the dye. They may, therefore, appear violet or not stain at all. The change is ascribed to complex formation between stainable acid mucosubstances and basic proteins. Sudan black stains a lipid component in normal azurophilic granules. The reaction does not overlap entirely with the peroxidase reaction, particularly in leukemic cells.

Lysosomal hydrolases (acid phosphatase, arylsulfatase, beta-glucuronidase, esterase and 5-nucleotidase), cationic proteins, and lysozyme can be identified by special stains. The iron-binding protein lactoferrin and alkaline phosphatase appear in myelocytes and are always present in later stages. They are present only in granulocytes and lactoferrin only in neutrophils. The exact localization of specific enzymes or other substances can only be studied satisfactorily with the electron microscope supplemented by biochemical studies after isolation of the organelles.

The cytochemical reaction in smears is useful for the differential diagnosis of certain diseases. The alkaline phosphatase is increased in polycythemia and leukemoid reaction, diminished in chronic myeloid leukemia and paroxysmal nocturnal hemoglobinuria, congenital hypophosphatemia and some other disease states (see p. 122).

Living Cells

The oval nucleus has a definite concavity corresponding to the convexity of the centrosome. The blocks of chromatin are small, but very clearly defined. The granules are refractile and numerous, at the limit of the power of resolution of the optical microscope, and cannot be distinguished from mitochondria. Normally, the myelocyte is capable of only very slow locomotion, if any.

Metamyelocyte (M₅)

Smears

The metamyelocyte has all the cytologic characteristics of the myelocyte and the PMN, except for the shape of the nucleus which is that of a kidney bean or a bent rod. The convexity is usually at the edge of the cell. The concavity corresponds to the periphery of the centrosome. The nucleus has a dense chromatin with numerous dark blocks sharply circumscribed and much more evident than in the myelocyte. The granules are generally beige, brownish, lilac or light violet. Their color depends on the pH of the staining solution and on the number of remaining primary granules.

Living Cells

Except for the shape of the nucleus, their appearance is that of mature PMN described below. It is in this state of maturation that locomotion begins.

Electron Microscopy and Cytochemistry

The main features have been described under M₃ and M₄.

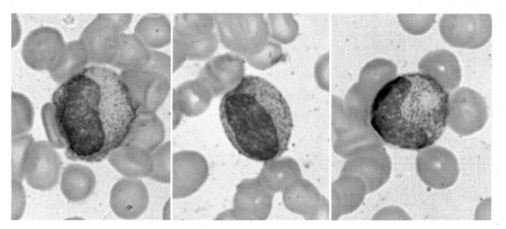

Figs. 1 and 2 – *Neutrophilic myelocytes (M₃ and M₄)*

Fig. 3 – *Neutrophilic metamyelocyte (M₅)*

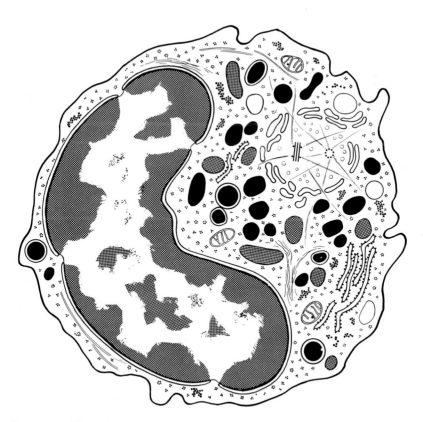

Fig. 4 – Both azurophilic and neutrophilic granules are present. At this stage, the Giemsa stain no longer permits a clear identification of azurophilic granules. They decrease in number, while neutrophilic granules increase

5 – M$_6$ and M$_7$ (Polymorphonuclear Neutrophils)

Smears

Mature cells of the granulocytic series are characterized by the shape of their nuclei which has earned them their name "polymorphonuclear". They belong to classes M$_6$ and M$_7$.

Their slightly acidophilic cytoplasm contains neutrophilic granules intermingled with a few azurophilic ones (see p. 112).

Nonsegmented PMN (Band Cells, M$_6$)

A certain number of PMN have a sausage-shaped nucleus or a nucleus with segments connected by broad bands rather than fine threads of chromatin (Fig. 1). Some hematologists attach considerable significance to changes in the percentage of these cells. They belong to Arneth's class I (see below).

Segmented PMN (M$_7$)

The nucleus is segmented into different lobes connected by filaments of chromatin. The individual lobes or segments have no particular relationship to each other and form an S, Y, Z, E or other shapes (Figs. 1 and 2, p. 119; Figs. 1 to 3, p. 123) depending solely on the manner in which the cell has been spread. The observation in living cells (at rest or in locomotion) shows that the lobes are always arranged in a circle around the centrosome which occupies the geometric center of the cell (see p. 117).

The degree of segmentation or polylobulation probably corresponds to the age of the granulocyte*. However, this has not been completely proven.

The chromatin is dense with dark blocks of chromatin separated by a clear network of lighter staining bands. Various nuclear appendages may be seen in normals and in pathologic states (see p. 120).

Arneth Counts[1-3]. The classification of Arneth, which contained numerous subgroups, has been simplified into the following six categories which are given with their respective percentages:

I	Nucleus kidney or sausage-shaped	0 to 5
II	Two lobes	10 to 30
III	Three lobes	40 to 50
IV	Four lobes	10 to 20
V	Five lobes	0 to 5
	More than five lobes	0

Under pathologic conditions, the peak frequencies change to the "left" or to the "right." Shift to the right occurs primarily in megaloblastic anemia and is one of the helpful signs in B-12 and folic acid deficiencies. It may be seen in CML, in some chronic diseases of the kidney and liver, and occurs in a rare hereditary anomaly (Undritz).

Shift to the left occurs primarily in acute and subacute infections, after hemorrhage and in toxemias. In a relatively frequent hereditary anomaly (Pelger-Huet) all mature granulocytes have two segments only.

Granules. *The number* of granules is variable. Even under physiologic conditions, one can see some granulocytes with few and others with numerous granules. Their *staining* also varies. The neutrophilic granules are at the limit of the resolving power of the optical microscope. A small number of primary granules remain, which may have lost part of their affinity for the azure. Very slight modifications in the staining technique and the functional state of the cell also influence the appearance of the granules (see pp. 118 and 122).

Although stained smears do not permit a more precise study of the granules, they play an important role in cell metabolism and in phagocytosis as has been demonstrated by microcinematography of living cells and electron microscopy (see p. 118 and 122).

Electron Microscopy

Nucleus. The blocks of chromatin adjacent to the nuclear membrane are interrupted by numerous nuclear pores; their persistence makes it likely that nuclear-cytoplasmic interchanges still occur. There is no remnant of a nucleolus and it is likely that the synthesis of proteins has ceased. The filaments which join the nuclear lobes are formed by dense chromatin and enveloped by the nuclear membrane but have no nuclear pores.

Cytoplasm. The number of ribosomes is small and there is little endoplasmic reticulum. The Golgi complex forms a small sphere in the center of the cell and contains the two centrioles from which numerous microtubules radiate. Few microfilaments are present. The mitochondria are small and often elongated. Vacuoles are present and include pinocytic vacuoles, contractile vacuoles, and fat globules. Numerous glycogen particles from 50 to 200 nm in diameter are present throughout the cytoplasm.

Granules. They are identical with those of the metamyelocyte (see p. 112); their number may vary in the course of phagocytosis or as a result of other functional changes (see pp. 118 and 122). Formation of new granules no longer occurs in the mature granulocyte.

---

* The normal segmentation of the nucleus is quite different from "radial segmentation"[4] due to a pathologic or toxic environment, e.g. storage in EDTA sodium oxalate, a culture medium or any other long in vitro storage. The radial segmentation is equivalent to the formation of Rieder cells in mononuclear cells (see p. 206)[4].

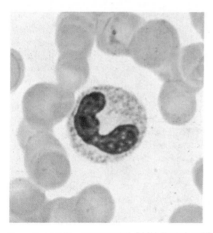 

Figs. 1 and 2 — *Neutrophil PMN* (band cell). The mixture of a large number of neutrophilic granules and few azurophilic ones shown in the corresponding electron microscopic section cannot be appreciated in the Giemsa stain

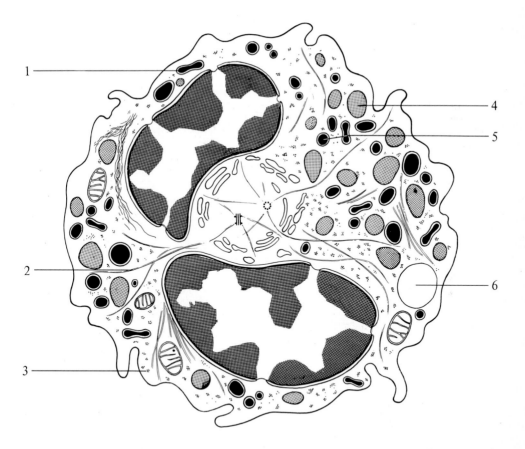

Fig. 3 — Note that the nuclear lobes which surround the centrosome are cut in such a way as to appear disconnected. The ribosomes have practically disappeared. Glycogen particles (3) are present in the cytoplasm. A contractile vacuole can be seen (6). (1, 5): neutrophilic granules; (2): microtubules; (4): azurophilic granules

115

Cell Movements

Between slide and coverslip, free-floating cells in thick preparations are round and the refractility of the granules does not permit the study of their internal structure. Slight compression gives a clear picture of the nucleus draped around the centrosome, often in horseshoe-shaped fashion (Fig. 3). More details can be gathered after strong compression although this usually causes death of the cell. The PMN are best observed when they have spontaneously spread out on the surface of the glass (Fig. 5).

Locomotion. The PMN can only move when they are attached to a solid surface. They extend a clear cytoplasmic projection (protopod) in the direction of locomotion, while the opposite end of the cell (uropod) is attached by a number of filaments to the support (the glass slide, the endothelium, or other surfaces, Fig. 2). The nucleus appears entirely passive and its consistency is more fluid than that of the central portions of the cytoplasm, but less so than that of the protopod. PMN progress at a speed between 19 and 40 µm per minute.

Adhesiveness[1]. Adhesiveness of PMN is the first step of spontaneous spreading on surfaces and of phagocytosis (see below and p. 118). The three properties are usually measured simultaneously. The degree of adhesivity of PMN to the wall of silicone test tubes or other materials used in extracorporeal circulation is of interest as an indication of the number of cells which may be damaged or lost in this fashion. In vivo, the marginating pool of leukocytes in the circulation is presumably due to the adherence of cells to the endothelium of small vessels.

Diapedesis. PMN constantly migrate into the tissue by diapedesis. Marginating PMN insert a pseudopod between endothelial cells and pass through the capillary wall.

Spontaneous Spreading[2,3]. Spreading, like adhesiveness, depends on intracellular and environmental factors. It is influenced by the milieu (spreading occurs more readily on wettable surfaces). Cells, first free in the surrounding liquid, begin to attach themselves to the surface. Cytoplasmic veils appear in different parts of the periphery. These veils attach themselves to the supporting surface and the cell becomes surrounded by a band of cytoplasm free of granules.

The diameter of the spread cell can reach 20 µm to 30 µm. The central portion of the cell remains unchanged during the process. In a second stage, the granulations extend closer to the edge of the cytoplasm, without entirely reaching the periphery. An "undulating membrane", 3 to 5 µm, in width, remains free of granules.

Under certain conditions, a spread cell can resume its original shape and start again to move in a definite direction. Spontaneous spreading is probably only a special form of ameboid movement: when the stimuli which elicit directional movement are equal in all directions, spreading rather than locomotion occurs. Spontaneous spreading has been compared to an attempt to phagocytize an object (the glass slide, for example) much larger than the cell itself.

The scanning electron microscope allows the study of the cell surface in the process of locomotion and spontaneous spreading (Figs. 1 to 5).

Intracytoplasmic Movements

Movements of the Centrosome[4]. The centrosome is a clear zone approximately 0.5 to 1.0 µm in diameter. The granulations surrounding it are often arranged in a radial fashion. Direct examination and particularly time lapse microcinematography has demonstrated that the centrosome undergoes to-and-fro movement with a periodicity of approximately 30 s and an amplitude of 5 to 10 µm. Since microtubules link the centrioles to the nucleus, the nucleus may participate in this movement (see p. 120). Probably these movements promote circulation of molecules and granules from the Golgi complex to other parts of the cytoplasm of the cell (see p. 16).

Movements of Granules. Granules may undergo a variety of movements other than those which are secondary to movements of the centrosome. Oscillations of less than 1 µm, i.e. Brownian movements, vary with the fluidity of the cytoplasm; they are absent in the part of the cytoplasm that has undergone gel formation, and they are increased in a hypotonic medium. The granules can also undergo rapid movement of a somewhat greater amplitude of 2 to 3 µm, either singly or more commonly in groups or as a single file of granules moving in one direction for a brief period, after which they resume their Brownian movement. The single file movements correspond to the cyclosis of the cytoplasm (see p. 16). During directional movements granules stream into the cytoplasm of the protopod a few seconds after its formation.

Chemotaxis[5–7]

Chemotactic factors attract PMN to the site of "inflammation". The methods for the measurement of chemotaxis are of great interest but have so far remained research methods only. Disorders of chemotaxis are still incompletely understood but are under active investigation. Both disorders of phagocytosis of certain organisms and of the killing of ingested bacteria have been identified in syndromes of recurrent infections, especially in children.

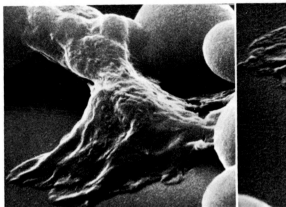

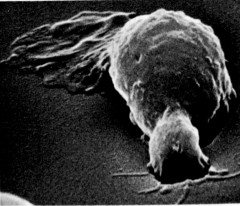

Fig. 1 – *Locomotion*. A PMN. A view of the proto-pod

Fig. 2 – *Locomotion*. Another PMN seen from the rear. The uropod adheres to the underlying glass by its cytoplasmic filaments

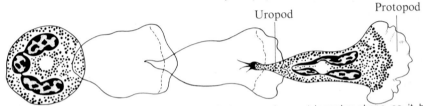

Uropod   Protopod

Fig. 3 – *Locomotion*. A PMN, spherical in the circulation, acquires a triangular shape as it begins to move

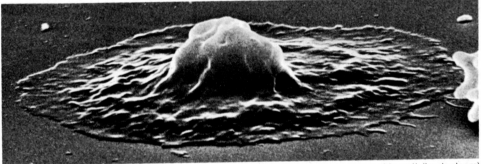

Fig. 4 – *Spreading of a PMN on a glass slide*. The cytoplasm spreads out into a thin veil (hyaloplasm) which contains no granulations. In the center, the nucleus, surrounding the centrosome, projects above the level of the peripheral undulating membrane. (This cannot be appreciated in the phase microscope) (see Fig. 5)

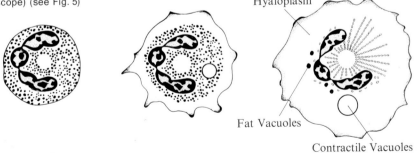

Hyaloplasm

Fat Vacuoles

Contractile Vacuoles

Fig. 5 – *Schematic representation of the stages of the spreading process*. Note the arrangement of nuclear lobes around the centrosome and that of granules aligned along the microtubules

7 – Phagocytosis

General Remarks

The ingestion and digestion of microorganisms is one of the functions of neutrophilic PMN. They have been referred to as microphages by Metchnikoff at the beginning of this century (see p. 12) in contrast to the macrophages which he thought were exclusively concerned with ingestion of large particles and cells. Actually, PMN can also phagocytize other cells and inorganic substances of considerable size (erythrocytes, leukocytes, crystals). The process by which either histiocytes (macrophages) or PMN (microphages) ingest foreign material is the identical process of phagocytosis. A general description of ingestion and digestion is given on pages 12 and 14 (autophagosomes of PMN are described on page 122).

Erythrophagocytosis

Living Cells[1]. Examination of the different stages of the phenomenon permits reinterpretation of the appearance in smears (see p. 164 and Fig. 4).

Smears. In hyperacute hemolytic syndromes erythrophagocytosis may be seen in smears of the peripheral blood or the bone marrow.

Accidental superimpositions of erythrocytes on PMN in the thick portions of the smear must not be mistaken for phagocytosis. It should also be noted that cytoplasmic "vacuoles" may represent the end stages of red cell phagocytosis and of hemolysis, which may take place either before or after ingestion (Fig. 2).

Phagocytosis of White Cells and Platelets[2]

Phagocytosis of white cells and platelets may be observed in autoimmune disease. It is quite comparable to that of red cells. The diagnosis of such phagocytosis in smears may be difficult if one does not keep in mind the various stages of the entire process. Very often the PMN ingest only a portion of a leukocyte which they have attacked (Fig. 5). When an entire leukocyte has been ingested, most frequently a lymphocyte, a tart cell results. Its appearance superficially resembles that of an LE cell described below. The sequence of events is, however, quite different, even if the leukocyte nucleus loses its structure and becomes homogeneous after phagocytosis. In the LE cell, only the nucleus is phagocytized and only after homogenization. In immunophagocytosis the entire cell is first ingested and the nucleus homogenized subsequently under the influence of enzymes liberated into the digestion vacuole.

Nucleophagocytosis (Formation of LE Cells)

Smears[3]. Although Hargraves discovered the LE (lupus erythematosus) cells in routine bone mar-

row preparations, they occur with regularity only in preparations in which the buffy coat has been previously incubated in one of several available techniques. LE cells are formed by granulocytes (very rarely monocytes) which have phagocytized the altered nucleus of another cell, most frequently a PMN nucleus or a lobe of a PMN nucleus. Thus the LE cell contains a large phagosome which stains pink in smears and gives a positive Feulgen reaction.

Living Cells[4]. In living preparations, it is possible to observe the entire sequence of events: homogenization of the nucleus of a PMN takes place within minutes of contact with the patient's serum; the normal chromatin structure is lost and lobes fuse, the volume of the nucleus increases; subsequently the altered nucleus becomes surrounded by a rosette of normal PMN attracted by chemotaxis; finally, an LE cell forms by phagocytosis of the homogenized nucleus. The cytoplasmic debris is not phagocytized.

LE cells are found in most cases of lupus erythematosus, although it is not an early sign of the disease. It can also be seen in other collagen disorders. It can be found in 10 to 15% of cases of rheumatoid arthritis if carefully searched for. It has also been described in myasthenia and some ill-defined allergic syndromes. It is occasionally seen after administration of certain drugs, perhaps also on an allergic basis.

Digestion[5–10]

During phagocytosis the cytoplasmic granules discharge their contents into the digestive vacuole. It has been suggested that the enzymes contained in the two different types of granules play different roles in the killing and digestion of bacteria. This is supported by the observation that in congenital disorders of bacterial phagocytosis some bacteria may be ingested and remain alive, while others are killed but not digested (chronic familial granulomatosis, deficiency of myeloperoxidase, deficiency of leukocyte G6PD, Chediak disease and others). These diseases are an important cause of susceptibility to infection in children.

Vacuoles. Vacuoles seen on smears may be the end stage of digestion of bacteria or of autodigestion or may represent fat vacuoles. The different types of vacuoles cannot be distinguished in stained smears. Phase microscopy or examination with the electron microscope is necessary to identify their origin.

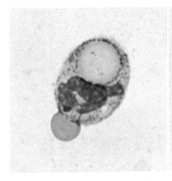 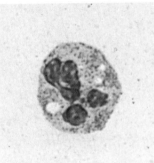 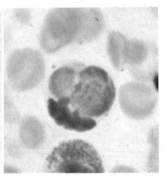

Fig. 1 – *Phagocytosis of a red cell by a PMN*. Note that the red cell adhering to the lower pole of the neutrophil has become spherocytic

Fig. 2 – *Remnants of phagocytized cells* appear as vacuoles

Fig. 3 – *Phagocytosis* of a swollen, rounded-off, and homogenized nuclear lobe of a PMN (L.E. cell)

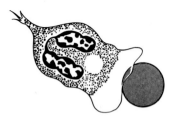 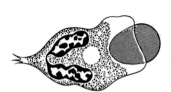 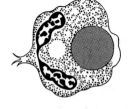

Fig. 4 – *Phagocytosis of a red cell* (cf. Fig. 1)

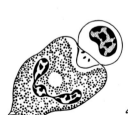   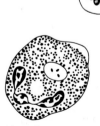

Fig. 5 – *Phagocytosis of a cytoplasmic portion of a lymphocyte*. The amputated nucleus assumes the same shape as in a Rieder cell and the phagocytized cytoplasm appears as a vacuole (cf. Fig. 2)

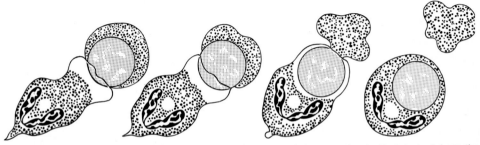

Fig. 6 – *Formation of a L.E. cell* (cf. Fig. 3). The cytoplasm of the granulocyte is detached from the homogenized nuclear segment and left behind in the process of engulfment

8 – Nuclear Appendages

The nucleus of the PMN frequently has appendages to which little attention has been paid, except for the sex chromatin. Although there is some controversy on this subject, recent work allows classification and reinterpretation of some of these appendages.

Certain misconceptions which have given rise to misinterpretation of the nuclear appendages must be recognized: 1) one must not confuse appendages with small nuclear lobes, 2) contrary to textbook statements, their size and shape are not specific. The different shapes (drumstick, golf club, sessile nodule) depend in part on the spreading of the nucleus in smears and on the degree of maturation or pathologic changes of the nucleus. Given these reservations, four types of appendages may be recognized with some assurance.

X Chromatin

It generally assumes the shape of a drumstick (which it resembles only remotely). Drumsticks are present in 1–5% of circulating granulocytes in women. At least 6 cells among 500 must have this feature for female sex determination. Like the Barr bodies (see p. 148), these appendages are the result of condensation of the "lyonized", inactive X chromosome. As long as the nucleus remains unsegmented, the W chromatin remains inside the nucleus. It becomes exteriorized during the segmentation of the nucleus.

Y Chromatin

It forms small clubs rather than X drumsticks. These clubs have a similarly low frequency of about 3.5% in circulating PMN of males. They can be identified with certainty only by their fluorescence after quinacrine staining (which also identifies the Y chromosome when it is not exteriorised).

Nuclear Extensions[4]

Smears. These are nuclear appendages which point towards the centriole (Figs. 2 and 3).

Electron Microscopy reveals that microtubules originating from the centriole are attached to the tips of individual appendages (Figs. 5 and 6). This represents an abnormal persistence of the connection between chromosomes and the microtubules of the mitotic spindle. The traction exerted by the microtubules on the tips of the appendages determine their shape and orientation.

Nuclear Clefts and Pockets[5–9]

They were first recognized by electron microscopy in leukemic cells (see p. 206). Subsequently they have also been seen in other pathologic conditions and occasionally in normals.

Smears. Nuclear pockets are often at the limit of visibility of the light microscope. They appear as unstained or slightly basophilic areas at the edge of the nucleus bordered by a barely visible rim of chromatin. Nuclear clefts are usually overlooked. They can occasionally be recognized as fine unstained cracks in the middle or at the edge of the nucleus (Fig. 1).

Electron Microscopy shows the nuclear pockets to consist of a thin leaf of chromatin covered on both sides by the nuclear envelope which forms a rooflike projection above a superficial invagination of the nucleus. The appearance of a pocket enclosing cytoplasm (which may contain glycogen, ribosomes, and even granules) results from cuts at right angles to the direction of the projection. The rooflike projections can become sufficiently exteriorized during maturation and segmentation of the nucleus to form actual nuclear appendages.

Pockets and clefts are due to anomalies of mitosis. The clefts may be due to persistence of fragments of the nuclear envelope which are not dissolved during mitosis and persist after reconstitution of the nucleus. The pockets present different disturbances in the development of the nuclear envelope and give rise to invaginations which develop into pockets (see figure below) or appendages during maturation and segmentation of the nucleus.

Occurrence[9–11]

Nuclear appendages and even nuclear pockets may occur normally but always in small numbers. An increase in the number of nuclear appendages has been described in chronic diseases with serious prognosis. One may observe them after irradiation or administration of a large number of toxic agents (e.g., acridine, nitrogen mustard, benzene and its derivatives, etc.). Congenital anomalies of granulocytes may be associated with multiple appendages (anomalies of sex chromatin, trisomy).

Stages of formation of a nuclear pocket. The appearance of an intranuclear vacuole results when the pocket is sectioned as indicated by the *arrow*

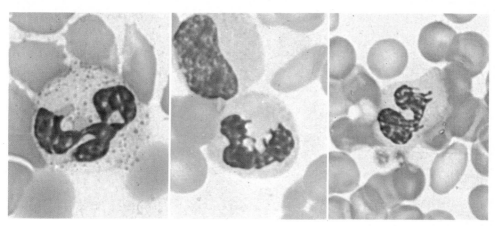

Fig. 1 – *Pockets and clefts* in a nucleus (cf. Fig. 4)

Fig. 2 – *Nuclear appendages* pointing towards the centriole

Fig. 3 – The nuclear *appendages* have become displaced during the preparation of the smear (see Figs. 5 and 6)

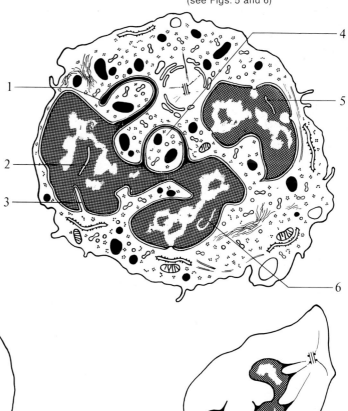

Fig. 4 – *Nuclear pockets (1, 4) and clefts (2, 3, 5 and 6).* Nuclear pockets frequently contain cytoplasm and its organelles (cf. Fig. 1)

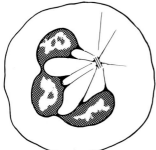

Figs. 5 and 6 – *Nuclear appendages* connected to the centriole by microtubules. On the *left,* as seen in the living cells with the phase microscope (see Fig. 2); on the *right,* after deformation during the preparation of the smear (see Fig. 3)

9 — Abnormalities of Cytoplasm and Granules

I shall not describe the gross alterations of organelles seen in some congenital diseases such as Chediak disease, Alder's anomaly, and Hurler's disease. I shall restrict myself to the description of some minor changes which are often overlooked in stained smears and which one can occasionally reinterpret.

Number and Content of Granules

Smears. Not all PMN have the same number of granules. In pathologic conditions, the granules can be greatly reduced in number or be entirely absent (Figs. 1 and 2). In that case, the cytoplasm is usually of uniform dusky rose-gray color. This appearance can also occur when neutrophilic granules have lost their normal affinity to dyes.

Cytochemistry allows the characterization of some anomalies as follows:

Lack of Myeloperoxidase[1, 2]. It may be due to the absence of azurophilic granules or the absence of the enzyme, although the granules look normal in the EM. In congenital disorders with this deficiency, the number of neutrophilic and azurophilic granules, the alkaline and acid phosphatase, PAS and Sudan black reactions are all normal.

Lack of Alkaline Phosphatase[3]. In the single case described, neutrophilic granules were absent.

Electron Microscopy is necessary to count the number of azurophilic and neutrophilic granules and to analyze their morphologic changes and the content of the different enzymes (see p. 112).

Vacuoles

Smears. In Giemsa stained smears vacuoles appear as holes in the cytoplasm.

Electron Microscopy[4]. Some vacuoles may be seen to contain lipids which can be identified in smears with Sudan black. Others contain remnants of phagocytized cells (see p. 118) or immune complexes, uric acid crystals (in the synovial fluid of gout)[5] or a variety of other substances dissolved during fixation and staining.
Vacuoles may represent the terminal stage of autophagocytosis. Granules may burst and their contents undergo autophagocytosis in a number of pathologic conditions.

Döhle Bodies

Smears. Döhle bodies are small areas of cytoplasm which stain a faint blue with Giemsa and are often located at the edge of the cell.

Electron Microscopy has demonstrated that they are formed from free ribosomes or more commonly from ribosomes attached to parallel sacs of endoplasmic reticulum (Fig. 5). They represent a special case of asynchronous maturation of the cytoplasm.
The basophilia persists in a few small areas of the cytoplasm. They were first described in scarlet fever, but can be found in a variety of other conditions such as erysipelas, diphtheria, tuberculosis, measles, burns, chemotherapy of leukemia, etc.

May-Hegglin Bodies

These inclusions resemble Döhle bodies and are frequently confused with them.

Electron Microscopy[6, 7] demonstrates that they are quite different from Döhle bodies: they consist of aggregates of very fine, filamentous structures sometimes in a paracrystalline arrangement. They are seen in 3–100% of platelets in the May-Hegglin anomaly. Occasionally similar inclusions may be seen in cells other than granulocytes in the peripheral blood of this disorder. Their significance is unknown.

"Toxic" Granulations

Smears[8]. In "toxic" states* one may see all manifestations of agony and death of granulocytes in smears. Frequently, cell death is preceded by the appearance of cytoplasmic vacuoles and "toxic" granules. These are large, irregular granules staining violet-purple with Giemsa. They can be visualized more readily by staining at pH 5.4, since normal neutrophilic granules do not stain at acid pH.

Electron Microscopy[9] has demonstrated that the "toxic" granulations are large azurophilic granules. Two or three such granules often fuse and form secondary lysosomes, either as a result of ingestion of bacteria in severe infections or of denatured serum proteins in rheumatoid arthritis or as a result of autophagocytosis (see p. 14). They contain peroxidase and acid hydrolases.
Toxic granules are almost always accompanied by other cellular abnormalities, particularly Döhle bodies and vacuoles. The number of neutrophilic granules is usually markedly diminished as a result of degranulation. The nucleus is often pyknotic.

---

* The term "toxic" is used to denote a general state of severe malfunction of many cells and organs which are seen in a variety of severe illnesses such as systemic infections, cancer, pneumonia, diabetic or liver coma, toxemia of pregnancy, etc.

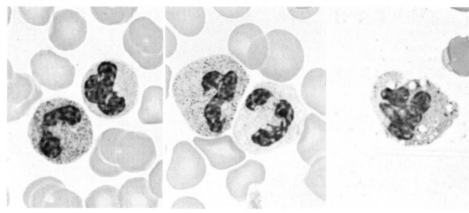

Figs. 1 and 2 — *Variability of staining of neutrophil granules*

Fig. 3 — *"Toxic" granulations and vacuoles* in PMN

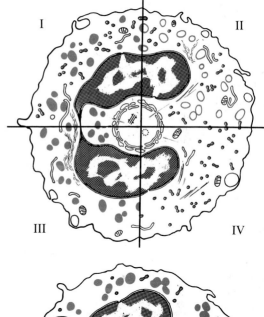

Fig. 4 — *Schematic representation of abnormal granules of PMN*
(I) Normal (azurophilic and specific) granules
(II) Azurophilic granules with negative peroxidase reaction
(III) Absence of neutrophilic granules
(IV) Absence of azurophilic granules

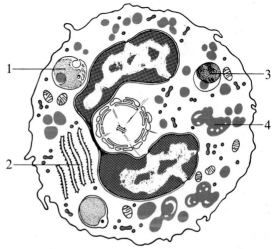

Fig. 5 — *A "toxic" PMN*
(1) Phagosome
(2) Remnants of RER (Döhle bodies)
(3) Autophagosome
(4) Fused azurophilic granules

Maturation and Amplification

Maturation of eosinophils parallels that of neutrophils, except that large eosinophilic granules take the place of neutrophilic granules in the myelocyte in which production of the secondary, specific granules starts and in all later stages. Whether myeloblasts and promyelocytes destined to become eosinophils are already identifiable morphologically is still uncertain.

Smears

The eosinophilic granules are larger than the neutrophilic granules. They stain at first purplish, then bluish, then orange, as the cell progresses through its maturation stages. The bluish purple granulation of immature eosinophils gives rise to the misconception that there are granulocytes which contain both eosinophilic and basophilic granules. No such cells exist, however, as cytochemistry and electron microscopy clearly show.

Electron Microscopy

The granules are formed in the Golgi complex in the same fashion as those of the neutrophils.

Life Span and Death

The fate of eosinophils is unknown. Some of them are phagocytized: granules can be found in phagosomes of macrophages. Others are probably eliminated, as are neutrophils, through the intestinal tract and the lungs. One may find them in nasal and bronchial secretions in large numbers, particularly in pathologic conditions.

Mature Eosinophils

Smears

The nucleus is usually bilobed and may be hidden by the granules. The granules are spherical or rhomboid, with a diameter of 0.5 to 1.5 µm but usually quite uniform. Normally, they fill the entire cytoplasm.

Cytochemical Reactions

The granules contain acid phosphatase, glycuronidase, cathepsins, ribonuclease, aryl-sulphatase, and other enzymes. Peroxidase is present. It is different from the myeloperoxidase of neutrophils. The eosinophilic granules contain phospholipids as well as a basic protein which is responsible for their staining characteristics.

Living Cells

The granules are so thick and refringent that halos tend to obscure details in the phase microscope.

After marked compression, their spherical, rhomboid or occasionally elongated shape can be appreciated. The behavior of the granules during phagocytosis is identical with that of neutrophilic granules.

Charcot-Leyden Crystals. These are seen in tissue in which eosinophils are present in large numbers. They can also form after eosinophils are crushed between slide and coverslip. Their composition is identical with that of the eosinophilic granules.

Electron Microscopy

The granules have a limiting membrane and are formed from two components: a homogeneous or granular matrix and an opaque crystal with a periodicity which can be visualized more or less clearly depending on the fixation, embedding, and staining. The periodicity differs from species to species. In man it is approximately 3 nm. Occasionally a concentric arrangement resembling myelin figures is seen. The acid phosphatase reaction is positive. The mitochondria of the eosinophils are more numerous and larger than those of neutrophils and the Golgi complex is better developed. Ribosomes and endoplasmic reticulum persist longer than in neutrophils. Glycogen particles are numerous.

Chemotaxis and Phagocytosis[1, 2]

The eosinophils respond to the same chemotactic stimuli as neutrophils, but particularly to soluble bacterial factors and antigen-antibody complexes. They play a particular role in allergy.

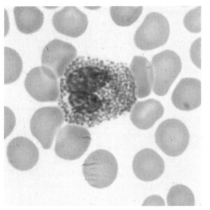

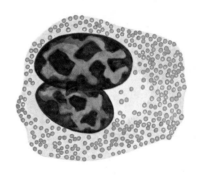

Figs. 1 and 2 — *An eosinophil*. The centrosome can be located by the lower concentration of granules

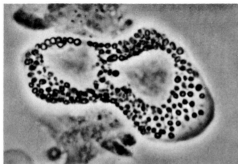

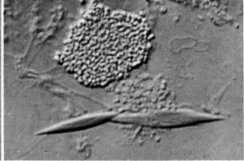

Fig. 3 — *Phase contrast*

Fig. 4 — *Charcot-Leyden crystals* (interference microscope)

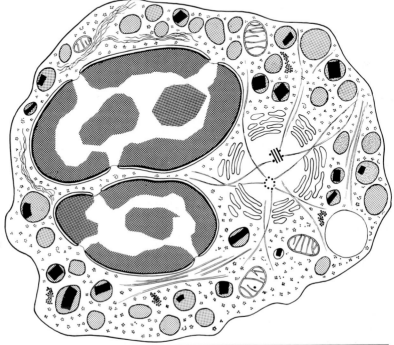

Fig. 5 — The most mature eosinophilic granules contain a crystal which can sometimes be seen at the light microscopic level. Azurophilic granules are rare

Maturation and Amplification

The maturation of basophils parallels that of neutrophils or eosinophils, except that metachromatic (erroneously called basophilic) granules take the place of the neutrophilic or eosinophilic granulations. Nothing is known about the amplification of the basophilic series.

Smears

Before becoming metachromatic, the granules are temporarily basophilic. They round off as they enlarge and may attain diameters of up to 2 μm.

Electron Microscopy

The earliest identifiable granules are formed in the Golgi complex during the promyelocyte stage. In the vesicles of the Golgi complex appear filaments, first in an apparently haphazard fashion. As the cells mature, the filaments form parallel lamellae 13 nm in diamter or form an angular honeycomb pattern with periodicity of 13 and 9 nm.

Life Span and Death

Nothing is known about these subjects.

Mature Basophils

Smears

The basophil has a diameter of 10 to 14 μm and is thus the smallest of granulocytes. The nucleus is usually masked by the numerous large and opaque granulations. The chromatin is homogeneous without clear cut chromatin blocks. The filaments joining the lobes, are quite short and not easily visualized.

The basophilic granules are round, oval or angular. They are stained a dark violet to purple with Giemsa. They measure 0.2 to 1.0 μm in diameter. Frequently, the granules appear to be contained in small vacuoles and occasionally cells are seen with a large number of vacuoles, of which only one or two contain a basophilic granule. This appearance is due either to the solubility of granules in water to which they are exposed during the staining process or to a smearing artifact. Whether the functional activity of the cell manifests itself by the extrusion of granules is controversial.

Acid mucopolysaccharides are responsible for the water solubility of the granules which may persist, however, after the fixation used in Giemsa or Wright's stains. These techniques frequently stain basophils poorly and the "basophilic" granules appear indistinct. The cytoplasm is rose or lavender colored or sometimes does not stain.

Cytochemistry

With toluidine blue, methyl violet or methylene blue, the granules stain a brick red. With neutral red, pyronine, safranine, and methylene azure, they stain yellowish. This metachromasia is due to the presence of acid mucopolysaccharides. The metachromatic granules of mast cells are morphologically quite different from those of basophils. However, both types of granules contain histamine and heparin in large quantities. Numerous enzymes have been identified in basophils: dehydrogenase, diaphorase, histidine-decarboxylase which converts histidine to histamine, and peroxidases. The PAS reaction is positive and appears localized entirely in the cytoplasm, the granules themselves being negative.

Living Cells

In phase contrast, the granules of the basophils are highly refractile and they appear as vacuoles due to the inversion of the phase contrast and are ringed by a dark border. Some are scattered irregularly through the cytoplasm, covering the nucleus. The movement of the basophil is ameboid and is similar to the movement of the eosinophil, though less active. The basophil has little phagocytic capability.

Basophilic Crystals[1,2]. Distilled water can produce intracellular crystals. The cell swells, the nucleus rounds off and becomes homogeneous, the basophilic granules increase in volume and a number of granules rupture. A small fusiform, highly refringent crystal forms and rapidly enlarges. It may attain 10 μm in length. Mastocytes do not form crystals when exposed to distilled water.

When numerous basophils are present and close to each other and are simultaneously destroyed in the preparation of the smear, larger crystals are formed which must not be confused with Charcot-Leyden crystals which they superficially resemble. They are distinct by origin and ultrastructure (see p. 124).

The electron microscope shows the crystals to have a periodicity of approximately 16 nm. The chemical composition of these crystals has not been determined.

Electron Microscopy[3]

Metachromatic granules vary in appearance in different animal species. In man they consist of a number of smaller particles, which are a uniform size in each granule, but which can vary considerably from granule to granule in the same cell. Some of the granules contain myelin figures; others have a floccular appearance and may contain fusiform crystals which appear hexagonal in transverse section with a periodicity of approximately 35 nm. These crystals must not be confused

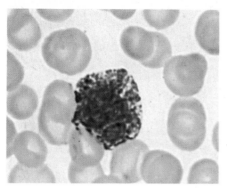

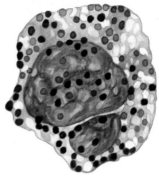

Fig. 1 — *A basophil*. Note the purple color of the metachromatic granules. Some have been emptied of their content during fixation and staining

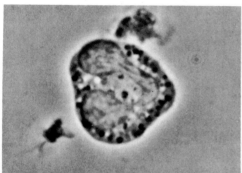

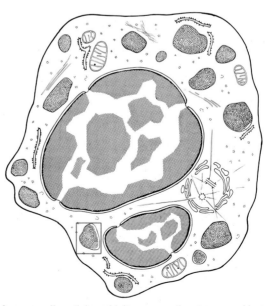

Fig. 2 — *Phase contrast*

Fig. 3 — *Crystal formation* after exposure to distilled water

Fig. 4 — The granules are made up of very small particles which are sometimes arranged in a concentric pattern. There is abundant glycogen

127

with the crystals of eosinophilic granules (see p. 124).

The Golgi complex is small. The centrioles, mitochondria, fibrils and microtubules are unremarkable. The ribosomes and the RER are rarely seen. The cytoplasm frequently contains glycogen particles, occasionally in large numbers (20 to 30 nm in diameter).

Chapter 4

Thrombocytic Series

General Remarks

Definition

The thrombocytic series consists of a succession of cells which starts with the basophilic megakaryocyte and ends with the thrombocyte or platelet*.

Origin[1-3]

The megakaryocyte is derived from a committed stem cell (see p. 18) susceptible to the action of thrombopoietin and in turn derived from a pluripotential cell.

Location

In the normal adult, megakaryocytes are localized primarily in the bone marrow. They can be found in small numbers in many other tissues (lungs, spleen, kidney, liver, etc.) as well as in the circulating blood (on the average 12 megakaryocytes/ml of blood). Concentration methods must be used to demonstrate them. The cytoplasm of the megakaryocytes gives rise to the platelets which pass into the circulation. The nucleus of the megakaryocyte is then phagocytized by the macrophages of the bone marrow. The platelets do not leave the circulation, except that 30% may be sequestered reversibly in the spleen. In splenectomized individuals, all platelets stay within the circulation.

Amplification and Maturation[2, 4]

In contrast to other cell lines in which multiplication (amplification) is accomplished by successive duplication of DNA accompanied by cell division, megakaryocytes multiply their DNA without cytoplasmic division (see p. 21).
Amplification thus consists of polyploidization of the cell. The cells enlarge during amplification, but maturation (characterized by the development of specific organelles) takes place almost exclusively after amplification is completed.
By light microscopy, maturation includes lobulation of the nucleus, increase in cytoplasm, appearance of granules and, later, of platelet territories.

Life Span and Death of Platelets[1-4]

It is still being debated whether platelets have a finite lifespan and normally die of old age, as do red cells, or whether they are randomly destroyed even in healthy individuals by continuously occurring minute events of intravascular clotting. Possibly a varying fraction of platelets is lost in this manner while most of them are removed from the circulation when they reach the end of their life span.
Depending on the degree of random loss associated with platelet survival in normal man, the finite lifespan of platelets has been estimated from 7 to 10 days, and the daily production from 35,000 to 70,000 per mm³.

---

* Purists will note that the name thrombocyte should be reserved for the nucleated cells which play a part in coagulation of blood of birds, fishes, and amphibia; while the name of platelets should be used for the corresponding non-nucleated counterparts in mammals. We will follow our rule to respect common usage and apply the names thrombocyte and platelet interchangeably.

1 — Thrombocytopoiesis

Amplification[1-3]

In normal adult man, the degree of amplification varies. The largest number of platelet forming megacaryocytes (76%) have a 16 N nucleus, corresponding to an amplification of 8. However, 8% of megakaryocytes liberate their platelets after a 4-fold amplification (at a ploidy of 8 N) and 16% after 16-fold amplification (32 N). Exceptionally one may find platelet-forming megakaryocytes at ploidy levels of 4 N and 64 N.

During amplification the intermitotic nucleus remains round. The lobes form during subsequent maturation. They do not correspond to normal 2C nuclei and the ploidy of megakaryocytes cannot be deduced from their lobe count as had been thought.

Maturation

Amplification, which leads to the formation of 8 N, 16 N, 32 N, and rarely 64 N (or even 128 N) nuclei is accompanied by an increase in nuclear and cell size without nuclear or cytoplasmic changes visible by light microscopy. The classification of the stages of maturation is not based on the size of the cell, but on the appearance of new characteristics, both nuclear and cytoplasmic. Any stage of maturation may be associated with any amplification (degree of polyploidization) and hence widely varying cell sizes.

Stages of Maturation[1-5]

It is difficult to separate individual stages of maturation sharply, because maturation is a continuous process, as in all blood cell series. The existence of a megakaryoblast can be postulated from the parallel development of the megakaryocytic and erythroid cell series. At present, however, the term cannot be applied to a clearly identifiable cell*. Occasionally, however, it is possible to identify a cell earlier than the basophilic megakaryocyte as belonging to the thrombopoietic series. In the rat this can be done by demonstration of acetylcholinesterase which does not occur in any other cell line in that species; and in man by fluorescent antiplatelet antibody or the presence of a special type of peroxidase in the nuclear cisterna (distinct from that seen in myeloblasts).

Ordinarily only four stages of megakaryocyte development can be recognized:
1) Basophilic megakaryocyte (MK_1)
2) Granular megakaryocyte (MK_2)
3) Platelet producing megakaryocyte (MK_3)
4) Platelets (and megakaryocytic nuclei)

In the rat, the total maturation time is estimated to be 34 h. Each amplification step (duplication of the genome) requires approximately 10 h. The total duration of amplification and maturation is thus about 3 days, assuming an average of four steps of DNA replication. In man, the total duration of maturation is estimated to be 5 days. The frequency of different stages encountered in normal bone marrow is:

MK_1: 10–15%
MK_2: 60–70%
MK_3: 10–20%

Number and Size of Platelets Formed by Megakaryocytes

This appears to depend partly on the degree of ploidy of the cell. It has been computed that the average 32 N megakaryocyte produces approximately 4000 platelets. It is probable that platelet production does not extend over several days but occurs at the end of the maturation of the megakaryocyte when it discharges all of its platelets in a relatively short time.

Regulation of Thrombocytopoiesis[1, 2, 5-8]

It has been postulated that thrombopoietin triggers the maturation of a committed stem cell similar to the action of erythropoietin in erythropoiesis: one can experimentally stimulate platelet production by injection of plasma of an animal made thrombocytopenic by bleeding or destruction of circulating platelets by antibody. Conversely, one can suppress platelet production by hypertransfusion of platelets.

The relationship between stimulation of thrombopoiesis, ploidy of megakaryocytes, and platelet size is still controversial. In all clinical conditions in which large or giant platelets have been observed, the megakaryocyte ploidy has been shown to be shifted to the left, including not only myelofibrosis, chronic granulocytic leukemia, and hemopoietic dysplasia, but also idiopathic thrombocytopenic purpura, which may be taken as a model of accelerated platelet regeneration. Since megakaryocytes of lower ploidy have in general also lesser amounts of cytoplasm, it has been postulated that the megakaryocyte of lower ploidy produces fewer platelets of larger size. This seems to be the general rule, the only exception being experimentally observed large platelets associated with increased ploidy in the acute phase of platelet regeneration. Conceivably in this particular circumstance the megakaryocyte increases granule production (in parallel to the increased synthesis of hemoglobin in stress erythropoiesis) while production of the DMS (see p. 132) lags behind, leading to larger platelet territories and larger circulating platelets.

---

* The megakaryoblast is a 2 N cell antedating polyploidization. Polyploid cells previously described as megakaryoblasts should be considered basophilic megakaryocytes.

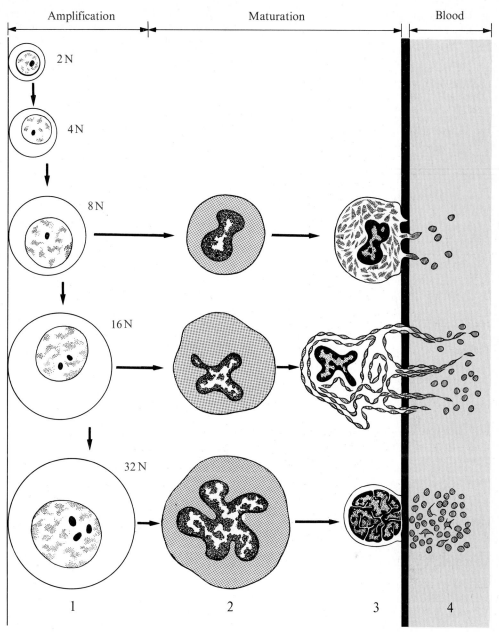

Different stages of thrombopoiesis. (1) MK₁, basophilic megakaryocyte (amplification = polyploidization 2 N to 32 N). (2) MK₂, granular megakaryocyte. (3) MK₃, platelet-forming megakaryocyte and liberation of platelets. (4) Platelets in the circulation. Note that maturation starts after amplification has been completed

2 – Basophilic Megakaryocytes (MK₁) and Granular Megakaryocytes (MK₂)

Basophilic Megakaryocytes

Smears

The basophilia of the cytoplasm is variable but in general intense. The nucleus, whatever its size and DNA content (4 N to 32 N or more), is always spherical. Next to the nucleus, and occasionally indenting it, is an uncolored or pale pink zone corresponding to the centrosome. On its surface appear numerous very fine azurophilic granules. They rapidly extend throughout the cytoplasm and transform the basophilic into a granular megakaryocyte.

Electron Microscopy[1−4]

Electron microscopy indicates that the generally spherical nucleus contains invaginations in which the cytoplasm already contains some organelles. Polyribosomes and ribosomes are very numerous, accounting for the basophilia. The Golgi complex synthesizes the first granules. The first demarcation membranes appear at this stage (see below).

Granular Megakaryocytes

Smears

The abundance of fine azurophilic granules gives the cytoplasm the wine-red color, which together with the large size of the cells (up to 100 µm and more) allows one to recognize the megakaryocyte immediately even under low magnification. The granular megakaryocytes represent the most numerous cells of the series, approximately 60 to 70%. A peripheral basophilic zone may occasionally persist and give the appearance of an exoplasm.

Whatever the degree of ploidy, the nucleus is no longer round but contains a variable number of lobes. In cells of 16 to 64 N, the shape of the nucleus is quite irregular; the lobes are of different size and joined by a major part of their surfaces or by very fine chromatin threads.

Mature megakaryocytes contain considerable amounts of PAS-positive material, for the most part glycogen. Its quantity and distribution varies markedly in pathologic states. The glycogen can be aggregated in conglomerations of various sizes which appear as empty zones in Giemsa-stained smears.

Presumed Phagocytic Role of Megakaryocytes. In smears, images of phagocytosis of red cells or leukocytes are simply appearances of superimposition. Megakaryocytes are not phagocytic. The electron microscope demonstrates, however, that they can ingest submicroscopic particles. Emperipolesis (see p. 150) of granulocytes (or lymphocytes) can produce pictures that are confusing in the light microscope.

Electron Microscopy[1−4]

The nucleus is polylobed and contains small nucleoli. The cytoplasm has developed a large number and a great variety of granules and membranes. The mitochondria are small and numerous; the number of ribosomes diminishes progressively.

Demarcation Membranes (DMS: demarcation membrane system). All along the cell surface invaginations appear and elongate into tubules which remain in communication with the extracellular space. These tubules extend continuously until they reach the immediate vicinity of the nucleus. They are quite tortuous and as they extend throughout the cytoplasm they give the cell a spongelike appearance. On the external surface of these membranes a glycoprotein accumulates. It constitutes the surface coat of future platelets.

Dense Canaliculi (DTS: dense tubular system). These appear to develop from the RER. The fact that a special peroxidase can be demonstrated after suitable fixation in the nuclear envelope, the RER, and the DTS favors this hypothesis. Tracer substances, such as ruthenium red or lanthanum nitrate, placed into the neighborhood of the living cell, penetrate the open canalicular system (DMS) but never the DTS, the RER or the cisternae of the Golgi.

Centrosome. The centrosome is relatively small in relation to the size of the cell. The sacs of the Golgi complex, which form the periphery of the centrosome, give rise to small vesicles which develop into lysosomes. Dense granules containing serotonin are rare or absent at this stage of development.

The centrosome contains a number of pairs of centrioles corresponding to the ploidy of the cell. The number of pairs thus generally varies between 4 and 32.

Microtubules. These are relatively scant in the megakaryocyte, except around the centrioles and during mitoses. However, one can demonstrate in the cytoplasm a number of isomers and dimers of tubulin ready to synthesize tubules, which explains the formation of a large number of marginal tubules found in platelets.

Microfilaments. Microfilaments are few in number in the usual preparation. Special techniques (glycerinization) make them appear. They constitute thrombosthenin, which represents 20% of platelet protein (see p. 138). Thrombosthenin contains both an actinlike and a myosinlike substance, thrombosthenin A and thrombosthenin B.

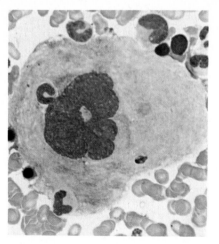
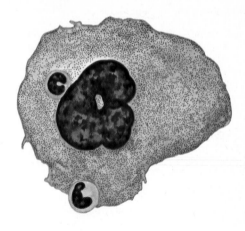

Fig. 1 – *A granular megakaryocyte* (MK₂). Note two neutrophils which appear to be within the cytoplasm, but which are in fact external to it. (They are not undergoing phagocytosis.) Because of the large size of the cell, the magnification here is 500 × instead of the usual 2000

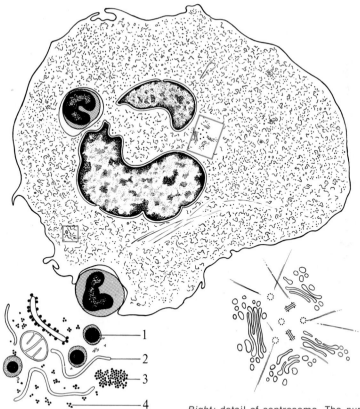

Fig. 2 – *Left:* details of cytoplasm. (1) azurophilic granule; (2) demarcation membrane; (3) glycogen; (4) ribosomes

Right: detail of centrosome. The number of centrioles corresponds to the ploidy. Here only six centrioles are present in the plane of section, presumably the cell contains eight

3 − Platelet-Forming Megakaryocytes (MK₃)

Smears

In this stage the azurophilic granules assemble in small groups of 10 to 12, separated by narrow channels of clear cytoplasm. This appearance is usually uniform throughout the cytoplasm, but can remain localized in an otherwise diffusely granular cytoplasm. Exceptionally a few such platelet territories may even be seen in a basophilic megakaryocyte.

The platelet-forming megakaryocytes represent 10 to 20% of all megakaryocytes. In addition to the megakaryocyte types enumerated, one can find naked megakaryocytic nuclei on the one hand and fragments of platelet producing cytoplasm on the other. These are artifacts in the preparation of the smears, although in pathologic circumstances fragments can result from an asynchronism of maturation (see Figs. 4 and 5, p. 137).

Living Cells

In the phase microscope, the megakaryocyte appears round and highly refractile, mimicking a mass of iron filings brightly illuminated. It is only on compression that one can distinguish the contours of the nucleus and of the future platelets.

Electron Microscopy[1−4]

The future platelets can be readily seen at this stage, more or less delineated by the demarcation membranes and containing a few organelles (granulations, mitochondria, ribosomes, etc.). The DMS, like the coat of platelets, contain glycoprotein. The Golgi complex is quite small and contains only two or three cisternae. It appears to fragment and to become distributed among the platelet territories.

The number of ribosomes is diminished, as is the production of granules. Large mitochondria remain within the perinuclear zone, while small mitochondria appear and are distributed throughout the cytoplasm of future platelets. The granulations are azurophilic. Serotonin-containing dense bodies are rare or absent (see p. 138).

The dense canalicular system is fragmented throughout the territory of future platelets. The canaliculi give a peroxidase reaction as does the perinuclear cisterna. The microtubules, although later abundant in platelets, are very seldom seen at this stage. However, vincristine leads to the appearance of numerous characteristic crystals, which demonstrate the existence of monomers of tubulin in megakaryocytes.

Thrombosthenin is abundant but exists in a monomeric form (in contrast to the platelets, see p. 138) and becomes visible in form of microfilaments only after glycerinization. The filaments are 5−12 nm in diameter.

Numerous vesicles, the nature of which is as yet undetermined, lysosomes and accumulations of glycogen may be seen. The nucleus becomes smaller and smaller; the granulations become larger and more closely aggregated in the center of the platelet territories.

Number of Platelet Territories

It has been suggested that the number of platelets formed by a megakaryocyte depends on the quantity of DMS synthesized which, in turn, depends on the ploidy of the cell. In this fashion, megakaryocytes of 64 N will form approximately 4000 platelets of normal volume, and megakaryocytes of 8 N only 500 to 1000, but of a large volume (macrothrombocytes). Not all authors agreed on this point (see p. 130). Under pathologic circumstances the number and volume of platelets formed vary considerably (see pp. 130 and 142).

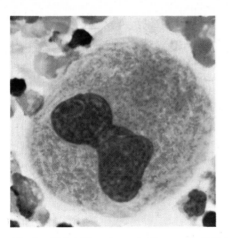
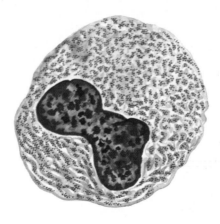

Fig. 1 — *Platelet-forming megakaryocyte (MK₃)*. The platelets have become individualized and can be seen clearly in some parts of the cytoplasm (500 ×)

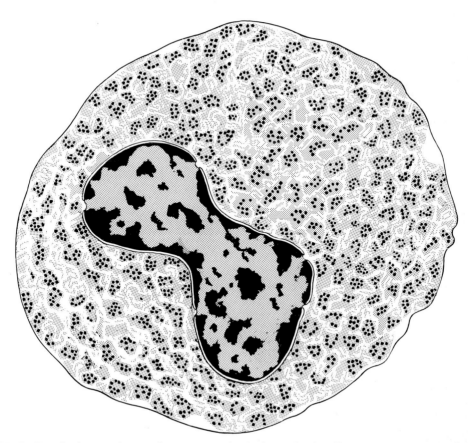

Fig. 2 — Note the demarcation membranes separating the future platelets. The centrosome is not included in the section

4 – Platelet Formation

Living Cells[1]

In 1906 Wright correctly deduced from the examination of excellent histologic preparations that platelets are preformed in large bone marrow cells (called megakaryocytes by Howell in 1890) as a result of cytoplasmic differentiation and that the differentiated part of the cytoplasm herniates across the wall of sinus capillaries and liberates platelets into the circulation by fragmentation of the cytoplasm. Even so, the origin of platelets became the subject of numerous controversies which have only recently been set to rest by the accumulation of more and more convincing evidence. The direct proof of Wright's hypothesis as well as description of the stages of liberation of platelets have been furnished by observation of the entire phenomenon in the living state between slide and coverslip.

The surface of megakaryocytes deforms. Cytoplasmic extrusions appear, become elongated and curve in various directions, giving the cell the appearance of an octopus. The major part of the cytoplasm is now contained in the arms or projections of the cell. They continue to become elongated, ramified, and filiform. At regular intervals these filaments have small bulges which correspond to the platelets. Within minutes the projections extend further. The platelets emit some dendrites and remain interconnected only by a fine thread of cytoplasm which finally breaks.

In accelerated time-lapse cinematography, one may see intermittent contraction of the megakaryocyte extensions. They shorten and thicken abruptly, then slowly resume their former shape and continue to elongate. If one taps the coverslip at the moment when the megakaryocytic filaments have completely developed, the currents provoked by this sudden tapping detach the platelets from the fine threads which connect them. Subsequently, they adhere to the glass and spread (see p. 140). The time required for the development just described depends on the stage of maturation of the cell. Some megakaryocytes may be transformed into platelets in less than 3 h, while others require 12 h and more.

Smears[2]

The octopuslike appearance obtainable in vitro is only rarely seen in direct smears of the marrow aspirate. It is probable that the cytoplasmic extensions encounter capillary sinuses in the course of their elongation; that they enter the sinuses and liberate rosaries of platelets, which are then carried off by the bloodstream and dissociated. Even though such rosaries of platelets are not seen normally, they are quite commonly seen in the blood of numerous thrombopathies, thrombocytopenic purpuras, and thrombocythemias.

Electron Microscopy[3, 4]

In the elongated arms of megakaryocytes, territories comprising several future platelets enter venous sinuses by diapedesis and so liberate the platelets. Sometimes the liberation of platelets appears to take place in a synchronous manner. After the megakaryocyte has liberated the platelets, the naked nucleus (actually always surrounded by a small amount of cytoplasm) becomes pyknotic and is phagocytized by neighboring macrophages.

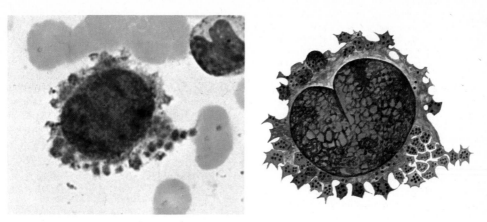

Fig. 1 — *Platelet-forming megakaryocyte*. In this cell only a narrow ring of cytoplasm remains bordering the nucleus

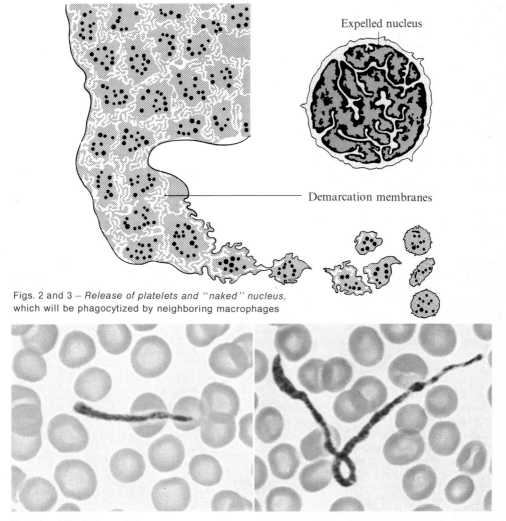

Expelled nucleus

Demarcation membranes

Figs. 2 and 3 — *Release of platelets and "naked" nucleus,* which will be phagocytized by neighboring macrophages

Figs. 4 and 5 — *Strings of platelets in the peripheral blood in thrombocytopenia*

5 – Platelets

Smears

Blood platelets aggregate almost immediately on contact with tissues, glass, or even air. A large number are rapidly destroyed in this manner. It is, therefore, necessary to collect the blood with a sufficiently large needle, in siliconed glassware without an anticoagulant, or an anticoagulant that affects platelets minimally, and to make one's preparations immediately.

In well-made smears, 80% of platelets appear as small round or oval bodies similar to their appearance in vivo when directly visualized within capillaries. They appear as a small mass of clear cytoplasm, slightly basophilic and containing azurophilic, reddish-violet granulations. The granules remain dispersed or form a crown around a central vacuole or are tightly packed in the center of the platelet and give the appearance of a nucleus.

A certain number of platelets will fail to display the discoidal shape just described even in well-made smears. In carelessly made smears, almost all platelets appear in the form of stars with three to a dozen points or mimic a comma, a cigar, a butterfly, etc. These appearances are due to the delay between the collection of the sample and the preparation of the smear, which allows the platelets to assume different dendritic forms (see p. 140). Before considering such appearances as pathologic, they must be verified in fresh preparations examined immediately by phase microscopy, in which case normal platelets are always discoid.

Platelets measure 2 to 5 μm in diameter, and their volume varies from 5 to 10 μm³. However, during active regeneration and in pathologic states the diameter may reach 10 and even 20 μm (macrothrombocytes, see p. 142).

Electron Microscopy[1-8]

Membrane and Outer Coat. Surrounding the membrane which borders the platelet and corresponds to the membrane of other cells, a coat approximately 15 nm in width exists. It consists primarily of glycoprotein and plays an important role in the physiology of the cell. It contains different adsorbed proteins which can also enter the interior of the cell through the system of open canaliculi. It contains a number of plasma factors involved in coagulation as well as a number of enzymes.

Microtubules. Microtubules are similar to those of other cells but are arranged in bundles of 3–10 and form a circumferential band at the periphery of the cell. The microtubules are probably responsible for the maintenance of the lens-shaped form of the circulating platelet.

The microtubules disappear in the cold at 4° C and reappear on reheating. Colchicine and vinca alkaloids lead to disappearance of microtubules. After washing, the tubules reappear, but fail to rearrange themselves in the usual form, and distribute themselves haphazardly in the cytoplasm.

Microfilaments. These are very numerous, particularly in the periphery of the platelet, close to the marginal ring of microtubules. The monomers of microfilaments represent approximately 20% of the proteins of platelets. They are polymerized only under the influence of physiologic stimuli, or experimentally by glycerine.

Granulations. Granules are of several types. The azurophilic granules seen by light microscopy correspond to the alpha granules of electron microscopists. Their content is still being debated. Hydrolases (arylsulfatase and acid phosphatase) have been demonstrated in lysosomal vesicles distinct from alpha granules. The "dense" granules are storage sites of serotonin (5-hydroxytryptamine) which is made in the enterochromaffin cells of the intestine and picked up by the circulating platelets from the plasma. The dense granules also contain calcium and probably nucleotides. Catalase has been identified in perioxisomes, which are also small vesicles. Platelets secrete all of these substances as well as platelet factor 4 and fibrinogen during the "release" reaction. However, the peroxidase localized to the dense canalicular system is not secreted.

The granules are always surrounded by a membrane. Their diameter varies from 150 to 300 nm, but occasionally much larger granules may be seen. They are generally round, but occasionally shaped like a rod, a golf club or a drumstick. Occasionally, the central zone is dense, giving the appearance of a bull's eye.

Reticulum, Vacuoles, and Canaliculi. Occasionally, one may see some cisternae of smooth or rough endoplasmic reticulum or free ribosomes. Numerous vesicles are present, of varying shapes and sizes. Their origin and function have not yet been adequately defined. Some of them are derived from the Golgi complex of the megakaryocyte, others from DMS, rarely from pinocytosis. The canaliculi, sometimes dilated, represent invagination of the plasma membrane and probably are continuous with the DMS. They have also been called the open canalicular system.

Mitochondria. These are smaller than in other cells (0.15 to 0.30 μm in diameter). One to five are usually found in each section of a platelet.

Inclusions. Inclusions of glycogen form granulations of 15 to 30 nm in diameter, usually with irregular outlines. Lipid inclusions are occasionally seen in normal platelets, but more frequently in degenerating forms.

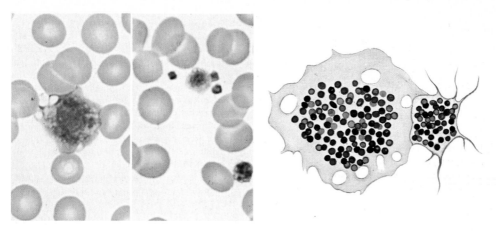

Fig. 1 — *Macrothrombocytes (giant platelets) and normal-sized platelets*

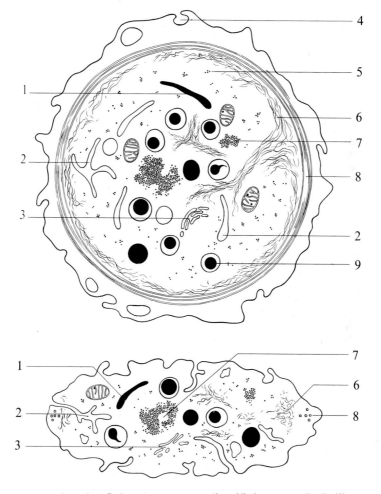

Fig. 2 — *Above: equatorial section. Below: transverse section.* (*1*) dense canaliculi; (*2*) open canaliculi; (*3*) remnants of Golgi; (*4*) pinocytosis; (*5*) ribosomes; (*6*) microfilaments; (*7*) glycogen; (*8*) microtubules; (*9*) azurophilic granule

6 – Movements of Platelets

Living Cells[1]

Circulating Forms. When the preparation is made rapidly and examined immediately, particularly from venous blood anticoagulated by sodium citrate or heparin, platelets are discoid or oval. They rapidly become spherical. The microtubules disappear if platelets are preserved at 0–4° C for some hours.

Dendritic Forms. They develop within minutes between slide and coverslip. The projections are at first short and gradually lenghten to 3.0 to 15.0 μm. They are broadly based but have filiform extensions (Fig. 1). The extensions change their shape and direction continuously. The phenomenon is too slow to be easily appreciated on direct observation, but becomes very apparent in time lapse cinematography.

Spread-out Forms. When the projections of pseudopods encounter the glass slide, they adhere, enlarge, and other dendrites join and spread similarly. The cell adheres entirely to the slide, becomes rounded off, free of dendrites and transparent, except in the center where the granules are concentrated (Fig. 3). Spread-out forms are reversible and must not be considered degenerative.
The hyaloplasmic border may undergo undulation and movements which carry along granulations and contractile vacuoles. Nevertheless, platelets are incapable of active locomotion.

Pinocytic Vacuoles. These may be seen to form in spread-out platelets after observation for long periods. The platelet projects a veil; its central portion becomes more transparent, while its borders become darker. Subsequently, two arms project on each side of the center portion which becomes a vacuole as the arms rejoin. Finally the vacuole moves to the center of the cell. Between slide and coverslip, platelets change after approximately 1 h; the vacuoles fuse and form bizarre patterns. One may observe shapes which resemble the vacuolated platelets in the circulating blood of certain thrombopathies.

Contractile Vacuoles. Contractile· vacuoles are generally found in the center of the cell, occasionally in the midst of the granules of the chromomere. They expand slowly over a period of 20 to 30 min at room temperature. In contrast, they contract rapidly in about 15 s. After fixation and staining, the contractile vacuoles appear as round holes.

Electron Microscopy of Spread-Out Forms[2, 3]

Spread-out platelets may attain diameters of 10 to 20 μm. The granules, mitochondria, and vacuoles are accumulated in the center. The periphery is formed by a very thin cytoplasmic veil.

Because electrons can readily pass through the hyaloplasm, platelets were among the first cells examined by electron microscopy when the techniques of thin sections were not yet perfected. With the then available primitive techniques the hyaloplasm appeared to contain fibers, particularly well seen after shadowing (Fig. 2). We can now identify some fibrils as microtubules, while others represent bundles of microfilaments (Fig. 2). Using a technique which allows one to cut cells parallel to the surface on which they have spread, the disposition of microtubules and fibrils in the course of spreading has been defined. The central zone of granules projects above the rest of the cell. Four to eight microtubules surround it concentrically. In sections closer to the bottom of the spread out cell, the hyaloplasm is seen to contain tubules and bundles of microfibrils which surround the centrally located granules and occasionally extend into the dendritic projections of the cytoplasm.

Functions of Platelet Organelles[4–8]

Surface Membrane. It plays a role in adhesion and aggregation. It is capable of endocytosis (pinocytosis, phagocytosis, micropinocytosis) and this permits the cell to accept and later reject a certain number of substances from the environment. The surface is continuous with the canaliculi which penetrate deeply into the interior of the cell; various substances can enter and leave the cell by this route.

Microtubules. They appear to give the platelet the shape it assumes in the circulation and, in rupturing, give rise to the dendritic projections.

Microfilaments. Microfilaments play a role in concentrating the granules in the center of the platelet, the contraction and the degranulation of the platelet, and the retraction of the clot.

Granulations. During the "release reaction", the contents of alpha granules and dense granules (as well as lysosomes and other storage sites) are discharged. The materials secreted or discharged include calcium, fibrinogen, nucleotides (ADP), hydrolytic enzymes, serotonin, and platelet factor 4. The release reaction and aggregation can take place without contact with glass or other foreign surfaces. For example, thrombin, other proteolytic enzymes, bacterial endotoxin, and collagen can trigger the release reaction and subsequent aggregation.
Under pathologic conditions, some of the granules or their contents may be deficient or the appropriate stimuli may fail to lead to their release or both (storage pool disease, aspirin effect). In thrombasthenia, normal-looking platelets undergo a normal release reaction without subsequent aggregation.

Fig. 1 — *A platelet with dendrites* (scanning electron microscope)

Fig. 2 — *A platelet spread on slide* (transmission electron microscope after shadow casting)

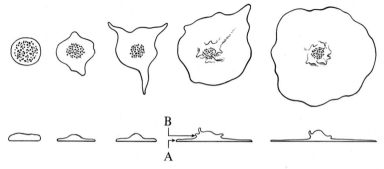

Fig. 3 — *Different stages in the spreading of a platelet on a glass slide. Above:* as seen on slide. *Below:* cross section. A and B sections at different levels illustrated in Figure 4

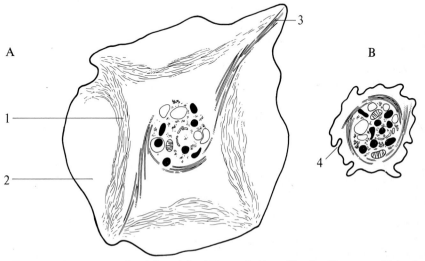

Fig. 4 — Electron microscopy at level A and B of Figure 3. Note (*1*) microfilaments; (*2*) hyaloplasm; (*3*) microtubules; (*4*) pinocytosis

General Pathology of Megakaryocytes

The pathologic changes of megakaryocytes may be provisionally classified as disorders of 1) amplification, 2) maturation, and 3) altered cytoplasmic organelles. These three categories are interrelated to various degrees.

Surprisingly, an association between specific changes in organelles of megakaryocytes and particular malfunctions or diseases of platelets has been documented only in a few instances. A variety of changes including vacuolization of granular megakaryocytes (Fig. 4) have been claimed to be characteristic of idiopathic thrombopenic purpura, but the consistency of such changes has been questioned. Electron-microscopic studies have indicated only moderate quantitative changes in large membrane complexes which cannot account for the vacuolization described in smears. Disordered maturation is even less frequently recognized, perhaps because increasing lobulation of the nucleus and increasing granularity of the cytoplasm are not highly correlated during normal maturation and consequently asynchrony of maturation is not readily recognizable, except in micromegakaryocytes (Figs. 1, 2, and 3).

In contrast, the degree of polyploidy or amplification is relatively easy to measure and a shift toward lesser degrees of ploidy has been well documented in myelogenous leukemia and preleukemic states, now more properly designated as hemopoietic dysplasias (see p. 203). In these conditions large or giant platelets, occasionally with giant granules ascribed to fusion of the azurophilic granules, have been noted. Reduced ploidy has also been documented in the majority of cases of idiopathic thrombopenic purpura in which platelets tend to be large (Fig. 5). Since idiopathic thrombopenic purpura is due to platelet destruction by an autoantibody and increased platelet size is known to be associated with recovery from thrombocytopenia, it was postulated that ploidy and platelet size are inversely correlated. However, experimental destruction of platelets by an isoantibody is temporarily followed by a parallel increase in ploidy and platelet size (see p. 130).

Small Megakaryocytes (Micromegakaryocytes)[1-5]

Smears. Although the ploidy of a megakaryocyte cannot be deduced directly from its size or from lack of lobulation, cells with single or double small nuclei are common in chronic granulocytic leukemia and hemopoietic dysplasias in which the ploidy is known to be reduced to 8 N or 4 N. The cytoplasm of these small cells is granular and sometimes considerably in excess of the nuclear volume, making the cell readily recognizable as a megakaryocyte. Sometimes in the same pathologic states, e.g., chronic myelogenous leukemia with impending blastic crisis, some micromegakaryocytes have a very scant basophilic cytoplasm which may contain a few localized areas of platelet territories, as may occasionally be observed in normal basophilic megakaryocytes (see Figs. 1 and 2, and p. 134). The basophilic micromegakaryocyte must not be confused with the megakaryoblast as defined on page 130. Some leukemias classified as lymphoblastic have been shown to be in fact megakaryoblastic leukemias by the presence of the special peroxidase in the perinuclear cisterna. These megakaryoblasts have nucleoli and do not have identifiable platelet formation.

Electron Microscopy. In some instances, electron microscopy demonstrates the existence of DMS in the cytoplasm and areas of platelet formation. Occasionally the appearance is so abnormal that their identification as megakaryocytes requires the identification of the characteristic peroxidase enzyme, which differs from myeloperoxidase.

General Pathology of Platelets

Pathologic platelets may differ from normal in size, shape, and staining characteristics. In practice, however, only increased size is readily seen in smears. It requires electronic volume measurements to appreciate the reduction in platelet size characteristic of Aldrich-Wiskott syndrome (thrombocytopenia and increased susceptibility to infections in male children, transmitted by their unaffected mothers).

Macrothrombocytes[6-8]

Normally platelet size varies considerably and it has been suggested that the larger platelets are those just released which then become progressively smaller during their 8–10-day lifespan in the circulation. This thesis has now been recognized as being incorrect. Nor has it been proven that larger platelets differ from smaller ones in density. Macrothrombocytosis is said to be present when more than 20% of the platelets measure more than 2.5 µm in diameter. Megathrombocytes* may occur which reach diameters of 20 µm.

Stress Macrothrombocytes[6-8]

Thrombocytopenia stimulates platelet production. When the stimulus is very marked, large platelets appear. These platelets are said to be more effective in hemostasis**. This is correct in the sense that a smaller number of these large platelets produce the same effect as a somewhat larger number of normal platelets. This is particularly evident during active regeneration after acute thrombocytopenia and is probably due to the fact

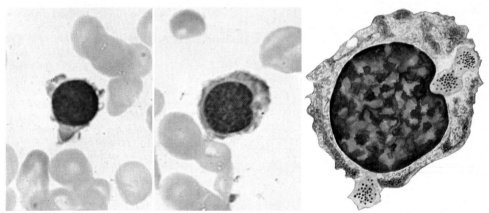

Figs. 1 and 2 – *Micromegakaryocytes* in the peripheral blood (may be mistaken for lymphocyte). In Figure 2 a few maturing platelets, one ready to be released

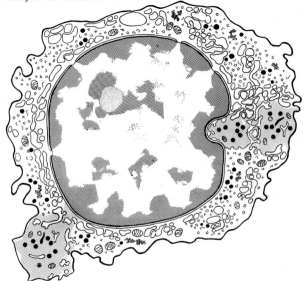

Fig. 3 – Micromegakaryocytes are often abnormal. In the cell shown here, the open canaliculi (demarcation membranes) are widened

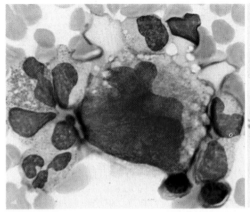

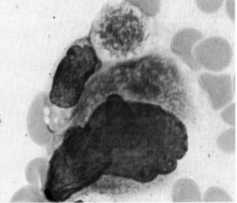

Fig. 4 – *Abnormal granular megakaryocytes* (700 ×). Note the numerous vacuoles

Fig. 5 – *Basophilic megakaryocyte and macro-thrombocyte*

143

that the total volume of platelets per unit of blood or plasma rather than the number of platelets determines the hemostatic effect. In many instances the production of large platelets has been found to be associated with a "shift to the left" and lesser ploidy as well as smaller size of megakaryocytes. This relationship has been well established in myeloproliferative disorders as will be discussed below and it is probably also true in some instances of idiopathic thrombocytopenic purpura. However, during the recovery from acute thrombocytopenia produced experimentally, a shift to the right with increased megakaryocyte ploidy has been reported (see p. 130).

Macrothrombocytes in Leukemia and Allied Disorders

Very large "giant" platelets may occur in these disorders and are frequently associated with abnormalities of cellular organelles when examined with the electron microscope. Giant granules, believed to be the result of fusion, can occasionally be identified in the light microscope. They may be as large as 2.5 µm in diameter, the size of an entire normal platelet. Such pathologic granules may be an early sign of hemopoietic dysplasia (preleukemia).

Abnormalities of Number of Granules

Although a decrease in the number of granules has been discussed frequently, no constant relation between number of the azurophilic granules seen in smears and any functional abnormality has been established. Similarly, vacuolization is not indicative of platelet pathology. Dense granules, which contain serotonin, ADP, and probably calcium are diminished in a rare thrombopathy, the Hermansky-Pudlak syndrome (see below). Functionally, HPS platelets are similar to storage pool disease platelets. Both types react poorly to aggregating agents as do normal platelets exposed to aspirin in vivo or in vitro. All are characterized by deficient serotonin release and absence of a second wave platelet aggregation. The reduction of dense bodies is seen in HPS platelets and is not visible in smears. A concentration of or conglomeration of the entire complement of granules in the center of the platelet can be induced by initiating the clotting process and making blood smears before the blood has actually clotted. Such appearances are thus usually seen as results of careless blood collection and can probably not be used for the diagnosis of intravascular coagulation which might conceivably induce them. (A single case of absence of granules, the so-called Gray Platelet Syndrome, has been described.)

Thrombo(cyto)pathies[9]

These are, by definition, functional platelet defects not due to thrombocytopenia itself, though platelet levels may be reduced. The classical disorder is thrombasthenia. It is now known to be due to a defect in the platelet coat which fails to become sticky in response to endogenous ADP (which is released normally) or when exposed to exogenous ADP. The platelets do not adhere normally to glass. No ultrastructural abnormalities have been identified. Bernard-Soulier syndrome is associated with giant platelets which, like thrombasthenic platelets, fail to adhere to glass or endothelium. There is coexistent thrombocytopenia. Again specific ultrastructural defects are missing.

---

* Giant platelets must not be confused with fragments of megakaryocytic cytoplasm, which one may find in the blood. As a rule, these fragments are not rounded off like platelets, but are elongated, filamentous, or helmet shaped (see pp. 136 and 137). Nor must giant platelets be confused with cytoplasmic debris of other cells, which is frequently seen in leukemia. Such cytoplasmic fragments are always round and, in general, basophilic. Their granules, if present, are azurophilic, and thus resemble those of platelets, but are distributed regularly throughout the cytoplasm rather than aggregated in the center.
** *Mediterranean Macrothrombocytosis.* An inverse relationship between the level of platelets and their size may be a general phenomenon. Recently a constitutionally lower level of platelets in Mediterranean people compensated by a larger volume of individual platelets has been reported. No studies of the ploidy of the megakaryocytes in the different populations are yet available.[10]

Chapter 5

Lymphocytic Series

General Remarks

Uncertainties of Present Knowledge

Our concepts of the origin and function of lymphocytic cells have undergone drastic changes in recent years, thanks to the introduction of new techniques and an enormous number of investigations on the role of lymphocytes in immunology. Older theories which considered the small lymphocytes as terminal cells have been entirely abandoned. It is now well established that most lymphocytes are, on the contrary, in a state of quiescence and have multiple possibilities for further evolution. It is also recognized that the term lymphocyte encompasses a large number of cells with different functions, and possibly even different origins.

Notwithstanding the immense quantity of research[1,2], which has taught us a large number of new facts, there remain large tracts of unexplored territory, and many interpretations remain uncertain. We shall, therefore, give only a broad outline of the major questions which are currently under discussion and will emphasize what is useful for the reinterpretation of blood smears.

Definition

The lymphocytic series consists of a succession of cells which, starting with the committed stem cell, leads to the production of the small lymphocyte. It comprises the lymphoblast, the large lymphocyte and the small lymphocyte. It also comprises a series of cells originating from the small lymphocytes which have the capacity of "rejuvenation" or transformation to use the term generally employed. The progeny of transformed lymphocytes may resume the appearance of the original small lymphocyte.

From a functional point of view, one may distinguish virgin lymphocytes and conditioned lymphocytes; cells with a short lifespan and others with a lifespan of many years; cells that are capable of recognizing an antigen, and memory cells. These different functions, however, do not correspond to morphologic characteristics that can be recognized in smears, except for a few special instances (see p. 156).

Origin

Only a few years ago it was thought that lymphocytes were produced in the "germinal centers" of the lymphoid tissues. It is now known that these are "reaction centers" (see p. 154), and that lymphocytes originate in the "central" lymphoid organs, i.e., the bone marrow, and perhaps the thymus.

Migration, Circulation, and Recirculation

Contrary to red blood cells and platelets, whose entire functional life takes place in the blood, and unlike mature granulocytes, which leave the blood vessels without reentering them, lymphocytes leave the circulation and return to it many times in the course of their life. They leave the bloodstream (circulation) predominantly in the lymphoid spaces of the tissues (migration), are taken up by the lymphatics, and after traversing one or more lymph nodes return to the bloodstream by way of the thoracic duct (recirculation). It appears that this recirculation permits a large number of lymphocytes to come into contact with antigens wherever these may have entered the organism. In the absence of recirculation, only local lymphocytes could come into contact with antigens, and consequently only a very small number of cells would be available to trigger immune reactions (see p. 152).

Life Span

The estimation of the life span of lymphocytes poses complex problems, because of their recirculation, because they can transform into other cells (see p. 154) and because we still do not know all of the possible ways of their eventual death or disappearance.

In general it is assumed that at least two populations of lymphocytes exist, as far as their life spans are concerned. One-third of circulating lymphocytes are believed to live 10 to 20 days, the rest may live several months or even years. A life span of 25 years has been documented for some human lymphocytes.

Death and Disappearance of Lymphocytes

Some lymphocytes disappear through transformation into other cells as a result of antigenic stimulation (see pp. 154 and 171). Another group of lymphocytes is lost through the digestive tract and the lungs. Finally, the largest number of lymphocytes are phagocytized and digested by histiocytes in the lymph nodes and lymphoid follicles of other tissues. This phagocytosis is extensive in cases of marked antigenic stimulation and gives rise to the "tingible bodies" of Flemming in macrophages (see p. 162), best seen in the germinal centers.

1 – The Lymphoblast and Large Lymphocyte

Lymphoblast

The description to follow is that of the classical lymphoblast, although it is uncertain that one can distinguish it from the first stages of a stimulated lymphocyte (see p. 154).

Smears

Lymphoblasts have a diameter of 15–20 μm. Their nucleus is round or oval, stains a light pink color, and has a high nuclear-cytoplasmic ratio. The chromatin if finely distributed. One or two nucleoli are always present and are characteristically a pale blue color or even colorless. The cytoplasm is sharply delineated and scanty. It forms either a uniform perinuclear band or is slightly more developed toward one side due to a larger centrosome. The cytoplasm is basophilic and often more deeply basophilic at the margins of the cell.

Electron Microscopy

The nucleus has finely dispersed chromatin and contains a large nucleolus, often surrounded by a narrow ring of chromatin or small blocks of chromatin. The cytoplasm contains ribosomes and polyribosomes, but little RER. The number of microtubules is fairly large. The Golgi complex is small. One or two lysosomes are usually seen. A Gall body and one or two compound vacuoles (see pp. 148 and 149) are frequent, but not unique features of lymphocytes. The surface of the lymphoblast has microvillosities and is actively micropinocytic.

The Large Lymphocyte

It is doubtful that the size of the lymphocyte, its basophilia, or the distribution of its chromatin can be used as criteria of its age and function. At present it is impossible to distinguish in smears such functionally different lymphocytes as virgin lymphocytes and conditioned lymphocytes, lymphocytes with a short or long life span or B- and T-lymphocytes, except under very special conditions (see p. 156).

Only two types of lymphocytes can be distinguished, large and small ones. This is an arbitrary classification, since lymphocytes of intermediate size are always present. They must be assigned to one or the other group. It is doubtful that this distinction between small and large lymphocytes has any physiologic or diagnostic significance. It should rather be considered a provisional "categorization" (see p. 187).

Smears[1]

Large lymphocytes measure approximately 9.0 to 15.0 μm, corresponding to a volume of 300 to 900 μm³. The nucleus is centrally located or slightly eccentric. It is smaller than the lymphoblast nucleus, darker, and with chromatin arranged in compact blocks, separated by lighter zones without sharp demarcation.

Nucleoli are only rarely seen in stained smears (in fact they are always present, but are obscured by the perinucleolar dense chromatin and can only be visualized with special stains, e.g. methylene borate). The cytoplasm is scant, moderately basophilic, or a light blue. It may be so scant as to be barely visible as a fine blue line surrounding the nucleus.

Azurophilic Granulations. Azurophilic granulations correspond, at least in part, to lysosomes. They are relatively large, 0.3 to 0.6 μm, and are occasionally contained within a small vacuole. They are more frequently seen in the circulating cells than in the aspirates of lymphoid organs. Normally they may be seen in 5–30% of lymphocytes. Their number varies usually from one to six and rarely exceeds a dozen.

Gall Body. This name refers to an uncolored vacuole (see p. 148).

Cytochemistry. Numerous attempts have been made to find cytochemical characteristics to separate lymphocytes from monocytes and to distinguish the different populations of lymphocytes that are known to exist. So far no absolutely specific characteristics have been found, with the exception of cells which secrete antibody (see pp. 148 and 149). The large lymphocyte presents cytochemical reactions similar to those of the small lymphocyte (see p. 148).

146

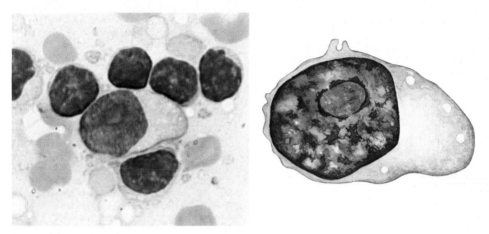

Fig. 1 — *A lymphoblast* (lymph node aspirate). Note the fragments of cytoplasm among the small lympho-cytes surrounding the lymphoblast. These originate from cells either spontaneously dead (see Fig., p. 23) or fragmented by the puncture

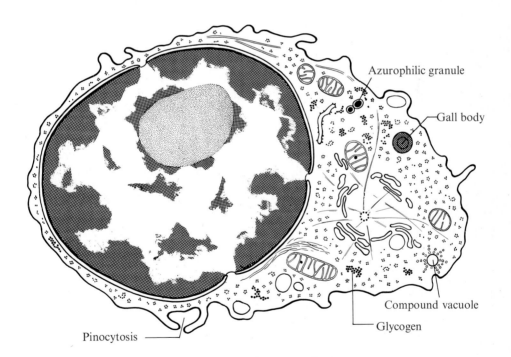

Azurophilic granule

Gall body

Compound vacuole

Glycogen

Pinocytosis

2 – The Small Lymphocyte

Smears

The small lymphocyte has a diameter of 6.0 to 9.0 µm, corresponding to a volume of 200 to 300 µm³. The nucleus is round or very slightly oval, rarely reniform, quite dense, a dark violet, with very heavily staining chromatin masses. The nucleolus cannot be seen except with special stains. The nucleus usually occupies nine-tenths of the area of the cell. The scanty cytoplasm is often barely visible and quite often seen only in part of the nuclear periphery. It may form a fine blue line around the nucleus or have the appearance of a pointed hat.

The female chromatin (Barr body) is usually obscured but may be seen in cells in culture. In contrast, the chromomere of the Y chromosome can be visualized in male lymphocytes by fluorescent techniques, even in the resting intermitotic nucleus. Binucleated lymphocytes may be seen with a frequency of one or two per 10,000.

Cytochemistry

The following cytochemical features are noteworthy:
1) The Gall body may be stained lightly with Sudan black indicating that it contains neutral fat.
2) Glycogen is present in only small quantities. During preparation of smears, it is generally condensed into small masses. It is seldom seen in form of PAS positive granules in normals, although it is not infrequently present in leukemic cells.
3) A large number of enzymes can be identified, particularly lysosomal ones.
4) The peroxidase and nonspecific esterase reactions are always negative, a useful characteristic to differentiate lymphocytes from cells of the granulocytic and monocytic series.

Electron Microscopy[1]

Except for the size and the configuration of the chromatin, large and small lymphocytes have the same complement of organelles and are quite similar.

The small lymphocyte has a nucleolus surrounded by compact masses of chromatin, which accounts for the fact that it cannot be seen in Giemsa-stained smears. The cytoplasm contains ribosomes and polyribosomes and some smooth or rough endoplasmic reticulum. It contains little glycogen, particularly compared with the PMN. It is always finely dispersed, which indicates that the small masses of PAS-positive material seen in smears are artifacts.

The Gall body, which is perfectly round, has a grey center and a dark periphery and consists primarily of lipids. One or more compound vacuoles or multi-vesicular bodies are usually present, a number of microtubules, and frequently some bundle of fibrils. Contrary to some texts, they are not characteristic of monocytes. The azurophilic granules, which are often too small to be seen in the optical microscope, have a dense center surrounded by a membrane.

Cytoplasmic Inclusions[2,3]. A variety of inclusions have been described which are too small to be seen with the light microscope. Some were first noted in abnormal lymphocytes and subsequently identified in a small number of cells in healthy individuals. They include rodlike, vermiform, or tubular structures or crystals. The crystals and rodlike structures probably represent secretions of lymphocytes, such as globulins (see p. 194). Spherical inclusions (vacuoles in smears) may be seen in storage diseases (see p. 168).

Surface of Lymphocytes. The lymphocyte surface sometimes extrudes veils (Fig. 1) and almost always presents microvillosities indicating that lymphocytes are capable of active micropinocytosis. They are readily visualized with the scanning electron microscope (Fig. 1, p. 151). It has been suggested that T-lymphocytes have a smooth surface, while the surface of the B-lymphocyte is studded by numerous villi. It is more likely that the appearance of the surface is an expression of the functional stage of the cell and the microenvironment in which it finds itself.

Hyperbasophilic ("Pyroninophilic") Lymphocytes

Some lymphocytes (large or small) may have a deeply basophilic cytoplasm. If stained with pyronin, the blue of the Giemsa stain is replaced by a vivid red. However, the significance lies always in the increased number of ribosomes, indicating increased synthesis of proteins. It is not known whether these pyroninophilic cells represent cells at the beginning of their transformation (see p. 154) or whether they are characteristic of a functional state or a specific subgroup of lymphocytes. There is nothing specific about the basophilia itself.

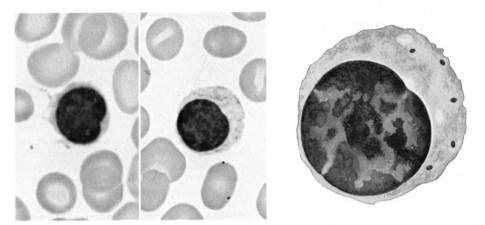

Figs. 1 and 2 — *Small lymphocytes*. Note the indentation of the nucleus caused by the centrosome, which is quite small. The granules, barely visible, stain red and represent lysosomes. Cytoplasmic veils project from the cell on the left

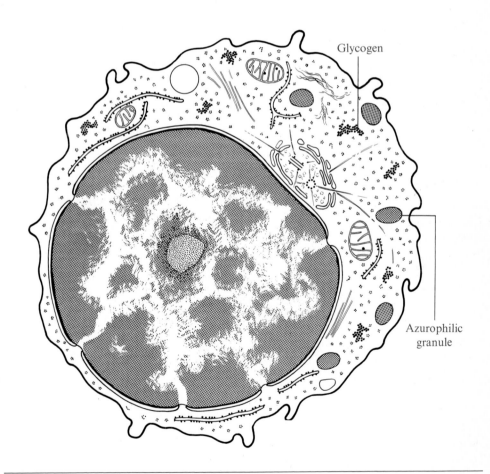

Glycogen

Azurophilic granule

3 — Movements of Lymphocytes

Locomotion

Lymphocytes are actively motile at 37° C. They move as fast as granulocytes, approximately 35 µm per minute.* During locomotion they usually assume the shape of a "hand mirror". The anterior portion of the mirror proper contains the nucleus and the centrosome. The handle corresponds to the uropod, which has the same appearance as uropods of other leukocytes with multiple ruffles and filaments attached to other cells, to debris or to the support on which the lymphocyte rests. It has been suggested that in lymphocytes the uropod plays a role in exchange of information with other cells during peripolesis and in the destruction of target cells. During locomotion, the anterior portion of the lymphocyte, the protopod, presents veils which are quite small compared to those of granulocytes and monocytes (see p. 16 and footnote, p. 160).

Intracytoplasmic Movements

Mitochondria, lysosomes, vacuoles and Gall bodies are carried along in active intracytoplasmic movements. The Gall body, which is seldom obvious in smears, is readily seen in phase contrast as a very small clear vacuole surrounded by a dark line. It increases in size if the cell survives for a few hours between slide and coverslip, but never exceeds 1 µm in diameter. It doesn't stain with either Janus green or neutral red. A small number of cells contain two or three Gall bodies. The centrosome moves rhythmically and in the process indents the nucleus as we have already described for granulocytes (see p. 116).

Chemotaxis and Spreading

Unlike granulocytes and monocytes, lymphocytes do not spread spontaneously. Nor are lymphocytes capable of phagocytosis, although this statement may require some modification (see below).
Although chemotaxis cannot be demonstrated in lymphocytes, some means of taxis must exist in the organism to account for the large accumulation of lymphocytes under pathologic conditions in unusual locations. Conceivably they are only retained in these locations under pathologic conditions, a form of pathologic homing (see p. 24).

Endocytosis and Exocytosis

While there is complete agreement that lymphocytes cannot phagocytize large particles, they can be made to phagocytize fragments of erythrocytes under very special experimental circumstances. In contrast, lymphocytes can very rapidly take up small particles less than 1.0 µm in diameter by micropinocytosis.
The exocytosis of products secreted by the lymphocyte has not yet been observed directly, but the Jerne method of hemolytic plaques demonstrates clearly that the lymphocytes can secrete the antibodies they produce. Lymphocytes are capable of clasmatosis, i.e., they shed fragments of cytoplasm, a process which some authors believe contributes to the origin of plasma proteins (see p. 14).

Capping[1, 2]

Particles and molecules which attach themselves to the surface of the lymphocyte (or other leukocytes) are transported and collected in a small portion of the cell surface, generally corresponding to the uropod. If fluorescent molecules are studied, the semicircle in which they collect gives the appearance of a cap covering the lymphocyte, hence the name "capping". The particles or molecules may subsequently be incorporated into the cytoplasm by micropinocytosis.

Rosette Formation[3]

In certain circumstances, a crown of erythrocytes attaches to a lymphocyte, thus forming a "rosette" (see p. 156). It has been shown by phase microscopy and time lapse cinematography, that during locomotion obtained by heating the slide to 37° C the rosetting erythrocytes move toward the uropod to which they remain attached. This is a special case of the general tendency of surface movements which result in the transport of particles toward the uropod.

Peripolesis and Emperipolesis

Peripolesis is a term given to a phenomenon seen frequently, though not exclusively, in lymphocytes. The lymphocytes attach themselves to another cell and surround it. It has been suggested that this attachment of lymphocytes to macrophages or epithelial cells plays an important role in the transmission of information from one cell to another. Emperipolesis implies the penetration into the interior of the cell or the passage of one cell through another. Both cells remain healthy and mitosis of lymphocytes has been observed in the interior of liver cells, cells of the ovary, the intestine, and even cancerous cells.

---

* The movements of lymphocytes can only be observed when the distance between slide and coverslip is sufficiently great to avoid compression of the cell. The PMN will move even in very thin preparations.

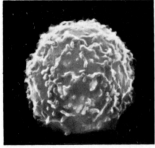

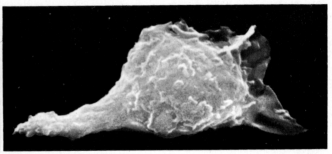

Fig. 1—*A small lymphocyte* as it exists in the circulation. The villosities of its surface may be short or long

Fig. 2 — *A large lymphocyte in motion*. Note the characteristic appearance of protopod and uropod

  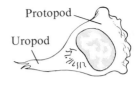

Protopod

Uropod

Fig. 3 — *Different stages of the locomotion of a lymphocyte*. Note the orientation of the centrosome toward the uropod

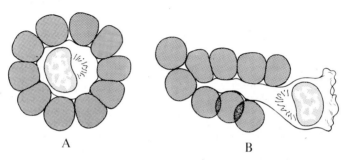

Fig. 4 — *Successive stages of "capping"*. (1) the fluorescent antibody is distributed over the entire cell surface; (2) and (3) it moves posteriorly; (4) it is localized at the uropod

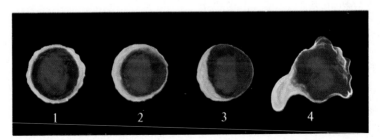

A   B

Fig. 5 — *Red cells forming a rosette* (A). When the cell moves, the red cells migrate toward the uropod (B) (from a filmed sequence)[2]

We shall describe the cytologic transformation and the movement of cell populations produced by the introduction of an antigen[1-4]. They are important for the recognition of certain appearances in smears of lymph node aspirates.

Progression From Local to Systemic Immunity

When an antigen penetrates into the tissues, it induces a local reaction. A small number of cells (lymphocytes, histiocytes) capture the antigen and carry it to the regional lymph node via the lymph that drains the tissues. In the node lymphocytes are transformed and proliferate and give birth to new cells. Some of them remain in the regional node, while others are transported to more central lymph nodes to proliferate there. This results in an increased number as well as dispersion of immunized cells.

A small number of cells, having traversed the entire chain of nodes, enter the circulation via the thoracic duct and recirculate (see p. 145) for months or years as memory cells, which can respond at any time to the new arrival of the same antigen. In this fashion the initially local immunity becomes generalized and protects the entire organism.

Regional Lymph Nodes

After contact with the antigen, lymphocytes "transform" (see p. 154) and multiply in the germinal centers. Subsequently, they can evolve in at least two directions: 1) they may develop into plasmocytes, remain in the medullary cords, and secrete humoral antibodies (see p. 174); 2) leave the node and colonize other more central nodes where they can again evolve in these two directions. A large number of lymphocytes leaves the circulation through postcapillary venules and thus reenter germinal centers.

Lymph

It is important to note that the population of lymphocytes in a stimulated lymph node is quite different from that in the lymph. 1) The histiocytes enter the node through the afferent lymph but are absent from the efferent lymph. Their fate is unknown. 2) The number of lymphocytes which enter the nodes represents approximately one-tenth the number which leaves the lymph node. 3) Plasmocytes appear to remain in the medullary cords. They are only very rarely seen in the efferent lymph.

Reinterpretation of Smears of Lymph Node Aspirates. The "adenogram" can give important diagnostic information, although in more than half the cases it must be supplemented by a biopsy, because lymph node puncture results in intermingling of aspirated cells and deprives one of the diagnostic value of the topography of the lesions.

The aspirate of a normal lymph node consists of small and large lymphocytes, a few lymphoblasts, histiocytes, and rare mastocytes and plasmocytes (Fig. 1, p. 155).

The aspirate of a normal lymph node consists p. 155) has a predominance of large lymphoid cells as well as small markedly basophilic cells which are difficult to identify in stained smears. In order to reinterpret these appearances, one must keep in mind the phenomena of cell transformation and traffic of different cell populations (Fig. 2, p. 155), of which we have just given a very simplified and abbreviated schematic representation.

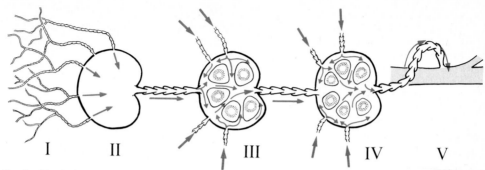

Fig. 1—*Circulation of lymphocytes after local antigenic stimulation.* (I) network of capillary lymphatics draining the local tissue; (II) local lymph node; (III) regional lymph node; (IV) central lymph node; (V) drainage into the venous circulation

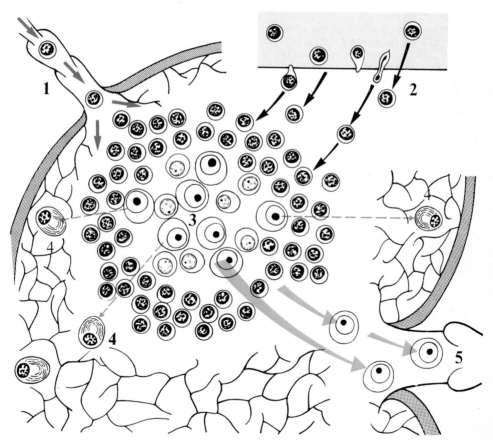

Fig. 2 — *Cell alterations in a stimulated lymph node.* (1) lymphocytes entering the node; (2) lymphocytes entering the general circulation after passing through the wall of a postcapillary venule; (3) "transformation" of lymphocytes in a germinal center; (4) migration of individual plasmocytes derived from the transformed lymphocytes toward the cortical zone of the node; (5) passage of "transformed" lymphocytes into efferent lymphatics and drainage into the regional node

Transformation in Vitro[1, 2]

It has been known for many years that the following phenomena take place in cultures of leukocytes of the peripheral blood: granulocytes degenerate within hours, monocytes spread on the glass surface and become macrophages, and the majority of lymphocytes remain alive. One or two percent of the surviving lymphocytes enlarge and within 3 to 10 days form large cells with prominent nucleoli. Some of them subsequently divide.

When one adds phytohemagglutinin (PHA) to the cultures, 70–90% of lymphocytes enlarge, acquiring the appearance of "blast cells" and are referred to as "transformed" lymphocytes. Two to five percent of the transformed cells divide, producing small lymphocytes while the rest may revert to their original size without division.

A large number of investigations have attempted to elucidate the mechanism and the significance of lymphocyte transformation. At present (1977) three types of transforming factors are recognized: 1) nonspecific mitogens (PHA), pokeweed mitogen (PWM), concanavallin A, periodate, lipopolysaccharides and many others; 2) allogeneic cells (histoincompatible); 3) antigens to which the donor of the lymphocytes has previously been sensitized (viruses, bacteria, and drugs).

Stages of the Transformation of Lymphocytes. Living Cells

Within 8 h of culture with PHA, production of RNA is demonstrable by isotope incorporation. Within 48 to 72 h the maximum number of cells have become transformed and at the end of this period mitoses appear. Many of the transformed cells die. They may exhibit marked alteration of the cytoplasm with numerous vacuoles or pyknosis of the nucleus.

The division of the large, transformed lymphocytes may in turn give rise to a new generation of small lymphocytes, which can again be transformed.

Smears

The completely transformed lymphocytes are cells 20 to 30 μm in diameter with a homogeneous, markedly basophilic cytoplasm, often with a somewhat more lightly stained cytoplasm immediately adjacent to the nucleus. Numerous vacuoles are often present. They can be shown to contain neutral fat by special stain. The nucleus is large, round, and has finely reticulated chromatin with several nucleoli which stain blue and are often of irregular shape.

All intermediate stages between small lymphocytes and completely transformed ones may be seen in smears. (The name blast cell should be avoided.)

After a number of days (the exact number of days depending on the origin of the cells and the conditions of culture), the transformed lymphocytes enter mitosis, of which all stages may be seen in smears.

Electron Microscopy[3]

Electron microscopy provides details of all of the previously described features. The nucleus becomes lighter, nucleoli hypertrophy, elongate or become irregularly branched. The mitochondria increase in size and develop more numerous, slightly packed, parallel cristae, which give them an unusual appearance (Fig. 4). The lamellae of the Golgi complex multiply rapidly and soon occupy a large portion of the center of the cell. The number of ribosomes similarly increases, and in a number of cells numerous long polyribosomes are formed, consisting each of 12 to 20 individual ribosomes. The endoplasmic reticulum develops, but remains relatively inconspicuous compared with that of plasma cells. Multiple vesicular bodies and fatty vacuoles are present in many cells. Azurophilic granules, possibly lysosomes, increase only slightly in number and size. Acid phosphatases and other hydrolases may be identified in the granules. The cell surface shows evidence of pinocytosis and micropinocytosis.

One can identify IgA, IgG, and IgM in the process of secretion, if the transformed cell is of B cell origin.

Transformation in Vivo

Reactive (Inflammatory) Lymph Node

Smears of lymph node aspirates show (Fig. 2) a large number of giant lymphocytes with lightly stained nuclei, several bluish nucleoli, and basophilic, often deeply basophilic cytoplasm. All intermediate stages between small lymphocytes and transformed cells may be seen. When the inflammation lasts several weeks, plasmoblasts, proplasmocytes, and plasmocytes appear.

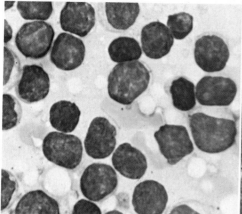 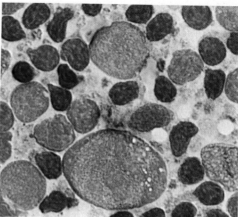

Fig. 1 — *Normal lymph node aspirate.* Presence of both large and small lymphocytes. A poorly stained preparation: the narrow rim of cytoplasm is barely stained and appears pale, rather than the usual blue, except in one or two cells in the upper left corner

Fig. 2 — *Lymph node aspirate in an infection.* All stages of transformation of lymphocytes are seen

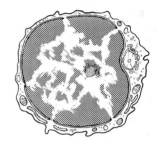

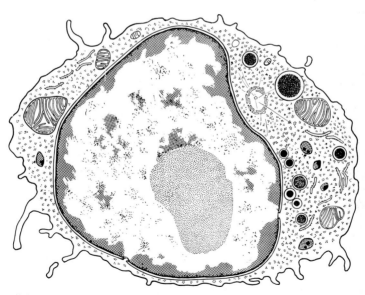

Fig. 3 — *A small lymphocyte and a transformed lymphocyte.* Note the hyperplasia of the nucleolus, the giant mitochondria, numerous lysosomes, and a large number of ribosomes, which are responsible for the intense basophilia of the cells in smears

6 – B- and T-Lymphocytes and Others

We have seen that different types of lymphocytes exist, based on differences in size, appearance or staining of the cytoplasm or nucleus, granulations or inclusions. These differences do not always correspond to differences in function which can now be distinguished thanks to special techniques.

B-, T-, and Null Lymphocytes[1-3]

As has already been noted, many different populations of lymphocytes exist. The most important classes are: 1) the T-lymphocytes, involved in cellular immunity (transplantation immunity, delayed hypersensitivity, viral and fungal immunity). They originate in the thymus or are controlled by it. 2) B-lymphocytes, involved in humoral immunity. In birds this population originates from the bursa of Fabricius, in mammals their origin is still in doubt.

The distinction between these two cell types is based on a number of reactions which comprise determination of membrane markers (receptors for Ig, for aggregated IgG, and for certain complement fractions, specific antigens for B and T cells, E and EAC rosettes); transformation in vitro by mitogens (PHA, PWM, Con A); and other reactions. .

The peripheral blood contains 15–20% B-lymphocytes and 70–80% T-lymphocytes, the rest being difficult to classify.

There remain many facts to understand or clarify: some monocytes may have similar markers, some lymphocytes have features of both B and T cells, some T-lymphocytes appear to be capable of transforming into B cells, etc. Thus the separation of B and T cells may not be as complete and precise as was originally thought.

Other lymphocytes exist which have neither B nor T cell markers and which have been called Null cells. They probably comprise several, perhaps quite different, cell types. It should be noted that in leukemias any cellular marker can be modified or supressed as a result of the disease (see p. 208).

Smears[4]

The morphology of B-lymphocytes is very variable, extending from the small lymphocyte to very large transformed cells. Colorless intracytoplasmic crystals may be present on rare occasions in normals and not infrequently in very old persons or in certain diseases. The crystals probably represent secretion of immunoglobulins (see p. 194).

The T- and other lymphocytes have no special features in Giemsa stains.

"Atypical" Lymphocytes

In normal individuals, a small number of mononuclear cells (1 in 10³) are large cells with markedly basophilic cytoplasm (see hyperbasophilic or pyroninophilic lymphocytes, p. 148). In order to see a sufficient number of such cells in a smear, one has to prepare buffy coats. Thymidine incorporation indicates that some of them are in S-phase. Examination with the electron microscope confirms the presence of large numbers of polyribosomes and a variable development of the RER.

These or similar cells are present in increased number in a variety of diseases, and since they occur in certain viral infections have been called virocytes. However, similar or identical cells can also be seen in a number of allergic reactions (particularly following administration of serum or of certain drugs) and sometimes in sterile or bacterial infections.

The terms atypical lymphocytes, infectious mononucleosis cells, virocytes, and stimulated lymphocytes have been applied to these cells more or less interchangeably and none of them is either fully descriptive or rational. The cells seen in the peripheral blood seldom have the prominent nucleoli of the lymphocytes transformed in culture by antigenic stimulation from which the term "stimulated" lymphocyte is derived. The term "atypical" is too broad to be useful. The cells in infectious mononucleosis have the additional feature of frequently horseshoe-shaped or irregularly indented nuclei, but these features are not unique and, therefore, do not warrent their designation as "infectious mononucleosis cells." Perhaps the term hyperbasophilic lymphocyte best expresses their common feature and that, at the moment, our knowledge of these cells is a purely descriptive one.

Occurrence. The principal diseases in which the number of markedly basophilic cells is increased are infectious mononucleosis, rubeola in which both the transformed lymphocytes and plasma cells are numerous, chicken pox, German measles, toxoplasmosis, brucellosis, acute viral hepatitis, viral pneumonia, herpes zoster, infection of the upper respiratory tract with adenovirus.

Chapter 6
Monohistiocytic Series

General Remarks [1-5]

The classical notions and definition of the mono-cytic series and its relationship with the so-called reticuloendothelial system have undergone drastic changes in recent years. New facts have been discovered thanks to new techniques. The old concepts founded on morphologic similarities and on indirect arguments based on histologic sections have been revised. Since it is important to link the present to the past, the principal stages of the evolution of our new ideas will be reviewed.

The Reticuloendothelial System. Metchnikoff, the great Russian zoologist working at the Pasteur Institute in Paris in 1892, was the first to recognize a group of cells which played a major role in the defense of the organism against a great variety of extraneous invaders. Metchnikoff named the white cells of the blood (now known as polymor-phonuclear) *microphages,* believing them to be concerned solely with the defense of the organism against small microorganisms, i.e., bacteria, and designated as *macrophages* certain cells in the tissues because they phagocytized large preys like parasites and other cells. This distinction is now known to be erroneous, since both cell types are capable of phagocytizing large and small particles. However, while the term microphage has fallen into disuse, the term macrophage has been retained for cells capable of ingesting a variety of particulate materials.

Ashoff (1913, 1924) found that a variety of cells distributed throughout the organism became stained when certain dyes were introduced into the living animal. Ashoff united all the cells so labeled under the name of RES, the reticuloendo-thelial system. This term has been justifiably criticized in recent years for a number of reasons: 1) a "system" according to the dictionary is an organic whole of interdependent parts such as the digestive or nervous system. There is no inter-dependence between reticular and endothelial cells; hence, the term "system" appears inappro-priate. 2) The term "reticulum cell" is now used (or should be used) only for cells that are (at least potentially) capable of forming reticulin fibrils and collagen. In Ashoff's time, however, the term reti-culum cell was applied indiscriminately to true reticulin-forming cells, to the histiocytes and mac-rophages which should be sharply separated from them, and sometimes even to presumed hemo-poietic precursor cells. 3) The endothelial cells are not truly phagocytic and only capable of micro-pinocytosis (see p. 12), i.e., the uptake of colloidal iron and similar materials, a common feature of

almost all cells. In some organs, however, macro-phagic histiocytes line sinuses and have thus re-placed the endothelial cells in that location. These are true histiocytes in endothelial positions and must not be considered as endothelial cells proper.

Thus the term reticuloendothelial system (RES) is a misnomer and another instance of nomencla-ture based on erroneous concepts blocking new ideas and engendering largely futile discus-sions.

We are nevertheless left, if not with a system, at least with an assortment of cells derived from monocytes in need of a name. A "system of mac-rophages" appears doubly undesirable, because the concept of a system appears inappropriate and because PMN are as much macrophages as monocytes. The term "monohistiocytic series" appears to me the most descriptive and I have adopted it in this book.

It has recently become customary for any hematol-ogist and immunologist who studies primarily the function of the free and stimulated cells under experimental conditions to use "macrophage" as a synonym for the histiocyte. I have bowed to this practice when feasible. However, we cannot elimi-nate the histiocyte altogether. We still talk of the malignancies derived from the histiocytic cell series as histiocytosis X and histiocytic lymphoma and not as macrophagosis X or macrophagic lym-phoma. In fact these malignant cells often have no tendency to phagocytosis. The term histiocyte also appears more appropriate for cells derived from monocytes but not actively phagocytic and for those cells the origin of which is not readily determined.

Definition and Origin

The monohistiocytic series comprises a variety of phagocytic cells, related by origin and function, which include the blood monocytes, alveolar mac-rophages, peritoneal macrophages, Kupffer cells, and free and fixed macrophages of the bone mar-row and other organs.

It is generally believed* that all of these cells are derived from bone marrow monoblasts and pro-monocytes, enter the blood as monocytes, and later enter the tissues to develop into "macro-phages".

---

* The classical theory according to which monocytes are derived from histiocytes (called reticulum cells by early investigators) is apparently in error and should be re-placed by the reverse hypothesis. Nevertheless, it is not established that *all* histiocytes are derived from mono-cytes, nor that *all* monocytes are derived from precursors in the bone marrow.

1 — Monocytes

The precursor cell of the monocyte in the bone marrow is unknown. Some believe it resembles a lymphocyte, others that it is related to the myeloblast. Although monoblasts and promonocytes undoubtedly exist, they cannot at present be identified with certainty in either Giemsa-stained smears or in the phase microscope.* It is, however, possible to do so with the electron microscope.

Smears

I will describe the two extremes of the appearance of monocytes, the large and the small monocyte, although all intermediate stages occur.

Large Monocytes. These have a diameter of 30 to 40 µm. Their nucleus is large, located centrally or eccentrically, usually irregular, kidney-shaped or multilobed. The chromatin appears pale and has a lacelike or reticular appearance without compact chromatin blocks. Occasionally, a nucleolus may be seen. The cytoplasm is ample, greyish-blue and has fine azurophilic granulations which are occasionally so numerous as to give the cytoplasm a pink color. Cytoplasmic vacuoles are quite common.

Small Monocytes. They measure 20 to 30 µm in diameter. The cytoplasm is a dusky blue. Occasionally it contains a few azurophilic granules; its nucleus is round to oval; it can be triangular or square, but this appearance is due to an artifact (see p. 224); the chromatin is pale and floccular. The nucleolus is generally not visible. The small monocyte is sometimes difficult to distinguish from a large lymphocyte by morphology alone; it is readily identified as a monocyte by its cytochemical reactions and phagocytic potential.

Cytochemistry [1-4]

The monocyte gives a positive reaction with peroxidases. The intensity of the reaction is always lower than in PMN and varies from cell to cell, which probably reflects differences in the number and composition of lysosomes. Although esterases are present in both monocytes and PMN, one can effectively differentiate the two cell lines if suitable substrates are used. Naphtol AS-D chloracetate is virtually "specific" for PMN, while the "nonspecific" esterase reacting with α-naphtyl butyrate is characteristic of monocytes. Both monocytes and granulocytes contain lysozyme, but only monocytes can continue to produce and excrete it. This permits their differentiation by lysis of bacteria in the surrounding medium in special tests. The monocytes fix immunoglobulins on their membrane, but do not produce them.

Electron Microscopy [5]

The nucleus contains one or two small nucleoli surrounded by perinucleolar chromatin. Nuclear projections and pockets may be seen even in normals.

The cytoplasm contains many ribosomes and polysomes, little glycogen and RER. The mitochondria are small and sometimes elongated. The Golgi is well developed and always situated in the indentation of the nucleus. Numerous microtubules are present. Microfibrils are better developed than in other blood cells. They are often present in bundles adjacent to the nucleus.

The granules, many of which are too small to be seen by the light microscope, are dense, homogeneous and limited by a membrane. Some of the monocyte granules are azurophilic, peroxidase positive, and contain lysosomal enzymes. After phagocytosis they fuse with phagosomes and form secondary lysosomes. There also exists a population of peroxidase-negative granules, the composition of which is as yet unknown. In young monocytes (promonocytes) the peroxidase can be seen to be present in the perinuclear cisterna, the RER, and the Golgi.

In mature monocytes, the Golgi is peroxidase negative. The peroxidase-positive and peroxidase-negative granules thus represent two successive populations of granules, similar to the successive production of azurophilic and specific granules in promyelocytes and myelocytes.

Life Span and Death

The sojourn of monocytes in the circulation has been estimated at 30 to 100 h. The cells leave the blood randomly and regardless of their age, by diapedesis, after they have become adherent to the endothelium. In sites of inflammation, they accumulate very rapidly.

The life span of macrophages (histiocytes) is long and can attain 75 days or more, as has been shown by direct observation of transparent chambers inserted into the ear of rabbits and observed continuously. The death of monocytes and histiocytes proceeds in an unknown manner, but it is known that the cells when damaged and particularly after intense phagocytosis can in turn be phagocytized by other histiocytes. This is frequently seen in the course of infection and severe immunologic reaction.

---

* A monocyte which contains a nucleolus visible in smears must not be identified as a precursor cell (monoblast): all monocytes have nucleoli, though they are usually obscured by perinucleolar chromatin (see p. 4, 146). Moreover, one of the first signs of the evolution of the monocyte toward the macrophage is the hypertrophy of the nucleolus.

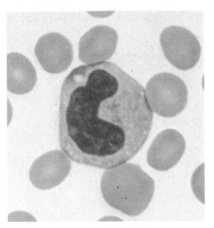

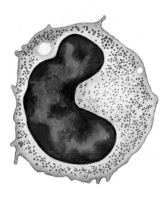

Fig. 1 — *A monocyte in the peripheral blood.* The azurophilic granules are at the limit of visibility and appear as fine reddish dust. A contractile vacuole is present

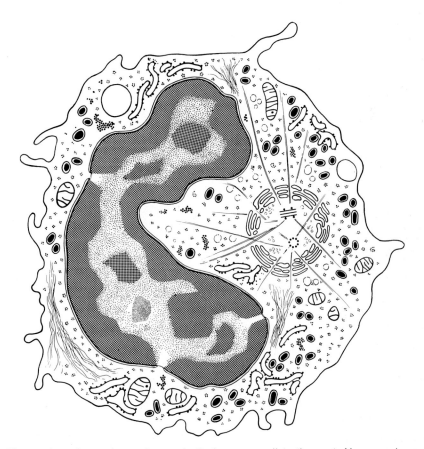

Fig. 2 — The number of granules varies markedly from one cell to the next. Many are beyond the resolving power of the light microscope

159

2 — Movements of Monocytes

Living Cells

In phase contrast the nucleus is pale, with finely divided and indistinctly reticulated chromatin, and often deeply indented. The indentation is due to the centrosome which undergoes oscillatory movements similar to those in other leukocytes (see pp. 116 and 150).

The cytoplasm is pale grey. The mitochondria are very fine. Occasionally they form a small juxtanuclear rosette encircling the centrosome. This appearance is not specific for monocytes (contrary to some textbooks). Azurophilic granules are variable in number. Many are at the limit of visibility. They sometimes give the appearance of a fine intracytoplasmic dust.

As all other types of leukocytes, monocytes contain several types of vacuoles, among which contractile vacuoles are particularly numerous. This is probably due to the fact that pinocytosis is very marked, which leads to the need to dispose of the ingested liquid.

Monocytes have undulating veils, particularly after a sojourn of several minutes in test tubes. The veils differ from those of granulocytes by their large size and their shape. Frequently they curve back towards the cell surface and form large pinocytic vacuoles in the process.

Locomotion

Monocytes engaged in locomotion generally assume a triangular shape similar to granulocytes. However, the undulating membranes forming the protopod are in general larger, the uropod shorter and thicker than in other leukocytes.

It has been claimed repeatedly that the different types of leukocytes can be distinguished by their mode of locomotion. This is only partially correct. The mode of locomotion of all leukocytes is basically the same, involving the formation of a protopod and a uropod (see p. 16), but the different nuclear-cytoplasmic ratios of different cell types result in different appearances during locomotion (triangular, hand-mirror shaped, etc.) which normally allow their identification. However, these criteria must be used critically in the identification of different cell types and cannot be directly applied to leukemic cells (see p. 210).

Chemotaxis of monocytes is quite marked, although the cells move more slowly than granulocytes.

Spreading

Monocytes have a pronounced tendency to spread on glass. The shape of the spread cell is characteristic: the center contains the nucleus and granules, the periphery abundant pale hyaloplasm with a fringed border (Figs. 4 and 5). Time lapse cinematography permits the observation of the quite typical movements of that border, the undulating membrane.

The scanning electron microscope shows the different stages of spreading and the undulating fringe of the cellular margin particularly well (Fig. 4).

Phagocytosis

The monocytes are very actively phagocytic. They phagocytize bacilli, viruses, antigen-antibody complexes, and a great variety of inorganic substances (iron, beryllium, plutonium, silica, dust particles, etc.). Probably because of their ample cytoplasm, they frequently phagocytize large objects, such as entire cells or large mineral particles (see the discussion about the term macrophage, p. 157). They ingest entire red blood cells much more frequently than granulocytes, which usually cut the red or white blood cells in two before phagocytizing them (see p. 94).

Pinocytosis is very marked (and since the veils of monocytes are particularly large, one might speak of "macropinocytosis"). Phagocytosis of erythrocytes may be observed in the peripheral blood, for instance in hemolytic disease of the newborn (erythroblastosis foetalis). It is a curious fact that small monocytes, which resemble lymphocytes, tend to phagocytize more actively than large monocytes.

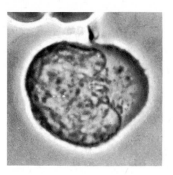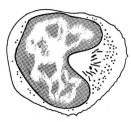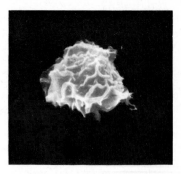

Figs. 1, 2 and 3 — *Monocytes in the peripheral blood* seen with phase contrast (Figs. 1 and 2) and with the scanning electron microscope (Fig. 3)

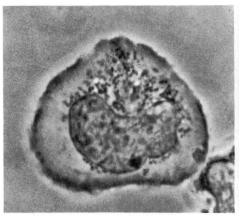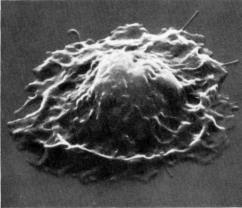

Fig. 4 — *Monocytes spread on glass* and seen with phase contrast (left) and scanning electron microscope (right). Note the undulating membrane

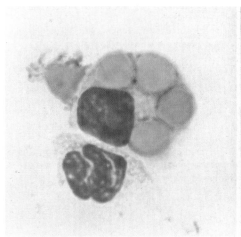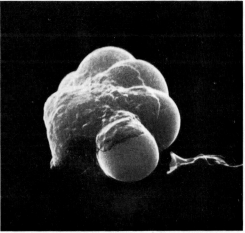

Fig. 5 — *Monocyte with four phagocytized red blood cells* in a smear (left). A similar cell seen with the scanning electron microscope (right)

3 – Histiocytes (Macrophages). Phagocytosis of Lymphocytes

Definition

As early as 1925, monocytes were shown to transform into histiocytes (macrophages) in vitro and to form multinucleated giant cells. These studies, forgotten for a time, have recently been resumed (see p. 157). It is now considered established that the histiocyte is derived from the monocyte and that these two cell forms can be transformed into each other reversibly.

Macrophages are actively phagocytic cells. They may ingest a variety of cells, including protozoa, bacilli, viruses, as well as antigen-antibody complexes, and a variety of inorganic substances (carbon, silica, beryllium, asbestos, etc.).

Location

The fixed macrophages are restricted to a permanent or semipermanent location. They may become specialized and have been classified, perhaps arbitrarily, according to their location: histiocytes of the serosa, of the lungs (alveolar macrophages), Kupffer cells of the liver, histiocytes of the spleen, bone marrow, and lymph nodes. Certain histiocytes are situated along capillaries and venous or lymphatic sinuses which provide them with ready access to material which they phagocytize.

Just as monocytes may be found in the tissues, so can histiocytes be seen in the blood in pathologic conditions. Their number is always small. One finds them in the buffy coat, particularly in subacute bacterial endocarditis, parasitic diseases, and tuberculosis.

Smears

Histiocytes measure approximately 30 to 40 μm in diameter. The nucleus is oval with a chromatin consisting of regular blocks in one or two nucleoli. The cytoplasm is ample, a pale grey with occasional pink or bluish zones. Azurophilic granules may be seen. Colorless vacuoles may be present, occasionally in large numbers, as well as all types of phagosomes (see below and pp. 164 and 166). It is not infrequent to see binucleated cells, particularly in smears from lymph node puncture (see pp. 163 and 167).

Cytochemistry

The cytochemical reactions are similar to those of the monocytes: positive nonspecific esterases and secretion and excretion of lysozyme.

All products of lysosomal origin are increased when the histiocytes are activated in the course of inflammation of immunologic reactions.

Living Cells

The cells have an undulating membrane. During phagocytosis and pinocytosis, they may interiorize up to 60% of their surface membrane in a few hours. In that case the veils retract and the cell becomes spherical. After several hours, the pinosomes and phagosomes are resorbed, new membrane is synthesized, and undulating veils reappear.

Like monocytes, histiocytes adhere and spread on glass (see p. 160).

Phase contrast microscopy shows mitochondria which are numerous and often elongated, lysosomes around the Golgi, contractile and pinocytic vacuoles, and highly refractile lipid inclusions (which appear as colorless vacuoles in Giemsa stained smears).

Electron Microscopy

The electron microscope reveals numerous ribosomes and polyribosomes, a large Golgi, manufacturing lysosomes, smooth and granular endoplasmic reticulum in variable, but generally moderate quantity, multivesicular bodies, vacuoles, tubules, and fibrils. Pinocytosis and phagocytosis with formation of secondary vacuoles of various types are common (see below and pp. 164 and 166).

Phagocytosis of Lymphocytes[1, 2]

In inflammatory lymph nodes one finds intense phagocytosis of lymphocytes and lymphoblasts. It is quite common to see the phagocytosis of several cells at one time, and the digestion of 10 to 20 phagocytized cells. One can follow the digestion of these cells in phagosomes into which lysosomes discharge their contents. They become strongly basophilic, diminish in volume and finally become dustlike particles next to still intact elements in the adjacent cytoplasm. The early cytologists called them "tingible bodies of Flemming."

The "Flemming cells" are simply histiocytes that have ingested other cells. Apart from lymphocytes, the phagosomes may contain red cells, granulocytes or even other histiocytes.

Histiocytes (Macrophages) and Immunity

The role of histiocytes in immunologic reactions has given rise to a large number of investigations (demonstration of receptors for gamma-globulins and for complement, role of these cells in viral immunity, delayed hypersensivity, interaction between histiocytes and lymphocytes, etc.). These important functions are not accompanied by any change in appearance visible in stained smears.

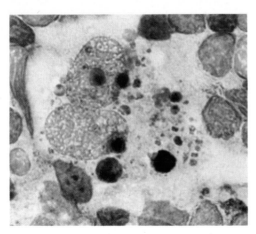 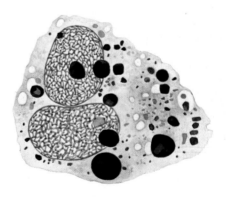

Fig. 1 – *A macrophage phagocytizing lymphocytes* (lymph node aspirate). The digestive vacuoles become more and more basophilic

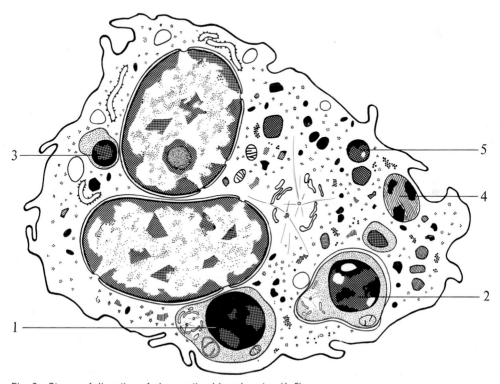

Fig. 2 – Stages of digestion of phagocytized lymphocytes (1–5)

163

4 – Histiocytes (Macrophages). Phagocytosis of Erythrocytes

General Remarks[1–6]

In a normal organism, only red cells which have reached the end of their normal life span of 120 days are removed from the circulation. They are recognized through an as yet unknown mechanism and phagocytized by histiocytes. The manner and site of this destruction has been a subject of controversy:

1) Earlier authors have been perplexed to find that the destruction in the organism of approximately three million red cells per second leaves almost no trace of the stages of their disappearance: actually, it is only the light microscope which fails to reveal the process.

2) Earlier isotopic studies have shown that the liver and spleen accumulate the largest quantity of injected labeled red cells; we now know that this applies only to pathologic red cells.

Normally it is almost exclusively in the bone marrow that erythrophagocytosis can be seen. It is particularly intense in the central histiocyte of the erythroblastic island (see p. 42).

In pathologic conditions red cells are destroyed at random and without regard to their age. Histiocytes, monocytes (see p. 161), and even granulocytes (see p. 118) actively phagocytize the cells or, in the case of intravascular hemolysis, their stroma. The site of this destruction depends on the quantity of damaged erythrocytes: if the number is very large, the liver destroys the majority, the spleen a lesser quantity, and the bone marrow the rest.

Living Cells

A single histiocyte (or monocyte, see p. 161, Fig. 5) is capable of phagocytizing many erythrocytes at the same time (Fig. 1). The study of living cells permits one to follow phagocytosis and digestion of red cells from start to finish (see pp. 118 and 160). When examining smears, one can only recognize those stages in which the ingested red cells still contain hemoglobin.

Phagocytosis of Red Cell Ghosts. When hemolysis of erythrocytes takes place in the blood, as may be the case in severe hemolytic states, histiocytes ingest the ghosts. Frequently a red cell will hemolyze at the very moment when it adheres to a histiocyte and before actual phagocytosis.

Digestion. After the engulfment, the erythrocytes or their fragments are progressively divided into smaller, still hemoglobinized spherules, then hemolysis supervenes and abruptly swells the digestive vacuole.

Subsequently the ghost fragments are divided into smaller and smaller particles. The most frequent end result is their complete disappearance. Occasionally, the cells extrude the residue of digestion.

Electron Microscopy

Electron microscopy allows one to follow all the stages of the digestion in histiocytes of the marrow, the spleen, the liver (and the skin in cases of ecchymoses). The content of the phagolysosomes depends on the degree of digestion of red cells. In the initial stages, one sees the cell membrane, myelin forms, ferritin molecules; the final stages are characterized by the appearance of different forms of hemosiderin (see p. 166) and ceroid substance (see p. 170), as a rule forming round or polylobulated bodies but occasionally resulting in elongated paracrystalline shapes.

The digestion is accompanied by an increase in the numbers and size of various cytoplasmic organelles (ribosomes, Golgi, lysosomes).

Smears

Early stages of red cell ingestions are readily identified (Fig. 1). Late stages of digestion or digestion of ghosts can only be seen in the form of vacuoles, the nature of which is difficult to ascertain in the absence of other visible signs of erythrocyte phagocytosis. The blue color of ceroid, when present, may give a clue to the origin of the vacuoles.

When a histiocyte has digested a large number of red cells (or other cell types, see p. 168) it degenerates rapidly*. Its nucleus becomes smaller in size, pyknotic; its chromatin loses its characteristic appearance, and its nucleolus becomes indistinct and disappears. Its cytoplasm fills with colorless vacuoles, resulting in a vacuolated histiocyte (foam cell). The moribund cells finally become lyzed and the debris phagocytized by other histiocytes.

Cytochemistry

In its first stages of development, the content of the vacuoles may give the reactions of hemoglobin (benzidine reaction, absorption in the Soret band) while in the later states it acquires the reactions of hemosiderin (positive PAS and Sudan stains—Perls' reaction: see p. 166).

Ceroid gives positive PAS and Sudan black reactions.

---

* The disorders of lysosomes or lysosomial diseases exhibit an abnormal and incomplete digestion of phagocytized erythrocytes and leukocytes, and will be described on pp. 168 and 170.

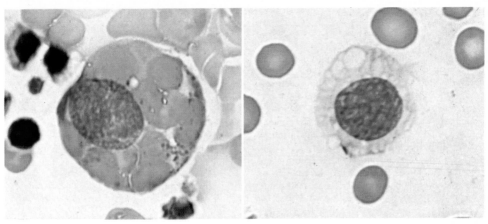

Fig. 1 – *A macrophage with several phagocytized erythrocytes*. The red cells have not yet been hemolyzed

Fig. 2 – *A macrophage with phagocytized erythrocytes*. The red cells have been hemolyzed and fragmented. (Bone marrow aspirate from patient with hereditary spherocytosis)

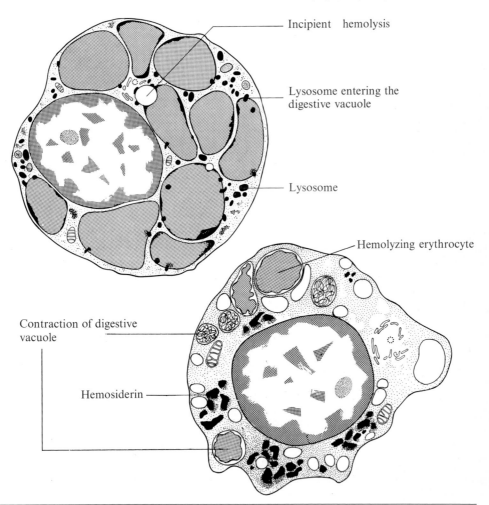

Incipient hemolysis

Lysosome entering the digestive vacuole

Lysosome

Hemolyzing erythrocyte

Contraction of digestive vacuole

Hemosiderin

5 – Histiocytes (Macrophages). Formation of Hemosiderin

Definition of Hemosiderin[1-9]

The term hemosiderin was introduced in 1888 to characterize the iron containing "ochre pigment." The chemist separating the iron from the cells found it to divide into a soluble portion, which was crystallized and named ferritin, and an insoluble portion called hemosiderin. On the assumption that it was a single entity, a large number of investigations were undertaken to analyze hemosiderin for its content of iron, lipids, glucides, proteins, copper, calcium, etc. Widely divergent results were obtained by different authors. Electron microscopic studies have elucidated the reason. Hemosiderin constitutes an extremely variable mixture. Its chemistry had been studied without the degree of homogeneity of the preparation being checked in the electron microscope. The chemists thus found themselves in a similar situation to that faced by early investigations on "microsomes," before electron microscopy made possible the identification and isolation of their constituents, such as ribosomes and endoplasmic reticulum.

Smears

When the cytoplasm of histiocytes contains a ochre or yellow pigment, this does not provide any assurance that one is dealing with hemosiderin. Only a positive Perls' reaction does. It should be noted that yellow pigment exists which is free of iron, but contains porphyrins.

Cytochemistry

Perls' reaction is positive in cellular inclusions which contain iron. It may also be weakly positive throughout the cytoplasm of a cell. This indicates the presence of a large amount of dispersed ferritin (see also p. 56). The PAS and Sudan Black reactions are positive in inclusions which correspond to erythro- and autophagosomes (see below).

Electron Microscopy

The iron is easily visualized when it is present in the form of ferritin molecules which have a characteristic appearance (see p. 28). Note that Perls' reaction can be applied to electron microscopic sections to confirm the iron content of a structure, if necessary.

When numerous ferritin molecules are present in the cytoplasm, Perls' reaction applied to smears gives the entire cell a sky-blue color (see Fig. 2 and p. 56).

Ferritin may appear in different forms:

Simple Siderosomes are masses of ferritin molecules in a protein substrate and generally surrounded by membrane.

Ferritin Crystals consist of ferritin molecules arranged in strips or lamellae with a periodic structure. Such crystals can attain considerable size (1 to 2 µm). It is not uncommon to observe empty places in the otherwise regular assembly of ferritin molecules: they correspond to molecules of apoferritin, the protein portion of the ferritin molecule.

Complex Siderosomes contain ferritin or masses of iron-containing micelles mixed or included in an amorphous substrate, granules or myelin forms.

Origin of Hemosiderin

Complex siderosomes are generally derived from phagocytosis of erythrocytes, occasionally from phagocytosis of other cells which contain iron such as erythroblasts and plasmocytes, occasionally from alteration of the histiocytic cytoplasm itself (autophagosomes) when it contains dispersed ferritin molecules.

Simple siderosomes and dispersed ferritin can also be the result of an external iron overload. These features are seen particularly prominently in hemochromatosis.

$\longrightarrow$

Fig. 3–(1) Digestion of the stroma of a red blood cell. Molecules of ferritin appear around myelin figures. (2) Siderosomes containing ferritin molecules. (3) Ferritin crystals. The intermediate enlargement shows the spatial arrangement of the molecules. Still higher magnification shows the presence of apoferritin (ferritin molecules without iron). (4) Peculiar appearance of siderosomes due to ferritin deposition in myelin forms. (5) Ferritin molecules dispersed in the cytoplasm. Note that the mitochondria in this heavily iron-laden cell contain no iron

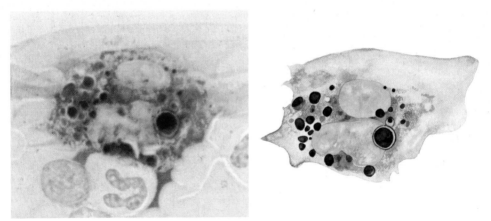

Figs. 1 and 2 – *A macrophage after digestion of a number of erythrocytes*. Perls' reaction demonstrates the iron in the form of hemosiderin granules or dispersed in the cytoplasm

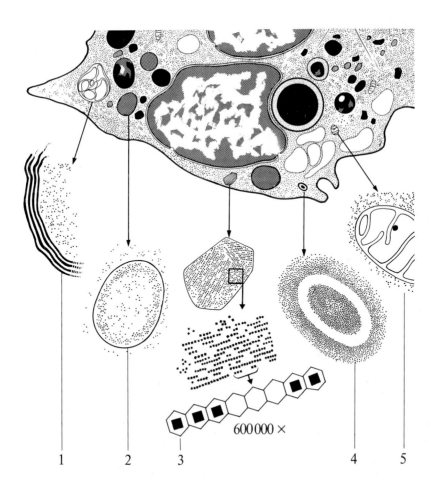

600 000 ×

1   2   3   4   5

General Remarks on Storage Cells

Definition

These are cells containing inclusions of phagocytized or pinocytic material that has not been completely digested and has led to the acccumulation of glycolipids and muco- and oligosaccharides. The phenomenon may be due to a congenital defect of an enzyme and may occur as a familial disease. It may be acquired and due to an overload of the phagocytized material that requires digestion: cells of this type are seen in numerous hematologic syndromes (see p. 172).

Familial Sphyngolipidosis[1]

These diseases have been known for a long time (Gaucher disease, Tay-Sachs, Fabry, Niemann-Pick, Hunter, Hurler, etc.). Each of these diseases may comprise one or several variants. Most of them have by now been characterized by the site in the metabolic chain which is blocked. The morphology, even with the electron microscope and cytochemistry, allows one to distinguish only three groups of storage diseases: Gaucher disease, Niemann-Pick disease and the ill-defined syndrome of sea-blue histiocytes. Only the identification of the deficient enzymes permits an identification of the individual diseases.

Gaucher Disease

Smears

Gaucher cells are seen in bone marrow or splenic aspirates. In smears they are seen either singly or in collections of 8−10 cells. These are very large cells with a diameter of 20−80 μm, round or polyhedral.

The cytoplasm is enlarged and distended by lipid inclusions, which have a varied aspect, depending on the stage of evolution of phagosome. The most characteristic appearance is the lamellar one: fine, elongated, empty spaces indicating lipids dissolved in the process of fixation, often arcuate and occasionally whorl-like, separated by a blue-stained cytoplasm. Less common are the sponge-like, finely or grossly vacuolated appearances of the cytoplasm.

Other inclusions represent the remnants of recognizable phagocytosis (debris of erythrocytes, leukocytes or hemosiderin).

The nucleus appears small in comparison with the hypertrophied cytoplasm. In cells that are only mildly affected, it resembles that of any other histiocyte, but it becomes altered with an increase in the cytoplasmic overload. Often the nucleus may appear to be strangled due to the filamentous structures that encircle it (Figs. 1 and 2). It loses its nucleolus and becomes more and more dense. Occasionally it disappears and only the lipid-filled cytoplasm is found in smears.

Cytochemistry

The inclusions of Gaucher cells are strongly PAS positive, faintly Sudan black positive and autofluorescent. In the cytoplasm one can stain the characteristic enzymes of the histiocytes (nonspecific esterase and a large quantity of acid phosphatase). Perls' reaction is strongly positive due to the presence of ferritin and hemosiderin. Cytochemistry does not permit, at present, the characterization of the storage material, which is a glucocerebroside.

Electron Microscopy[2−5]

The electron microscope allows the identification of all stages of formation of the inclusions. Storage becomes more and more marked with the prolonged duration of the life or the cell. In fully developed cells, the inclusions are elongated vacuoles filled with tubules of 50 nm in diameter, tightly packed against each other. High magnification reveals that the tubules have a characteristic twisted appearance (Fig. 3). The cell border shows numerous microvilli and evidence of micropinocytosis.

It appears that the stored substances are derived from the incomplete digestion of red and white blood cells (glucosphingolipids). In some cells one can observe all the stages of the digestion of the red blood cells and the formation of large quantities of ferritin and hemosiderin. Iron can also acculumate directly from plasma transferrin. The iron apparently cannot be reutilized for hemoglobin synthesis. A hypochromic anemia may develop in the presence of Gaucher cells filled with iron.

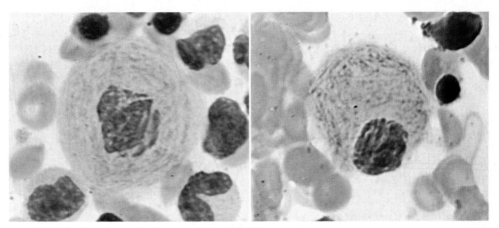

Figs. 1 and 2 – *Gaucher cells* (bone marrow aspirate)

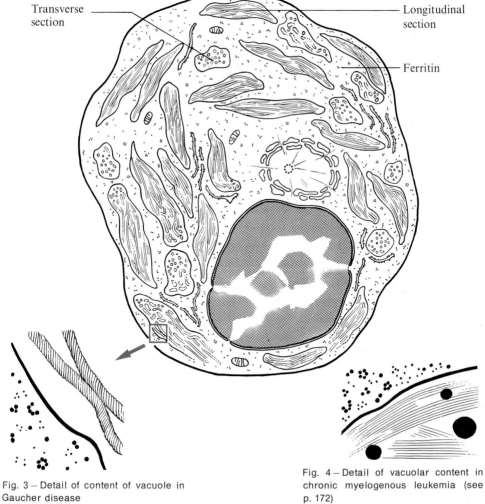

Transverse section

Longitudinal section

Ferritin

Fig. 3 – Detail of content of vacuole in Gaucher disease

Fig. 4 – Detail of vacuolar content in chronic myelogenous leukemia (see p. 172)

7 – Niemann-Pick Disease and Other Storage Diseases

Niemann-Pick Disease

Smears

In contrast to the lamellar appearance of Gaucher cells, the storage cells in Niemann-Pick disease and most other lipidoses contain only spherical inclusions. They resemble each other in smears, though they differ in the nature of the stored lipids. In fully evolved cells, the inclusions are dissolved during fixation and all that remains as a rule are very thin cytoplasmic bridges, which border the large number of colorless vacuoles, among which one finds a generally pyknotic nucleus, giving the appearance of a foam cell.

Before attaining this stage, histiocytes in all stages of phagocytosis and progressive accumulation of degradation products in their phagosomes may be observed. In the first stages the content of the vacuoles is not dissolved and is stained blue by Giemsa or Wright stains (see below).

Cytochemistry

The inclusions give positive reactions with PAS and Sudan black. Enzymes characteristic of histiocytes are found in the cytoplasm.

Living Cells

In phase contrast the cytoplasm is filled with highly refringent spheres. In polarized light, Maltese crosses may be seen. The cells are autofluorescent.

Electron Microscopy[1]

Vacuoles of all sizes are seen filled with concentric lamellae of myelin figures mixed with amorphous material and remnants of phagocytosis.

The different organelles present the same changes as those found in Gaucher cells (see p. 168).

Other Storage Diseases

Numerous syndromes or diseases with cells of similar appearance have been described. As already noted, the definition of these diseases is only possible by identification of the stored material or the deficient enzyme. However, two syndromes of slightly different appearance on smears may be noted.

Mucopolysaccharidosis[2-4]

This is also described as Hurler's syndrome of which numerous variants exist. The inclusions have been described as PAS positive, azurophilic Reilly bodies in histiocytes and fibroblasts. This disorder coexists frequently with Alder's anomaly, in which the PMN have abnormal granules.

The Syndrome of Sea-blue Histiocytes[5-8]

In this disorder a large number of histiocytes are present with small, tightly packed inclusions, which stain a bluish green with Giemsa or Wright stains. Electron microscopy reveals appearances similar to that of Niemann Pick disease.

Sea-blue histiocytes occur in various pathologic states (see p. 172) as an apparently secondary phenomenon, but there also exists a primary familial form of the disease. In one such family, a marked deficiency of sphyngomyelinase was demonstrated. Excess of sphyngomyelin concentration in tissues of other affected families has been described and is compatible with such an enzyme deficiency. However, the question of the unitary and specific nature of the syndrome remains to be resolved.

It should be noted that the blue color of the inclusions is not specific (see p. 172). Ceroid, for example, sometimes stains ultramarine blue. This pigment is closely related to lipofuchsin. It is the result of the oxidation and polymerization of unsaturated lipids. Many histiocytes which have undergone intense phagocytosis contain ceroid. The diagnosis of "sea-blue histiocyte syndrome" cannot be based on this feature alone, but must be based on biochemical and clinical findings.

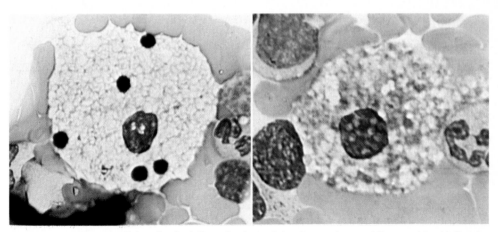

Figs. 1 and 2 – *Niemann-Pick cells. On left:* early stage of development resembling sea-blue histiocyte. *On right:* fully developed stage. The lymphocytes are superimposed, not phagocytized

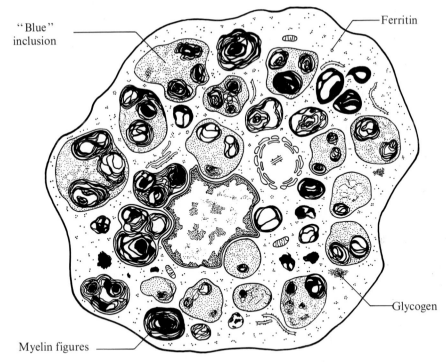

Fig. 3 – Different stages in the development of storage with appearance of many myelin figures

8 – Storage Cells Due to Overload

Cells similar to Gaucher or Niemann-Pick cells or sea-blue histiocytes or intermediate forms between them can be found in the bone marrow or lymph node aspirates in many diseases of hemopoietic organs and occasionally even in normal individuals. The accumulation of undigested products in histiocytes may be due to an acquired diminution of cellular enzymes (perhaps after administration of certain drugs) or due to an enormous increase in the products requiring phagocytosis as in leukemias or hemolytic anemias (see below).

Cells that manifest this phenomenon are virtually always histiocytes. Nevertheless, the same inclusions may be seen in small number in monocytes, granulocytes, lymphocytes, or even other cell types, such as fibroblasts. It appears that the frequency of overload in histiocytic cells is due to a relative long life span, which allows the accumulation of a large amount of undigested material.

Pseudo-Gaucher and Pseudo-Niemann-Pick Cells[1-7]

In chronic myelogenous leukemia and even some acute leukemias, the destruction of large numbers of pathologic leukocytes results in the accumulation of material awaiting degradation, and pseudo-Gaucher cells form, even though the enzyme that is deficient in Gaucher disease is present in normal amounts. Electron microscopy permits the separation of true Gaucher cells, in which the inclusions form tubules, from pseudo-Gaucher cells, which contain nontwisted fibers of 10 nm in diameter.

In thalassemia, the congenital dyserythropoieses, and idiopathic thrombocytopenic purpura, a similar need to dispose of large numbers of cells exists because of the high turnover of red cells and platelets in these diseases. The overloaded macrophages may acquire an appearance indistinguishable from Niemann-Pick cells even by electron microscopy. Occasionally, however, fibrillar formations resembling those in pseudo-Gaucher cells are present.

The development of pseudo-Gaucher and pseudo-Niemann-Pick cells is accompanied by the appearance of macrophages in all stages of phagocytosis and digestion, leading to the formation of storage vacuoles. It appears that excessive cytolysis results in the accumulation of incompletely digested material by exhaustion of the capacity of hydrolytic enzymes or reveals latent enzyme deficiencies.

Sea-blue Histiocytes[8-11]

Occasionally histiocytes with blue-colored inclusions may be seen in smears of normal or pathologic marrows. These inclusions are found frequently adjacent to vacuoles containing undigested debris and entirely colorless vacuoles, the content of which has been dissolved by fixation and staining. These cells may resemble the first stages of development of Gaucher cells, or more frequently those of other storage diseases. One may see them when there is an increase in the number of cell deaths (granulocytic leukemias, thrombocytopenic purpura, Hermansky-Pudlack disease, lymphomas). These cells have also been described after the administration of tranquilizers and antirheumatic, antileukemic, and anorexic medications.

Chapter 7

Plasmocytic Series

General Remarks

Definition

The plasmocytic* series begins with the committed stem cell and ends with the production of a mature plasma cell which secretes antibodies.

Origin

The origin of plasmocytes has been a matter of lengthy discussion in the past. Recently, it has been established beyond doubt that B-lymphocytes can give rise to plasmocytes (see p. 152). It has even been suggested that plasmocytes represent only a functional state of lymphocytes and not a separate cell series. However, other cells cannot be entirely excluded as being capable of developing an RER and of assuming the appearance of plasmocytes: morphologically intermediate stages between histiocytes and developing plasmocytes have been observed. Although it appears necessary to record this fact, it is well known that little reliance can be placed on morphologic transitions between cell types.

Life Span and Death

Little is known about the fate of plasmocytes. It is possible that some of them die immediately after they have secreted their antibody. It is also possible that the cells that have ceased to produce antibodies can be transformed into cells indistinguishable from small lymphocytes, the memory cells which in turn can redevelop into plasmocytes. Alternatively, plasmocytes retain the appearance of mature cells. It is noteworthy that, notwithstanding their mature appearance, they are capable of division. Dead plasmocytes are phagocytized by histiocytes of the marrow and lymph nodes. The process can be readily identified by electron microscopy by the presence of abundant RER in the digestive vacuoles.

Location [1, 2]

Plasmocytes are tissue cells. They are seen wherever lymphocytes and histiocytes are found, which means that they are almost ubiquitous. Lymph nodes and particularly their medullary cords are normally very rich in plasmocytes (see p. 153, Fig. 2). They are always present in the spleen, the thymus, the skin, the bone marrow, and the intestine.

Plasmocytes frequently form a crown around histiocytes and around the capillaries of lymphoid and hemopoietic organs. As a result, they may be seen lined up along capillaries in appropriate portions of marrow aspirates, particularly when plasmocytes are present in increased numbers.

Normally plasmocytes do not enter the circulation. Occasional ones may be found in smears of buffy coats. Their number increases as a result of antigenic stimulation. In that case plasmocytes originate in the lymph nodes and enter the circulation by way of the thoracic duct.

In the lymph, plasmocytes are always very few in number, even when they are very numerous in the lymph nodes.

Stages of Maturation and Amplification

Whatever the origins of the plasmocyte, it is obvious that one may see plasmocytic cells with signs of immaturity. For instance, the nucleus may have finely distributed chromatin and contain nucleoli. Electron microscopy indicates that these cells have a less developed RER than the mature plasmocytes. Immunochemical investigations have identified that they secrete globulins. They have therefore been called plasmoblasts or proplasmocytes (see p. 174).

Nothing is known about amplification of plasmocytes.

---

* Since plasma cells and plasmocytes are used interchangeably in many hematologic texts, we have selected plasmocyte as more consistent with the general rules discussed under Nomenclature.

1 – Proplasmocytes

Definition

Proplasmocytes can be defined as immature plasmocytes (see p. 173). A large number of synonyms
can be found in the literature: plasmacytoid lymphocytes, intermediate forms, virocytes, Turk
cells, plasmoblasts, etc.

Smears

Proplasmocytes measure 15 to 20 µm in diameter.
Their principal characteristic is the marked cytoplasmic basophilia and the presence of the nucleolus, generally hidden by the nucleolus-asso-
ciated chromatin. The nucleus is only slightly eccentric.

The preceding description makes it obvious that
these cells are sometimes quite difficult to distinguish from the hyperbasophilic lymphocytes
(see p. 154).

Electron Microscopy[1, 2, 3]

Electron microscopy clearly demonstrates the progressive development of the RER. Special techniques have made it possible to follow the elaboration of antibodies in these cells, as well as in
the B-lymphocytes from which they are derived.
Antibodies appear at first in the perinuclear space,
subsequently in the RER and pass through the
Golgi complex before being secreted (see Fig.
p. 178).

The secretion of the products synthesized by the
plasmocyte probably takes place by means of extrusion of small granules derived from the Golgi.
Another mode of secretion is clasmatosis and possibly the lysis of the entire cell. However, it is
uncertain that this mode of secretion is physiologic.

Some cytoplasmic vacuoles are of nuclear origin
(as can be demonstrated by microcinematography) and their contents can be diverted to the
external milieu. Their appearance is usually colorless, and it is possible that they represent invaginations of the nuclear envelope containing products of secretion (see p. 196).

Presence in the Blood

Proplasmocytes and plasmocytes can be found in
the blood as a result of antigenic stimulation or
certain infectious diseases. They are particularly
numerous in rubeola, where they may represent
15% of all leukocytes, and in infectious mononucleosis, toxoplasmosis, syphilis and tuberculosis.
(It should be noted that abnormal plasmocytes can
be found in the blood in multiple myeloma and
plasmocytic leukemia.)

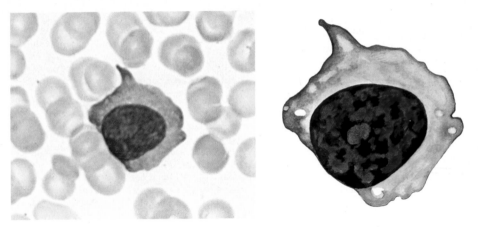

Fig. 1 – *A proplasmocyte*. The cell may be seen in the blood in virus diseases, for instance in rubeola

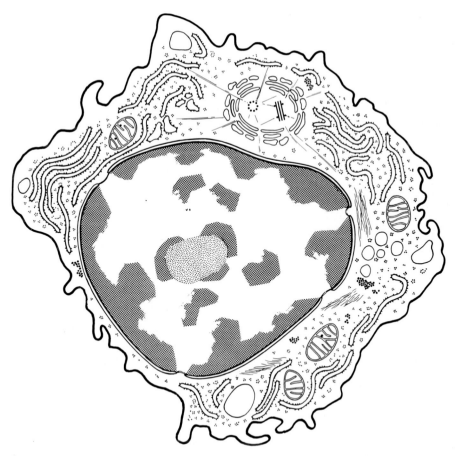

Fig. 2 – Note the abundance of RER which is beginning to be organized in parallel sacs

2 – Plasmocytes

Smears

Plasmocytes are generally oval with a greater diameter of 12–20 μm and a lesser diameter of 7–12 μm. The nucleus is almost always situated at one pole, and is likewise oval. However, its longer axis is at right angles to the longer axis of the cell: this is a unique feature of plasmocytes*.

The arrangement of the chromatin is also characteristic. The chromocenters form seven to nine large blocks of approximately polygonal outline, resembling a tortoise shell.

It is not unusual to find several nuclear lobes. One may also find giant plasmocytes with tetra- or even octoploid nuclei.

The cytoplasm is intensely basophilic and its ultramarine color identifies it immediately. Occasionally, the periphery of the cells appears multilayered, corresponding to the lamellae or sacs of RER (see below).

The centrosome is colorless or only faintly stained. It is always clearly visible in plasmocytes. Although normally juxtanuclear in position, the centrosome may occasionally be located in a cytoplasmic extension.

Surrounding the centrosome there are always numerous round or oval, occasionally vermiform, cellular spaces, which correspond to the negative images of mitochondria. These can be visualized with special dyes and are occasionally stained reddish by the azure of the Giemsa.

Living Cells

In the phase microscope, plasmocytes have a very dark cytoplasm which resembles that of polychromatophilic erythroblasts (E_5). Unless the cell is compressed between slide and coverslip, the nucleus is poorly visible because of the high refractive index of the cytoplasm. The cell is readily distinguished from E_5 by its centrosome and the presence of large mitochondria surrounding it.

When the cell is sufficiently compressed, the RER may be seen in forms of lamellae, arranged in parallel with the margin of the cell. The lamellae surround the centrosome almost entirely. The nucleus and its blocks of chromatin become clearly visible. The nucleus is pushed aside by the more rigid centrosome (see p. 6).

Locomotion. The plasmocytes form very small hyaloplasmic veils and their directional mobility is due to protopods, first clear, and subsequently filled with dark cytoplasm. The centrosome is usually situated behind the nucleus, occasionally in the uropod.

Cell Death. The agonal stage is characteristic. The RER becomes vacuolated, its sacs dilate more and more, fragment and round off, giving the cytoplasm an alveolar structure. At this stage, the cells resemble Mott cells (see p. 180), but the dilated sacs are not filled with secretions but reflect the vacuolisation due to the death of the organelles (see p. 22).

Electron Microscopy

The RER occupies almost the entire cytoplasm, except for the center of the cell, which is occupied by the centrosome (which corresponds to the clear "paranuclear zone" or "hof"). The RER is formed of flattened parallel sacs, which surround the nucleus concentrically. The sacs are studded with polyribosomes, each containing 5 to 15 elements in small chains or spirals. This structure is particularly well seen in tangential cuts (see p. 234).

The nucleus comprises gross blocks of chromatin which adhere to the nuclear membrane. A central block contains the remnants of the nucleolus.

The Golgi complex is particularly well developed. It contains neither endoplasmic reticulum nor free ribosomes, which explains the absence of basophilia from this zone. The Golgi apparatus is rigid and pushes the nucleus into an eccentric position.

In the thin layers of cytoplasm which remain between the sacs of the RER, large elongated mitochondria are seen, particularly around the Golgi complex. A small number of lysosomes is also present. Micropinocytosis is frequently seen.

---

* The eccentric position of the nucleus is due to the presence of the large centrosome. After division and reconstruction of the daughter nuclei but before the cytoplasmic separation, the two nuclei may be seen one next to the other at one pole of the cell, with their large axes at an angle of approximately 30 to 40 degrees to each other, again as a result of displacement by the centrosome. This appearance is quite frequent in aspirates of lymph nodes or bone marrow whenever a slight plasmocytosis is present.

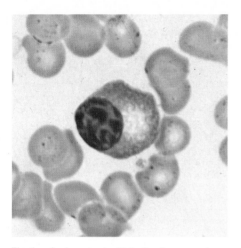

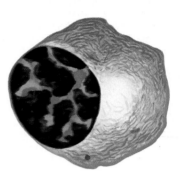

Fig. 1 — *A plasmocyte*. Note the large centrosome, the barely visible negative images of mitochondria and the parallel lamellae of the RER

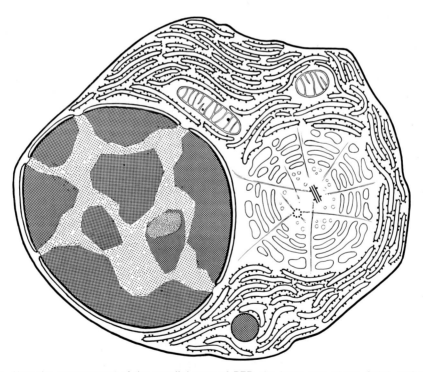

Fig. 2 — Note the arrangement of the parallel sacs of RER, the large size of the Golgi, and a large lysosome at the lower pole of the cell

3 — Flaming Plasmocytes

Smears

The cytoplasm of these plasmocytes, instead of the usual ultramarine, is purple tending to red, particularly at the edges. Flaming plasmocytes are usually seen in pathologic states, but may be present in normal bone marrow.

Cytochemistry[1, 2]

Immunofluorescence and immunologic labels such as ferritin and peroxidase have shown that the cytoplasm has an increased content of globulins compared to normal plasmocytes. The globulins, which appear to be responsible for the flaming color, may be of any of the known subtypes. (Although flaming plasmocytes were first thought to be associated with a particular type of globulin in multiple myeloma, this was not proven to be the case).

Electron Microscopy[3]

The sacs of the RER are dilated by globulins, which allows the reinterpretation of the images seen in living cells or stained smears.

Living Cells[4]

Phase microscopy allows one to visualize the dilated sacs of the RER, particularly after slight compression of the cell. Their content is greyish. One must not confuse these appearances with various vacuoles, including contractile vacuoles and vacuoles of nuclear origin, all of which may be present in plasmocytes. Nor must they be confused with the rounded-off, altered RER and Golgi complex, which form vacuoles preceding death of the cell (see pp. 22 and 176).

Clasmatosis. This particular mode of secretion is particularly frequent in flaming plasmocytes. One sees first a cytoplasmic projection, which grows rapidly, becomes isolated, except for a thin peduncle. Sometimes the volume of this cytoplasmic mass exceeds a quarter or half of the entire cell. Finally, the cytoplasmic projection, which may move independently of the rest of the cell, becomes detached and entirely separated from the cell. It swells; the sacs of the RER become visible and rupture, liberating their contents into the external milieu.

In smears, or in sections, it is impossible to be certain of the presence of clasmatosis. Such observations must be made on living cells.

Reinterpretation of Smears[5]

The reddish color of the cytoplasm of flaming plasmocytes is due to the superimposition of the bluish color due to ribosomes and the red color of glycoprotein present in the sacs of RER. This explains the different shades of blue to red of the different cells.

In the intact cell, the basophilic and acidophilic portions of the cell can be separated by ultracentrifugation. The basophilic portion moves with the centrifugal force. At the other pole of the cell a brick red zone appears which represents the contents of the sacs of RER of lesser density.

Occasionally, a similar separation of the two zones actually takes place during the preparation of the smear and the gamma globulins become displaced toward the periphery of the cytoplasm (Fig. 2).

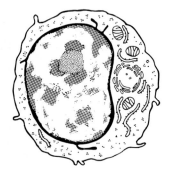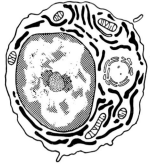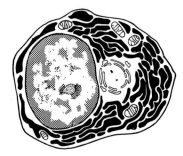

Stages of antibody secretion (here an antiperoxidase antibody after injection of horseradish peroxidase into an animal)

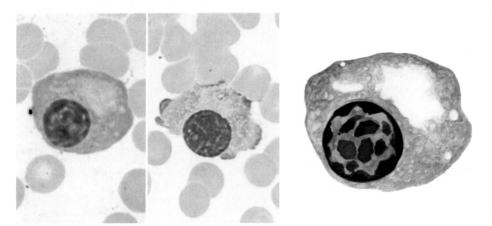

Figs. 1 and 2 — *"Flaming" plasmocytes*. The purple color of the cytoplasm is due to the superimposition of blue-staining ribosomes and red-staining globulins. During the preparation of the smears, these globulins may be displaced toward the periphery. The unstained mitochondria become also clearly visible (Fig. 2)

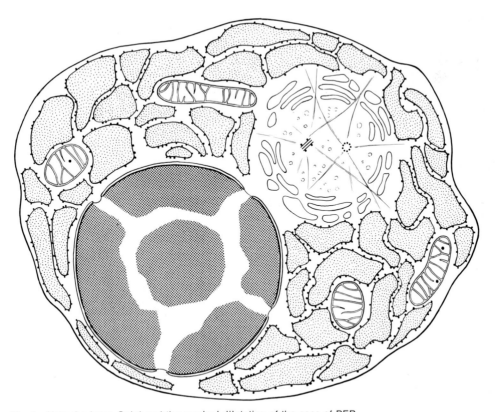

Fig. 3 — Note the large Golgi and the marked dilatation of the sacs of RER

4 – Plasmocytes with Inclusions

Mott Cells

Definition [1, 2]

Mott cells (or grape cells) are plasmocytes filled with "Russell bodies"*.

Smears

Russell bodies are spheres of variable size. They may reach 2 to 3 µm in diameter. They stain a blue-violet or pink and may be dissolved by the usual fixation and staining sequences. In that case, the plasma cell is more-or-less filled with colorless vacuoles (Figs. 1 to 3). In smears of the bone marrow and lymph nodes with a plasmocyte hyperplasia one can readily see intermediate stages between healthy plasmocytes and Mott cells. A few spherules appear in some of the cells. They are initially small, of blue or pink color, round, sharply circumscribed. Gradually the volume and number of the spherules increase until the entire cytoplasm of the cell is filled. The cytoplasm constitutes only a filiform basophilic network between the spherules and the nucleus becomes pyknotic (see p. 196).

Occasionally, the contents of a Russell body assume a crystalline appearance. In that case the plasmocytes contain fine needle or fusiform or sickle-shaped structures.

Cytochemistry

Different cytochemical reactions have identified mucopolysaccharides, absence of lipids, of nucleic acids and of glycogen in these inclusions. Immunofluorescence has demonstrated the presence of globulins.

Living Cells [3]

The cells have active directional movements. The RER swells and rounds off, commensurate with the accumulation of secretions in its sacs until the entire cell is filled with them. Mott cells must not be confused with degenerated plasmocytes or with those cells in which the RER is vacuolated (see p. 178).

Electron Microscopy

Russell bodies are due to accumulation of different types of material in the sacs of the RER, SER or Golgi complex. The Golgi complex is frequently abnormal. Its sacs are dilated, vacuolated, and may contain small Russell bodies. In fully developed Mott cells the nucleus and Golgi complex are compressed by the enormous sacs containing the Russell bodies. The cytoplasm itself has almost entirely disappeared (Fig. 4).

It is probable that Mott cells represent the result of an obstruction of the progression of substances secreted by the RER and the Golgi complex. They are generally cells destined for an early death. Nevertheless, some Mott cells have been observed to incorporate thymidine.

Crystals. Crystals contained in the RER or SER or the cisterns of the Golgi bodies may have variable periodicities. These crystals have the same cytochemical and immunochemical reactions as the Russell bodies.

Frequency

Plasmocytes that contain Russell bodies and crystalline inclusions represent pathologic alteration of the cell. They can, however, be observed in smears from entirely normal bone marrow or lymph nodes. Mott cells are most numerous in diseases which are associated with chronic plasmocyte hyperplasia, such as hypergammaglobulinemias and cryoglobulinemias; parasitosis, such as malaria, kala-azar, and trypanosomiasis; and malignant tumors. In a rare form of multiple myeloma, virtually all cells are Mott cells.

Other Cellular Inclusions

Apart from Russell bodies, a large number of different inclusions have been described in the nucleus and cytoplasm of normal and pathologic plasmocytes.

PAS-Positive Inclusions [4]

These have been described in Hurler's disease and more generally in storage diseases (see pp. 168 and 170).

Hemosiderin Inclusions

These are frequent in hemochromatosis and hemosiderosis. They have been ascribed to micropinocytosis of ferritin or to the digestion of phagocytized erythrocytes. Thus one must consider the possibility that plasmocytes are phagocytic or that histiocytes which have ingested red cells can be transformed into plasmocytes, as some have claimed.

Snapper-Schneid Inclusions [5]

These are round or oval azurophilic granulations about 0.5 µm that appear after treatment of myeloma with Stibamidine, a therapy no longer in use. They remain of interest because similar granulations may be seen normally or after hyperimmunization. They probably represent a nonspecific reaction of lysosomal origin.

---

* They were described in 1890 as cellular parasites, particularly seen in cancers. This is an example of an error of earlier interpretation.

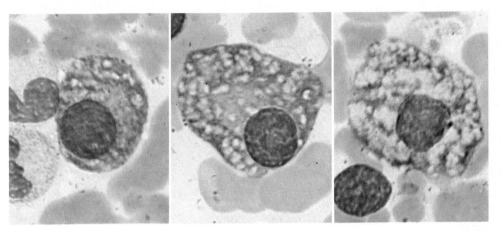

Figs. 1, 2 and 3 — *Stages of development of Mott cells*. The fixative and stain have dissolved the contents of the vacuoles

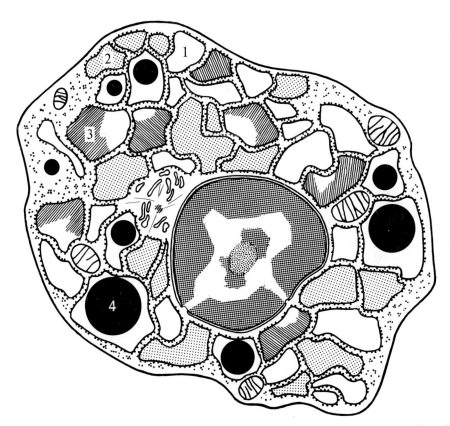

Fig. 4 — Note the extreme dilatation of the sacs of the RER and the progressive organization of the secreted globulins into compact masses and crystals (*1 to 4*)

Needle-like crystals resembling Auer bodies seen on rare occasion in plasmocytes and lymphocytes may be of similar origin and must be distinguished from crystals of immunoglobulin including Russell bodies.

Nuclear Inclusions

These inclusions described on page 196 have been particularly well studied in myeloma and macroglobulinemia, but may be seen in other non-malignant disease.

Amyloid Inclusions[6]

These may be identified by immunofluorescence.

Chapter 8

Mastocytic Series

General Remarks

Definition

Mastocytes* are mononuclear cells, which contain metachromatically staining granulations. They differ from basophils, although their granulations are also metachromatic (see p. 126).

Origin[1, 2]

The origin of mastocytes is still in dispute. Reticular cells, fibroblasts and lymphocytes have been implicated as precursors of mastoblasts, none of them convincingly.

Location

Mastocytes are widely distributed in the body. They are present in connective tissues, particularly around blood vessels, nerves, glands, and in the skin. They are always present in the bone marrow, lymph nodes, the thymus, and the spleen. They do not normally appear in the peripheral blood. They are very numerous in the peritoneum of rats and mice.

Life Span

Their life span is unknown. Mitoses are very rare. They have been observed in culture or in animals after exposure to compounds which cause degranulation (see p. 184).

---

* The term mastocyte has been justly criticized, because it is made up of a German and a Greek root. However, we have already dealt with the problems of hemopoietic nomenclature on page XV.
As to the choice between mast cell and mastocyte, mastocyte appears preferable, more consistent with the nomenclature of the other blood cell series.

1 – Mastoblasts and Mastocytes

They are normally present in small numbers in smears of aspirates of any hemopoietic organ. They must not be confused with developing basophils (see p. 126).

Mastoblasts

Smears

Mastoblasts have a diameter of 20 to 25 µm, and they are usually round or occasionally elongated. The nucleus is oval, and has a very finely distributed chromatin with only a few clumps. The cytoplasm, which is itself faintly basophilic, contains a mixture of azurophilic and metachromatic granules. The latter are of two types, round and oval. Giemsa stains these granules a purple violet. Frequently, the staining is less pronounced in the center, which may be entirely unstained. Toluidine blue colors the granules orange (see below).

Electron Microscopy

Electron microscopy shows the formation of the granules. It begins with elaboration of dense granules approximately 30 nm in diameter in the vacuoles of the Golgi body. These granules fuse and filaments appear, which intertwine into small balls. The endoplasmic reticulum and the Golgi body are very well developed, numerous polyribosomes are present, and the surface of the cell is studded with microvillosities.

Mastocytes

Smears

In Giemsa stains the cells are of variable sizes, 5.0 to 25.0 µm in diameter. They are generally round and occasionally polygonal or elongated. The nucleus is usually central, round to oval, and rarely bilobed. It stains uniformly; its size is small and rarely exceeds 7.0 to 10.0 µm. The cytoplasm is pale or entirely unstained. The granules stain variably from violet to red-purple and are 0.3 to 1.5 µm in diameter.

Cells which contain large granules usually have a pyknotic nucleus. The granules are sometimes small and contained within large vacuoles; occasionally, vacuoles are present without granulations. This is usually due to the solubilization of the granules during washing after staining. In some instances, however, the disappearance corresponds to a physiologic event (see below).

It is not uncommon to see a naked nucleus surrounded by granules without visible cytoplasm. This is an artifact produced during the preparation of the smear.

Cytochemistry[3]

The physicochemical nature of the metachromatic reaction is complex. In mastocyte granules it is due to the presence of sulfated mucopolysaccharides. They contain heparin, histamine, and a large number of enzymes which can be demonstrated by specific reactions. The granulations are quite different from the azurophilic granulations. They do not contain myeloperoxidase. They also differ from the basophilic granulations in the granulocytic series by being positive in the chloracetate esterase reaction. In rodents, mastocytes contain a large amount of biologically active amines (serotonin, dopamine).

Living Cells[4]

Phase microscopy gives particularly clear pictures of the mastocytes because of the high refractility of their granules. They are round or oval, either entirely black or with a clear center. The cells are capable of ameboid movement, which is not very pronounced, and of phagocytosis. Toluidine blue in very low concentration can be used as a vital dye to stain the mastocyte granules an intense red-orange.

Electron Microscopy[5]

The cell outline varies. In the connective tissue, the cell is usually elongated and presents very long pseudopods. In the hemopoietic organs their appearance is more frequently oval with microvillosities on the surface. The endoplasmic reticulum and the Golgi are not well developed. The mitochondria are not numerous.

The internal structure of the granules varies with the animal species and in the same species with the state of maturation. The granules are surrounded by a membrane. Depending on the state of maturation, the amorphous or finely granular material is organized in parallel lamellar structures or forms multilayered cylinders when the lamellae curl up on themselves. They are composed of individual particles which are very dense and approximately 5.0 nm in diameter. The appearances are quite different from the granules of basophils in the blood (see p. 126).

The scanning electron microscope shows the granulations in relief and numerous microvillosities of the surface.

Release and Discharge of Granules[5–7]

It is generally believed that mastocytes perform their physiologic function by releasing their granules extracellularly. They may act in the immediate environment of the cell or may reach the bloodstream by way of the lymphatics. The discharge of the granules has been observed with the light microscope in the living animal and, in

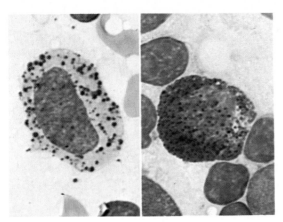

Fig. 1 – *A mastoblast* (lymph node aspirate)

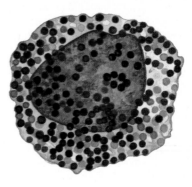

Fig. 2 – *A mastocyte* (lymph node aspirate). The zone of the centrosome is a little lighter

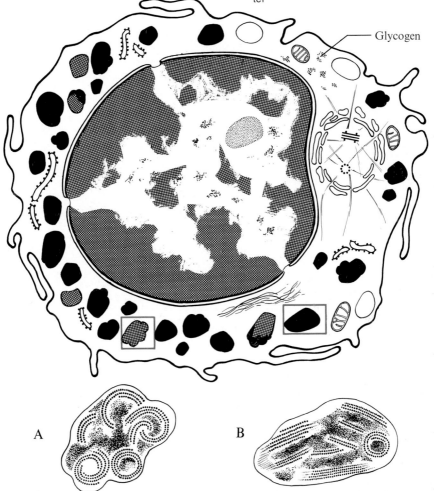

Fig. 3 – Note the various arrangements of the subunits, which differ greatly from the basophilic granules. They are shown both in transverse (A) and longitudinal sections (B)

parallel observations after fixation, with the transmission and scanning electron microscope.

Many compounds can lead to degranulation. Among those which appear to have physiologic significance are ACTH, dextran, certain venoms, vitamin A, protamine sulfate, antigens to which the body is responsive, and antigen-antibody complexes. Degranulation liberates histamine, heparin, serotonin, hyaluronic acid, and different enzymes. It produces local edema and permits fixation and inactivation of certain toxic agents. The mastocytes, therefore, play an important role in inflammatory reactions.

The released granules may be phagocytized by histiocytes, or neutrophils and eosinophils. Such cells subsequently contain basophilic granules and must be distinguished from actual mastocytes.

Chapter 9

Leukemias

A — Reinterpretation of "Classification" of Leukemias

At least twenty varieties of leukemia and a dozen related syndromes have been described. In addition a certain number of disease states occur in which the diagnosis is uncertain (see unclassifiable leukemias and preleukemic states, pp. 202 and 203). This is due partly to our ignorance of the causes of leukemias and partly to our insufficient knowledge of the pathophysiology of leukemias. These uncertainties prevent a true classification of leukemias.

Classification and Categorization[1, 2]

There are two types of classifications: natural or true classifications, which are based on the sum total of all characteristics, and all other classifications which depend on a single characteristic and are more appropriately called categorizations. A true classification rests on the knowledge of an inherent ordering principle or law. Hence a natural classification may predict the existence of hitherto unknown members of a set. Thus the periodic table predicted the existence of elements unknown at the time it was conceived. Intriguingly, the underlying principle may at first be only intuitively conceived as was the case when Mendeleiev proposed the arrangement of elements in the periodic table.

The laws which govern differentiation of cells are not known even for normal cells, let alone leukemic ones. Our classification of leukemias is thus merely a categorization. It recalls the elaborate classification of tuberculous lesions based on anatomic or clinical characteristics a hundred years ago. It was still thought useful when the cause of tuberculosis was already known but was discarded when effective treatment became available. We can expect that our present classification of leukemia may be similarly discarded when the etiology and cure of the disease will be known.

In deference to common usage, I shall continue to use classification and categorization interchangeably hoping the reader will keep in mind the preceding considerations.

Why Classify Leukemias in the First Place?[3−9]

The present classification of leukemias, though provisional, is useful in the prognosis and treatment of the disease and facilitates communication among hematologists.

"Classification" of Leukemias

Basis of Present Classification of Leukemias[10]

In general, leukemias are classified by their resemblance to normal cell lines (granulocytic, lymphocytic, monocytic, etc.) and, within a given cell line, by the stage of maturation of the prevalent leukemic cell (myeloblastic, promyelocytic, etc.). The wish to classify all leukemias, even those in which cells are difficult to associate with a given cell line or its subcategories (see pp. 190–203) has led to a multiplicity of "objective" methods of categorization using electron microscopy, immunocytochemistry, in vitro culture, cellular kinetics, etc., in addition to Giemsa stains.

The classifications are based on several assumptions:

1) Leukemias represent proliferations of cells belonging to a single cell line.
2) Within that cell line, a single type (frequently called a clone) constitutes the leukemic population.
3) The "clonal" population remains identical from the beginning to the end of the disease, disappears during remission and reappears during relapse.

The assumptions imply that the so-called unclassifiable cells and leukemias remain unclassified only for lack of appropriate means of study. Refinements of technics should permit the characterization of the cell line of origin in every leukemia.

The basic assumptions, however, are only a first approximation to the actual state of affairs. In effect: (see pp. 205–214)

1) All cell lines may be affected.
2) While one cell population may have outgrown all others at a given stage of the disease, new subpopulations may emerge continuously.
3) The population most resistant to the regulatory forces of the organism (or to treatment) may replace the original populations.
4) Disordered maturation may alter the morphology and physiology of leukemic cells to such a degree that they no longer retain any specific characteristics and no refinement of technic can help to identify the cell line of origin.

Nevertheless, the correlation between the current classification and current treatment and prognosis is good, though many exceptions can be found.

One day the present classification of leukemias, though useful at the moment, will probably appear as bizarre as an ancient Chinese classification of animals into 13 categories including:

1) Those belonging to the emperor
2) Tame animals
3) Four-footed animals
4) Those resembling flies
5) Embalmed animals
6) Mythologic animals
7) Those not included in the foregoing classes ...

As illogical as this classification appears to us, it was very useful: it would have been deadly not to recognize an animal belonging to the emperor (as it would be fatal not to recognize a promyelocytic leukemia which requires immediate treatment). It was no doubt beneficial to distinguish wild animals from tame ones (as it is beneficial to distinguish a hairy cell leukemia, which at present (1977) should not be treated by chemotherapy, from a leukemia known to respond favorably in at least a percentage of cases). Nevertheless, one must not forget that categorizations, though useful, contain no underlying principle that could stimulate new concepts. On the contrary, they may block new ideas for a long time.

B — Reinterpretation of Selected Types of Leukemias

The reinterpretation of Giemsa-stained smears is based on new knowledge brought to light by new cytologic techniques. In many leukemias, including some of the most frequent ones, new cytologic techniques have not added significantly to our ability to reinterpret ordinary smears. There are, however, some well-defined, though less common leukemias with very specific morphologic changes which allow their identification. These changes are often minor and escape the eye of the casual observer. Nevertheless, it is exactly the appropriate evaluation of these minor signs in the light of knowledge adduced by other techniques that permits the correct diagnosis. This chapter presents the reinterpretation of smears of some of these leukemias.

Here even more than in other chapters of this book, the reinterpretations are offered primarily as examples to induce the student to engage in his own reinterpretation as he looks at the cells of any other leukemia he may have the opportunity to observe.

Important Reminders

1) The cells described in the following pages characterize particular types of leukemia because they represent predominant cell types. Similar cells may, however, occur in small numbers in other leukemias or even in the absence of leukemia.
2) The diagnosis of leukemia in general and of specific types of leukemia is based on the presence of a population of abnormal cells. When this population is very small, the diagnosis becomes impossible for the time being (see p. 203). The more pronounced or specific the abnormalities are, the fewer cells are necessary to arrive at a diagnosis.
3) The disturbance of normal maturation of one or more cell lines, with or without morphologic abnormalities, is an important adjunct in the diagnosis of leukemia, but insufficient to establish it (see p. 208).

1 — Pseudolymphoblastic Leukemias

Definition

Rare leukemias (approximately 5% of all cases) in which the majority of the cells resemble lymphoblasts at first sight. A more detailed examination and recourse to cytochemical and electron-microscopic technics permits at times the identification of their true origin. These leukemias must not be confused with unclassifiable leukemias (see p. 202).

Pseudo-agranular Myeloblastic Leukemia

Definition

Leukemias in which the predominant myeloblastic cells contain "azurophilic" granules that are no longer stainable by Giemsa.

Smears

Most cells resemble lymphoblasts, though a number of minor signs may point to their myeloblastic nature: there may be numerous, small nucleoli; rare cells may contain azurophilic granules while all others, though of otherwise identical appearance, contain none.

Cytochemistry

Peroxidase and Sudan black stains may be positive and establish the diagnosis. Occasionally these reactions may be negative because the cell though still capable of making granules, has lost the capacity to synthesize sufficient peroxidases or lipids to give a positive reaction visible with the light microscope. One may even see Auer rods (by phase contrast microscopy) that give a negative peroxidase reaction.

Electron Microscopy

It may reveal granules or enzymatic reactions typical of the granulocytic series.

Agranular Myeloblastic Leukemias

Stained Smears

The appearance of cells is identical with that of pseudo-agranular myeloblastic leukemia.

Cytochemistry

The typical reactions of the granulocytic series, and particularly the peroxidase reaction, appear negative.

Electron Microscopy[1, 2]

Granulations are absent. However, the peroxidase reaction is positive in the perinuclear cisterna and in some of the sacs of the RER, indicating that the cells belong to the granulocytic series. The presence of enzymes or other chemical constituents (alkaline phosphatase, lactoferrin) may allow the recognition of cells as simplified myeloblasts (see p. 208). Theoretically, they could also be monoblasts or some other cells that have acquired the capacity to manufacture constituents such as peroxidase by derepression (see p. 28).

Acute Stages of Chronic Myelogenous Leukemias[5—7]

The progression of chronic granulocytic leukemias to a form often indistinguishable from acute myelogenous leukemia is discussed below (p. 212). Sometimes these myeloblasts have lost part or all of their granules and have thus become pseudo-lymphoblasts. Recently, several authors have drawn attention to the presence of terminal deoxynucleotidyl transferase in these cells which is considered specific for lymphoblasts. It has been suggested that one is dealing with an actual conversion to a lymphoblastic leukemia and not only with a simplification of the cells with a loss of all signs of differentiation (or the appearance of a previously repressed enzyme). This interesting hypothesis needs to be verified.

Apparently Lymphoblastic Cells[3, 4]

Almost all leukemias of whatever type contain pseudolymphoblastic cells. These cells can actually be myeloblasts, monoblasts or even megakaryoblasts (see p. 142).* They are usually disregarded by the hematologists when the diagnosis has been made and the minor population of pseudolymphoblasts can indeed be neglected. They are, nevertheless, of great importance for the understanding of the physiopathology of leukemic cells (see pp. 210 and 212).

---

* These are, however, pathologic cells and must not be mistaken for precursors of the normal megakaryocytic series (see p. 130) nor for micromegakaryocytes (see p. 143) which can be recognized in smears by the presence of platelet granules.

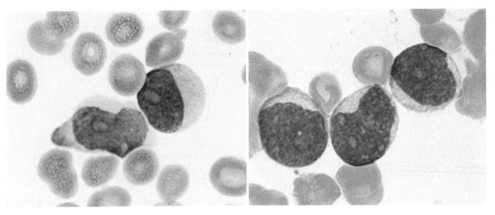

Figs. 1 and 2 — *Two cases of leukemia*, in which the cells are devoid of granulations

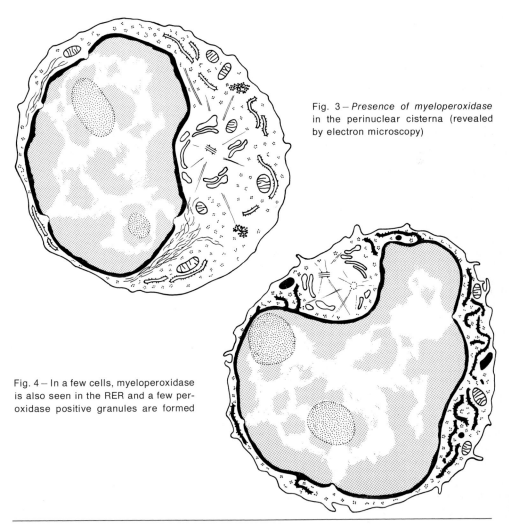

Fig. 3 — *Presence of myeloperoxidase* in the perinuclear cisterna (revealed by electron microscopy)

Fig. 4 — In a few cells, myeloperoxidase is also seen in the RER and a few peroxidase positive granules are formed

2 – Promyelocytic Leukemia (Syndrome of Jean Bernard)

Definition[1-4]

This syndrome links a specific cytologic appearance to a severe hemorrhagic tendency. The cytologic diagnosis becomes a matter of urgency because adequate treatment, if given immediately, can save the life of the patient and induce prolonged remissions.

Two important points should be emphasized:
1) The "promyelocytes" have specific abnormalities which identify them clearly: the abundance and excess size of granules, and their abnormal shape.
2) Leukemias with such abnormal promyelocytes do not necessarily have a severe hemorrhagic tendency, but the association is found in nine of ten patients. Conversely, non-thrombocytopenic coagulation defects and a hemorrhagic syndrome have occasionally been described in acute leukemias other than those of the specific promyelocytic type.

Smears

The number of leukocytes may vary from 2000 to 300,000. Most authors require 50% of cells to be abnormal for diagnosis.

The typical cells are large, 15 to 20 μm, with basophilic cytoplasm and abundant, often very large azurophilic granules. They may be so numerous as to entirely mask the basophilic cytoplasm and the nucleus. Some of the cells and occasionally a majority contain multiple Auer bodies of unusual appearance. They are very thin, needle-like and piled up so as to recall a carelessly tossed pile of firewood, in contrast to the Auer bodies in myeloblastic leukemia which are usually single and very rarely exceed two or three per cell.

In contrast, the granules may be very small and dustlike and be present in such large numbers that the cytoplasm appears diffusely red.

Cytochemistry[3, 6]

The peroxidase reaction is strongly positive both in the granules and in the Auer bodies. Due to the abundance of granules, the reaction colors the cell diffusely and obscures the nucleus. Nonspecific esterase and acid phosphatase are, in general, markedly positive. The PAS reaction may be positive or negative. This is also true of some other reactions, such as toluidine blue and lysozyme. These reactions vary from case to case and in the same patient at different stages of the disease.

Electron Microscopy[3-8]

In addition to disorders of organelles seen in other acute leukemias, the syndrome of Jean Bernard may be associated with some additional features:

1) The azurophilic granules may have a periodicity (see p. 204). Many fuse and form voluminous inclusions.
2) The ultrastructure of the Auer bodies appears different from that seen in classic myeloblastic leukemia. They have a periodicity of 25 nm, instead of 6 to 12 nm, and they appear to be composed of tubular elements.
3) Fibrillar formation in the RER of promyelocytes may occur, particularly in the clinical syndrome of defibrination due to disseminated intravascular coagulation.[9]

Acute Basophilic Leukemia[10]

It may lead to confusion with the syndrome of Jean Bernard, because of its numerous large granules of red-wine color. In principle, the metachromatic staining with toluidine blue or astra blue allows their differentiation. In practice, it needs to be stressed, however, that 1) the azurophilic granules of promyelocytes may, under certain conditions, be metachromatic (see p. 112); 2) the granules of leukemic cells may be highly abnormal and lose some of their normal cytochemical characteristics (see pp. 206 and 208) so that leukemic basophils may lose their metachromasia. Hence, the diagnosis of a basophilic leukemia may be difficult and should be based on more than one characteristic.

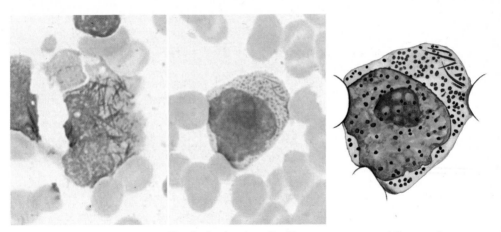

Fig. 1 — *Leukemic promyelo-cyte* containing bundles of Auer bodies

Fig. 2 — *Leukemic cell with numerous azurophilic granules,* among them some very thin Auer bodies

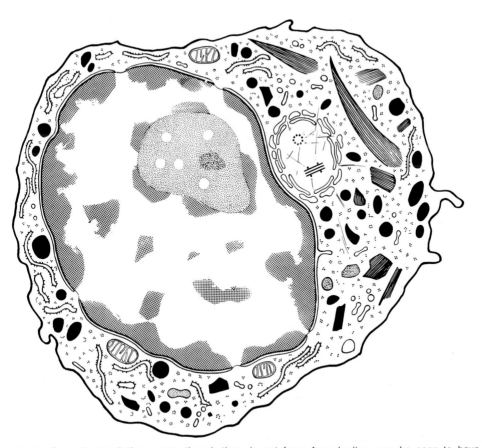

Fig. 3 — Several granulations, even though they do not form Auer bodies, can be seen to have a periodic structure

3 − Chronic Lymphocytic Leukemias − Special Types

Many attempts have been made to categorize the different leukemias of the lymphocytic series, either by their appearance in stained smears or by the result of cytochemical stains, and most recently by different functional tests. The discovery of surface markers considered specific for B- and T-lymphocytes has led to the interesting observation that most chronic lymphocytic leukemias are derived from B-cells. Keeping in mind that leukemic cells may lose old or acquire new antigenic determinants (see pp. 205 to 214), the observation must be interpreted with caution.

Notwithstanding these difficulties of subclassification, I shall present four particular types of lymphocytic leukemia, in which careful examination of the stained smears can point to a specific subtype. They are: the lymphocytic leukemia with crystals, the macroglobulinemia of Waldenström (p. 196), the Sézary syndrome (p. 198) and "hairy cell" (tricholeukocytic) leukemia (p. 200).

Lymphocytic Leukemia with Crystals

The presence of crystalline inclusions, which fail to stain with Wright's or Giemsa, characterize a particular form of CLL though not necessarily a special clinical course. Usually, the crystals are present in a percentage of cells, which varies little during the entire duration of the disease. The inclusions are due to the retention by the cell of its secretion, in general macroglobulins. They are, therefore, related by the mechanism of their formation to Russel bodies (see p. 180).

Smears[1]

As in all chronic lymphocytic leukemias, the cells are in general of a single type, which varies from patient to patient. This uniformity is often so pronounced that one can recognize the cells as belonging to a particular patient. In general, the lymphocytes are small, but occasionally they are of medium or large size. At times the abnormal lymphocytes have some typical pathologic features: a large centrosome, abnormally shaped non-basophilic nucleoli, etc. Frequently, however, the cells do not present any morphologic abnormality.

Careful examination of the smears reveals the presence of unstained rodlike structures* in the cytoplasm of the lymphocytes. They remain entirely unstained and can often be seen only with difficulty, except in markedly basophilic parts of the cytoplasm. When the cell has been crushed in the preparation of the smear and the nucleus is reduced to a shadow, one may see the negative images of the rods as they become superimposed on the nuclear chromatin.

Cytochemistry[2−4]

Leukemic lymphocytes contain much more glycogen than normal lymphocytes, hence the frequently more strongly positive PAS reaction, distributed in relatively coarse packets. The conglomerates seen in the optical microscope are, however, an artifact produced during preparation of the smears.

The rods do not stain with Sudan black, PAS, or Feulgen's. They give a positive reaction with antiglobulin antibody. This indicates that we are dealing with secretions of the cell which fail to be extruded.

Living Cells

The rodlike structures can be seen very clearly in the phase microscope. They appear grey, often abutting on the nucleus and the adjacent mitochondria.

When the cell is squashed or begins to lyse, they become darker and are clearly resistant to lysis. They persist as long as 24 and 48 h between slide and coverslip in entirely lysed cells.

Electron Microscopy[6−10]

The pathologic cells are lymphocytes, which may present one or several "leukemic" features (see pp. 206 to 214). In particular, nuclei are often riederiform and contain nuclear pockets; cytoplasm contains abnormal ribosomes and autophagosomes. The rods have a rectangular shape; their length varies between 1 and 10 μm, their width between 0.1 and 0.5 μm. They are occasionally surrounded by a membrane formed from RER. They are made up of parallel fibrils which have a periodicity of 10 nm.

These paracrystalline structures are quite different from Auer bodies; not only do they not stain with azure, they are rectangular, rather than rhomboid or needle-shaped. They are found in cells of the lymphocytic rather than the granulocytic series, and they are composed of immunoglobulins.

---

* Individual cases of lymphoid leukemia have been described in which a large percentage of cells contains a variety of inclusions characteristic of an individual patient: spherical inclusions, which fail to stain, large azurophilic granules, or azurophilic crystals resembling Auer bodies. The nature and significance of these inclusions remain to be defined.

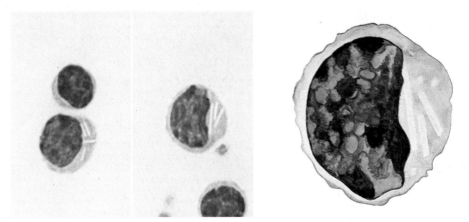

Figs. 1 and 2 – *Leukemic lymphocytes* containing globulin crystals not stainable by Giemsa. One can infer their presence from their negative imprints in the blue cytoplasm

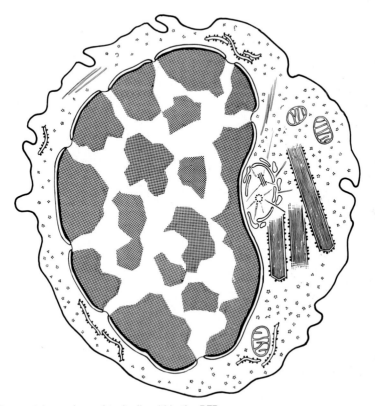

Fig. 3 – The crystals are formed typically within the RER

195

4 – Macroglobulinemias (Waldenström's)

Definition[1−3]

Waldenström's syndrome is characterized by macroglobulinemia associated with infiltration by neoplastic lymphocytes and/or plasmocytes in the bone marrow, liver, spleen, lymph nodes, and, occasionally, peripheral blood.

Smears

Blood smears are occasionally difficult to prepare, because of the marked rouleaux formation due to the hyperglobulinemia (see p. 48). The number of white cells is normal, occasionally low, and sometimes frankly leukemic.

The pathologic cells can be of one or several types. Generally, one may see all intermediate stages between lymphocytes and plasma cells. Several subpopulations recognizable by some particular pathological feature may be present. The most common cytologic forms are those with predominance of lymphocytes, while those that consist exclusively of plasma cells are quite rare. Subclassification of Waldenström's syndrome into primarily lymphocytic, mixed, or primarily plasmacytic types has not been shown to be of any utility.

Some of the lymphocytes or plasma cells contain nuclear vacuoles. They are often difficult to recognize in stained slides (Figs. 1 and 2). Careful examination, particularly with the aid of phase contrast (see p. 176), allows one to recognize them, or at least to suspect them. Cytochemistry, immunocytochemistry and electron microscopy are best suited to reveal their presence and nature.

Nuclear vacuoles are not pathognomonic of Waldenström's, but occur in this disease quite frequently.

Cytochemistry and Immunocytochemistry[6, 7]

The PAS reaction is sometimes positive in the cytoplasm, and more or less positive in the nuclear and cytoplasmic inclusions. It persists after digestion with amylase, which excludes its being due to glycogen and indicates the presence of neutral mucopolysaccharides.

Immunocytochemistry has shown the presence of macroglobulins in the inclusions and in the cytoplasm. The macroglobulins are generally identical in all cells and such cases are said to be monoclonal (erroneously so, see below). In rare instances, two or more macroglobulins may be present in different malignant cells (such cases are called "polyclonal") or even in the same cell.

In rare forms of the disease non-secretory cells have been found.

Electron Microscopy[4, 5]

In the cytoplasm of most cells, one may observe free ribosomes, vacuoles, lipid inclusions, numerous mitochondria, poorly developed Golgi complex and RER. In contrast, in a few cells RER may be highly developed. Occasionally it may be dilated and contain cytoplasmic inclusions visible with the optical microscope.

Nuclear inclusions of various sizes may be seen, some so small that they cannot be seen with the optical microscope. They are usually situated at the edge of the nucleus and are due either to an invagination of the cytoplasm, or a dilatation of the perinuclear cisterna, which herniates into the interior of the nucleus (Fig. 3 and 4).

Multiple Populations[7]

The pathologic cells may be all of the same type or multiple populations may exist.

The cells may reflect all types of pathology of neoplastic cells. The most important points are:

a) The abnormality of the RER and of the Golgi complex may be such as to prevent the elimination of the macroglobulins produced and give rise to the formation of intranuclear and intracytoplasmic inclusions.

b) The cells may retain all of the synthesized globulins in the cytoplasm and excrete none into the surrounding plasma.

c) The cells (or some of them) may be so severely affected as to prevent the synthesis of macroglobulins entirely.

Note that multiple populations may thus exist even if only a single type of macroglobulin appears in the plasma or on the surface of abnormal cells. Hence, such cases are not truly "monoclonal." In rare instances, however, subpopulations may indeed produce different macroglobulins.

Progression[8]

As in acute leukemias, the emergence of new populations less subject to regulation by the organism (more autonomous) leads to a progression of the disease. This phenomenon is rare, as in all lymphoid leukemias, compared with its frequency in myeloid leukemias (see p. 212).

When such progression occurs, the cells usually no longer secrete globulins and frequently resemble histiocytes (without being true histiocytes) or blasts.

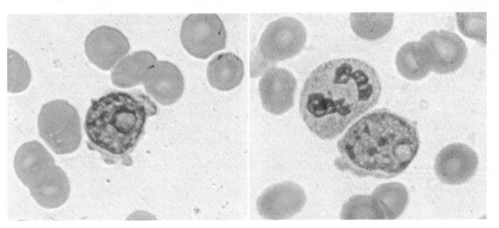

Figs. 1 and 2 – *Leukemic lymphocytes* in the bone marrow. Note the pseudonucleoli which correspond to nuclear inclusions (see Figs. 3 and 4)

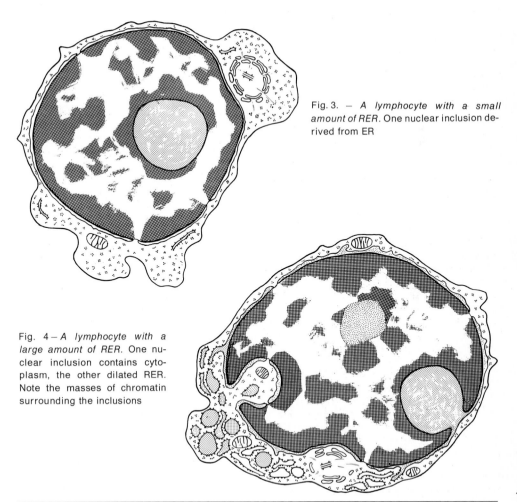

Fig. 3. – *A lymphocyte with a small amount of RER*. One nuclear inclusion derived from ER

Fig. 4 – *A lymphocyte with a large amount of RER*. One nuclear inclusion contains cytoplasm, the other dilated RER. Note the masses of chromatin surrounding the inclusions

197

5 – Sézary's Syndrome

Definition[1-5]

Sézary's syndrome is characterized by the association of a dermatosis, usually an erythroderma, with the proliferation of abnormal leukocytes in skin, lymph nodes, blood, and bone marrow.

Initially the abnormal cells were described as "monstrous" histiocytes and monocytes. More recently, they have been considered pathologic lymphocytes (see below).

The nature of the Sézary syndrome is not clear. It is possible that it comprises different diseases (e.g., cutaneous forms of leukemia, which may be monocytic or lymphocytic). Some authors believe that Sézary's syndrome is only a special variant of mycosis fungoides with circulating abnormal cells.

Smears

The so-called Sézary cells represent as a rule only a small percentage of circulating cells (1.0–20%).

They are generally large cells, 15–20 µm in diameter, though on occasion they may be of the diameter of a small lymphocyte. The basophilic cytoplasm represents a narrow band and occasionally contains multiple vacuoles. The nucleus stains deeply and occupies three-fourths or more of the cell. It is lobulated, the lobes usually overlapping, creating folds (Fig. 1). These are often difficult to see, particularly if the smear is poor, and the objective inadequate or inadequate magnification is used. It should be noted that 1) the abnormal cells vary considerably from patient to patient; 2) undoubtedly there exist pathologic cells in this syndrome which fail to present the characteristic folds of the nucleus and then resemble the cells of lymphocytic leukemias; 3) in some cases the pathologic cells are or resemble monocytes.

Cytochemistry[6, 7, 9, 10]

In most recently studied cases, the cells have the same cytochemical reactions as normal lymphocytes. In some cells, particularly those with folded nuclei, there is a significant increase of DNA content. The chromosome number is between 46 and 99 without any modal peak.

This large quantity of DNA in a relatively small cell explains perhaps the tendency of the nucleus to fold on itself and to give the heavily stained appearance in smears already described.

Cells resembling normal lymphocytes with a normal DNA content may also be present and may contain a normal chromosome number but abnormal chromosomes.

The cases in which the appearance of the cells in smears was that of pathologic monocytes have not yet been studied cytochemically.

Electron Microscopy[7-9]

The small cells do not differ from those of chronic lymphocytic leukemia.

The large cells have a characteristic nucleus. They appear cerebriform or serpentiform in sections. Measurements of the contours have shown that the surface is larger than that of normal or even of leukemic lymphocytes.

The cytoplasm contains many microfilaments in bundles, small granules of as yet undisclosed nature and masses of glycogen, which correspond to the cytoplasmic vacuoles seen in smears.

The histio-monocytic forms have not yet been studied with the electron microscope.

Origin of the Sézary Cell[11-14]

Recent studies indicate that the Sézary cells just described are derived from lymphocytes and not from monocytes. They do not adhere to glass and are nonphagocytic. They respond to PHA, which suggests that they are T-lymphocytes and, indeed, they do not have immunoglobulins on their surface, nor complement receptors, and they do form E-rosettes. Again, these results need to be confirmed and a larger number of cases need to be studied.

The so-called Sézary cell is not pathognomonic. It suggests but does not define the Sézary syndrome. A small number of similar cells can be found in lymphocytic leukemia and in unclassified leukemias.

Finally cases of monocytic leukemia and/or malignant histiocytosis with skin infiltration have been described, some by Sézary himself. It is impossible to determine at present whether they are a special form of the Sézary syndrome.

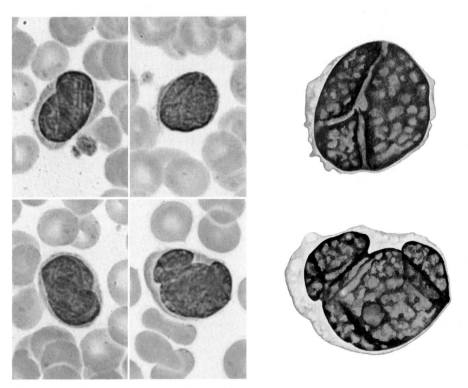

Figs. 1 to 4 — *Sézary cells.* Note the folded nucleus

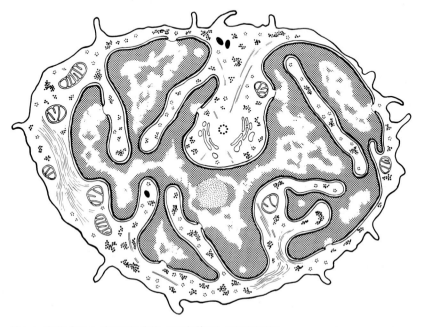

Fig. 5 — The nucleus has a characteristic cerebriform appearance

199

6 – Tricholeukocytic Leukemia (Hairy Cell Leukemia)

These are leukemias characterized by the association of very special cells with the clinical syndrome of pancytopenia with splenomegaly*.
Tricholeukocytic leukemia may serve as an example of a leukemia which can be easily categorized yet not easily classified (see p. 187). In fact, at this time (June 1977), there is still uncertainty whether the affected cells are lymphocytes, monocytes or another cell line.

Smears

The cells are of variable size, 10–25 µm. The nucleus is centrally located and round. The cytoplasm is faintly basophilic, its borders irregular and indistinct. It contains no granules.
In certain parts of the smear, the cells have dendrites and veils, but in other parts, and in particular the feather edge of the smear, the outline of the cytoplasm is sharp and circular. This is due to an artifact in spreading the cells. Examination in the living state shows that virtually all pathologic cells have ruffles (veils) and filiform extensions. In certain cells one may see a rodlike inclusion or a "basophilic double line". The electron microscope permits one to analyze these formations.

Living Cells[7, 8]

These are best examined after concentrating the white cells, because the number of leukocytes is often quite low. The cells are somewhat larger than large lymphocytes. The nucleus is round and clear. The cytoplasm contains many mitochondria. The surface is covered with spike-like and filiform extensions, which separate them clearly from other leukemic cells. Occasionally the characteristic rodlike inclusions may be seen, particularly after compression of the cell between slide and coverslip.
Some of the cells are capable of necrotaxis (a normal feature of monocytes, never seen in normal lymphocytes, see p. 24).
The scanning electron microscope shows the expansion of the surface of the cell very clearly.

Cytochemistry[9–11]

The peroxidase reaction is negative, but the PAS and nonspecific esterase reactions vary from cell to cell, and the overall percentage of positive cells from case to case. The acid phosphatase is positive in a considerable number of cells. Exposure to acid tartrate fails to inhibit the reaction as is the case for lymphocytes, plasma cells and monocytes and, in fact, may enhance it. Thus the cytochemical reactions do not give a clear answer to the question of the monocytic vs. lymphocytic nature of the cells.

Electron Microscopy

The examination confirms the presence of numerous projections of the cell surface and allows the recognition of a marked pinocytosis. The nucleus and cell organelles (Golgi, mitochondria, lysozomes) have no special features.

Tubular Inclusions[12, 13]

In a small percentage of cells, tubular inclusions 0.5 to 3.0 nm in length may be seen; these appear to consist of rolled up strips studded with ribosomes. In transverse sections they are round or ellipsoid and occasionally surrounded by a sac of RER. Although these inclusions are quite peculiar, they are nevertheless seen in other neoplasias, such as lymphocytic and myeloblastic leukemias. They have even been observed in normal cells. Thus they are not characteristic but merely frequent in hairy cell leukemia.

Nature of Tricholeukocytes[14–17]

This has not yet been established definitively. Results of different tests (fixation of gamma globulins and complement, adherence to glass, possible phagocytosis, enzyme characteristics) vary according to different authors and, perhaps, with different patients. Some investigators lean toward the monocytic, others toward the lymphocytic origin of the cell.
Certainly many of the cells have characteristics thought to be specific for one or the other of these two cell lines. Possibly, although the cells originate from one cell line, they may acquire features of another cell line due to derepression.

---

* Synonyms, e.g. lymphoreticular leukemia, leukemic rericuloendotheliosis, reflect our uncertainty about the nature of these pathologic cells.[1–6]

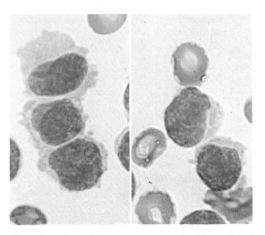
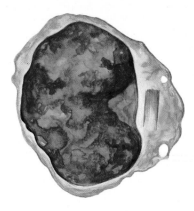

Figs. 1 and 2 — *Different aspects of tricholeukocytes*. The process of spreading the cell on a slide results in the disappearance of a number of cytoplasmic processes. At the feather end of the smear the cells appear rounded off and their borders are smooth. Note a peculiar cylindrical formation in the cytoplasm (Fig. 2)

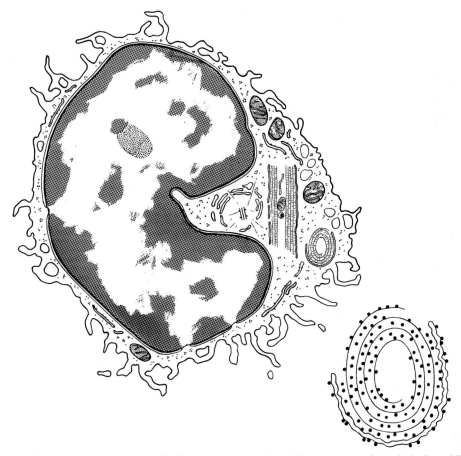

Fig. 3 — Note the numerous microvilli, the micropinocytosis and the presence of two inclusions (ribosome-lamella complexes) cut longitudinally and transversally (*detail on the right*)

7 – Unclassifiable Leukemias

Hamlet: Do you see yonder cloud that's almost in shape of a camel?
Polonius: By the mass, and 'tis like a camel, indeed.
Hamlet: Methinks it is like a weasel.
Polonius: It is backed like a weasel.
Hamlet: Or like a whale?
Polonius: Very like a whale.

Leukemic Cells Falsely Considered Unclassifiable[1, 2]

These are cells which appear unclassifiable as seen in smears*. A more thorough examination, be it cytochemical or with the electron microscope, may reveal features which allow identification with a given cell line. The so-called pseudolymphoblastic leukemias, which are in fact myeloblastic in origin, may serve as an example (see p. 190). Others are leukemias with undifferentiated blasts in which the immunofluorescence leads to the identification of globulins (see p. 190, 196).
Another example may be found in the Friend leukemia of mice, for a long time thought to be a reticulum sarcoma. Since it became possible to culture these cells, it has become obvious that they can synthesize hemoglobin and, consequently, belong to the erythrocytic series.
One can always hope that new techniques may be found which will allow the determination of the very earliest signs of commitment to a given cell line, but there is also reason to believe that certain leukemias are truly unclassifiable and will remain so.

Unclassifiable Cells and Unclassifiable Leukemias[1]

In any leukemia there exist cells that cannot be classified with certainty as belonging to a given normal cell line. The clinician is not so much concerned with these cells, once the diagnosis is made. However, occasionally these cells make up the majority of leukemic cells, either to begin with or as a result of progression. In that case we deal with an unclassifiable leukemia.
Unclassifiable leukemias can belong to one of three categories:
a) The cells have no signs of differentiation whatsoever.
b) Different cells appear to belong to different cell lines. This phenomenon is often due to pathology of organelles (see p. 206), to mosaicism and progression (see p. 212). For example, it is not too uncommon for a proliferation of abnormal erythroid cells to precede or coexist with a myelogenous leukemia, both erythroid and granulocytic cells originating presumably from a single affected stem cell. Very rarely leukemias actually derived from two cell lines coexist or follow each other.

c) The cells resemble the cells of one line, but in addition have features of another cell line (e.g., tricholeukocytes, p. 200).

Treatment of Unclassifiable Leukemias[1]

The clinicians are forever anxious to classify unclassifiable leukemias and to discover their true origins, for they believe that prognosis and indication for treatment can be inferred from such knowledge. That appears to be correct for a good number of classifiable leukemias. Indeed, a particular treatment may be effective because it corrects specific problems derived from the biochemistry of the causative cell. An eloquent example is given by the promyelocytic leukemia (see p. 192). The correction of the bleeding tendency caused by the intravascular coagulation promoted by the granules of the promyelocytes is effective. Here, or in the plasmaphoresis treatment of macroglobulinemia, we treat successfully the secondary effects due to the pathologic products of the cell line involved. In general, however, the correlation between successful treatment and the particular line of origin of the leukemic cell is not as clear. In fact, it is well known that leukemias become more or less rapidly resistant to a treatment which was originally effective, even when their appearance in smears, in the electron microscope, and with any other means of diagnosis remain the same.

Conclusion

In difficult cases the cytologic diagnosis of the cell line affected may occasionally be possible by additional investigations, be it by cytochemistry, by light microscopy, electron microscopy, chromosome study or other means.
However, there remain unclassified leukemias either because we do not, as yet, have adequate techniques to characterize the stem cell of lymphocytes and monocytes, or because the cells in question are simplified in such a manner that no sign of differentiation remains. Finally, cells may have undergone pathologic changes and have acquired the apparent characteristic of a different line (see pp. 205 and 212).

---

* Some hematologists have reported that they have submitted a number of slides to a group of highly competent investigators. They found that not only did the cytologic diagnosis vary from investigator to investigator, but that the same slides, having been submitted to the same investigators after a lapse of time and with different labels were now identified as a different type of leukemia by the same person. Perhaps the explanation lies in the anxiety to make a specific diagnosis, even in an unclassifiable leukemia.

Preleukemic States

The expression "preleukemic state" or "pre-leukemia" is ambiguous. It has been applied to quite different conditions:

1) The period of incubation of a leukemia, in which the characteristic signs of the disease have not yet fully developed and a diagnosis is not possible.
2) Syndromes that appear to favor the development of leukemia, such as anemia of Fanconi, anemia of Bloom, paroxysmal nocturnal hemoglobinuria, karyotypic abnormalities, in which leukemia develops in a small percentage of cases.
3) Hyperplastic, hypoplastic or normocellular marrows in which some cells present all the alterations seen in leukemia: these are "leukemia without leukemia."

A variety of names have been used to designate the last one of these disorders: preleukemia; pancytopenia, sideroblastic preleukemic anemia; refractory anemia with excess myeloblasts; preleukemic anemia, thrombocytopenia, leukopenia, leukocytosis, thrombocytosis or polycythemia. The term preleukemia has been severely criticized, because the evolution into a true leukemia is uncertain, and because the patient may conclude that he is suffering from a fatal disease. The prognosis is by no means comparable to that of acute leukemia. The disease evolves very slowly and may be completely arrested in its development and possibly heal spontaneously.

For these reasons, the name hemopoietic dysplasias has been adopted for such conditions. The term dysplasia has been used quite generally for pathologic lesions with a high likelihood of developing into cancers. The term hemopoietic has been chosen to indicate that lymphoid as well as myeloid cells can be involved. Most commonly, however, only the three major lines of the bone marrow—megakaryocytes, granulocytes and erythrocyte precursors—are affected.

Hemopoietic Dysplasias

Definition[1–6]

In view of our inability to predict whether a severe hemopoietic dysplasia will develop into leukemia or not, hemopoietic dysplasias must necessarily include all conditions in which cytologic changes suggest the possibility of a leukemic evolution.

Hematologic Examination[7]

The blood most frequently shows signs of a refractory anemia resistant to the presently known medications, or thrombocytopenia or neutropenia.
The bone marrow is, as a rule, hyperplastic, but may be hypoplastic.

Myeloblasts may be present in somewhat increased numbers of approximately 10%, with some abnormal forms. Frequently, pathologic sideroblasts are present, or abnormalities of megakaryocytes. All of these signs are nonspecific (see below).

Both karyotypic abnormalities and abnormal results of in vitro cultures are important and may precede progression of the disease by many months. These results together with those of morphologic examination are similar to those found in leukemia. Some authors, indeed, maintain that the dysplastic stages are in reality already leukemias (see pp. 214 to 216).

Smears[8–10]

Many minor signs, which may go unnoticed by the untutored eye, allow one to make the diagnosis. Nevertheless, these signs are not constant, and they can be partially or totally absent.

Granulocytic Series. The polymorphs may have abnormal nuclei with irregular lobes, and chromatin which is too dense, or, on the contrary, too lightly stained. The granulations may be entirely absent or present in markedly reduced numbers. One may often find granulocytic precursors with minor abnormalities (myeloblasts or promyelocytes with pockets or nuclear indentations, abnormalities of the azurophilic granules which may be of giant size or needle-shaped.

Erythroid Series. In the erythroid series, pathologic sideroblasts are the frequent and predominant lesion. Megaloblastic forms may be present, but are in general not very numerous, and only occasionally can the entire gamut of abnormalities and all stages of the megaloblastic series be seen. The erythroblasts may exhibit numerous anomalies, nuclear malformations, fragmentation of the nucleus, multiple Howell-Jolly bodies, abnormal mitoses, and punctate basophilia.

Abnormal mature red cells are frequently seen in the peripheral blood and include poikilocytes, hypochromic or normochromic macrocytes and other malformations.

Thrombocytic Series. Megakaryocytes may be small and mono- or binucleated. Platelets may be of giant size with abnormally large fused granules or lacking granules altogether.

A variety of abnormalities of the enzyme reactions or functional abnormalities may be noted in all three cell lines and particularly changes in blood group antigens.

Electron Microscopy[11–15]

Electron microscopy confirms the existence of pathology of organelles in all three cell lines in cases in which they were doubtful on examination with the light microscope. They are indistinguishable from those in fully developed leukemias.

Granulocytic Series. All cells of the series may present nucleolar abnormalities, asynchrony or anarchy of development of the organelles.

The azurophilic granules are often unusually large: 0.5 to 0.8 µm. Some of them have a lamellar structure, which is not found in the normal promyelocytic granules and which resembles the structure of Auer bodies. Perhaps these granules represent precursors of Auer bodies. Some of them may be elongated into rodshaped forms. The secondary granules are less numerous and sometimes absent. The granules may be deficient in myeloperoxidase and their ability to destroy microorganisms may be defective.

Erythrocytic Series. Pathologic sideroblasts are commonly seen (see p. 54). Mitochondria contain iron; the polyribosomes are replaced by monoribosomes; autophagosomes are present; occasionally lipid vacuoles are present. Erythroblastic islands consisting entirely of normal cells may be seen side by side with other islands, in which all of the cells are abnormal (double population of erythroblasts).

Thrombocytic Series. The megakaryocytes may have abnormalities of their ploidy, and platelets may have abnormal granulations due to the fusion or absence of individual granules.

The number of histiocytic macrophages is increased and they may actively phagocytize erythroblasts, granulocytes and platelets.

In summary, all of the signs, though minor in themselves, point to an involvement of a common stem cell of all three myeloid lines, at least in the majority of cases. As in true leukemia, one finds multiple populations of pathological cells, but the number of cells in each may be very small.

C — Reinterpretation of the Cytology of Leukemias

This third section constitutes an attempt to find general laws which govern the behavior of the leukemic cell population. I will discuss successively the pathology of cellular organelles (p. 206), the anarchy and arrest of maturation (p. 208), mosaicism (p. 210) and progression (p. 212).

Admittedly this separates artificially the different cytologic signs of leukemia which normally coexist.

However, I hope that this attempt to consider the morphologic signs, which are well-known, from a new point of view will stimulate some readers to rethink in their own fashion some of the fundamental problems of the cytology of human leukemia. I have kept in mind the admonition of Karl Popper:

"Unjustified speculations and hypotheses constitute our only means to interpret nature. Those who refuse to expose their ideas to risks of refutation, do not truly participate in the game of science."

1 – Pathology of Cell Organelles

In most leukemias the majority of cells contain altered cell organelles as seen in stained smears or made evident by cytochemical reaction. At times abnormalities can be found by electron microscopy or physiologic tests. These abnormalities are extremely variable. Nevertheless, when present in a sufficiently high percentage, they may characterize a cytological or a clinical form of the disease (see pp. 189 and 200).

It is important to note that occasionally leukemic cells can appear quite normal and indistinguishable from their normal counterparts and, yet, be no less malignant. This signifies that we see only the secondary effect, frequent but not constant, of the actual lesion which makes the cell leukemic.

Since alterations of the organelles exist, one would wish to classify them by their causes or, at least, establish relationships between the different lesions (for example: modification of certain chromosomes and changes in their granulations or the absence of certain enzymes). Unfortunately, in most cases this is, as yet, impossible. We will, therefore, attempt to give a brief catalogue of the changes of the cellular organelles, and we once more remind the reader that these changes are neither necessary nor sufficient for the diagnosis of leukemia.

Changes in the Number of Organelles[1,2]

In a leukemic cell, all the cellular organelles, whether normal or pathologically altered, can be present in reduced or increased numbers. A few examples:

The number of ribosomes may be markedly diminished, the cytoplasm of the cell assumes in the Giemsa-Wright stains a grey appearance. This may lead, erroneously, to the identification of the cell as monocytic. In other cases, the number of ribosomes is markedly increased and gives the cytoplasm a dark blue color, and may lead one to identify the cell, again erroneously, as a proerythroblast.

The synthesis of granules of the myeloblast or promyelocyte may be inhibited, and the cell may assume the appearance of a lymphoblast (see pp. 206 and 208).

The number of chromosomes may be increased or reduced, and this may also apply to centrioles, to various types of granules, to Pamellae of the Golgi, to microtubules, and to fibrils.

These changes in the number of organelles cannot, in general, be recognized in smears. Some of them, however, can result in modifications of the shape or stain characteristics of the cells and lead to errors in identification, as already exemplified by the influence of the concentration of ribosomes.

Pathologic Alterations of the Organelles[1,2]

Nucleus. The nucleus can be abnormally large or abnormally small. It is frequently polylobed, resembling that of a normal monocyte. It can exhibit constrictions usually due to agonal or adverse environmental changes. Such cells are called Rieder cells. It may have slits, or fissures, pockets, appendices, or other expressions of abnormalities of chromosomes or the mechanism of mitosis (see p. 120). It may contain cytoplasmic inclusions, so-called pseudo-nucleoli, lipid and other vacuoles.

Nucleoli. They are very often hyperplastic and are present in larger numbers than normal; they are more commonly seen at the edge of the nucleus than in normal cells.

Michochondria. Mitochondria may be swollen, more or less dense than normal. The cristae may be more widely spaced than normal.

Granulations. The granulations may be of giant size or very small, fused as in the Auer bodies, or they may contain abnormal enzymes, degenerate or rupture within the cytoplasm leading to the formation of autophagosomes.

Endoplasmic reticulum. This may be dilated by the products of its secretion, which accumulate and/or crystallize; on the contrary, they may fail to function and remain flat. The ribosomes may fail to become associated with each other and remain in the cytoplasm as nonfunctional monoribosomes.

Various inclusions. Inclusions may be formed by nuclear membranes (annulate lamellae) or by fibrils, by masses of glycogen, etc.

Chromosomes. Although numerous abnormalities have been found, leukemias exist in which the karyotype is entirely normal.

Alterations Due to the Necrosis of Cell Organelles or of the Entire Cell[3]

Organelles and cells undergoing necrosis must not be confused with leukemic alterations.

Many leukemic cells are so changed as to have developed functional disorders, which in turn lead to the death of some organelles; autophagic vacuoles frequently result. Many leukemic cells contain granulations (or lysosomes) which have ruptured or have fused and can be recognized in smears. These various alterations can lead to the agony and death of the cell, phenomena which are often difficult to recognize in smears, but which can be well seen in the phase contrast or electron microscope. In some smears of leukemias, one may see many basket cells (Gumprecht shadows). They represent dead cells which have ruptured during preparation of the smear (see p. 22).

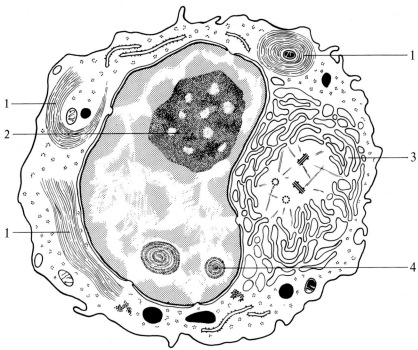

Fig. 1 — *In this diagram a number of abnormal features have been depicted artificially in a single cell.* (1) fibrillar body; (2) hyperplasia of the nucleus; (3) alterations of the sacs of the Golgi complex; (4) pseudonucleoli

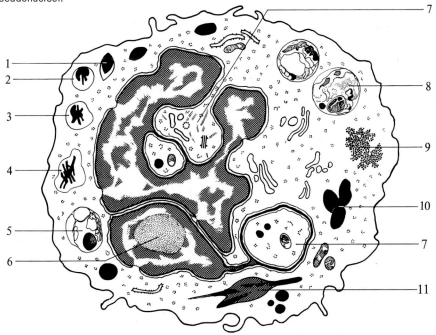

Fig. 2 — *In this diagram a number of abnormal features have been depicted artificially in a single cell.* (1 to 5) stages in the degeneration of a granulation; (6) nucleolus persisting in a polylobed nucleus with dense chromatin; (7) cytoplasmic inclusion in the nucleus; (8) autophagosome; (9) aggregate of glycogen; (10) fused granulations; (11) Auer bodies

2 – Anarchy and Arrest of Maturation

Superimposed on the alteration of cell organelles already described are the arrest and the anarchy of maturation. These changes lead to morphologic, biochemical or physiologic alterations. While these different manifestations of pathologic alterations are linked, their appreciation depends on our means of observation: a morphologic pathology indicates necessarily that the biochemistry and physiology are altered in some fashion. The reverse is not always true—at least, at present. Functional disorders of locomotion, digestion, etc., are not always expressed in morphologic changes that we can appreciate today.

Maturation Arrest[1-3]

This is one of the most frequent, but by no means constant, changes of leukemic cells. It is easy to recognize when the cell affected is the youngest of the line, for instance the myeloblast. In this case there exists a stage of the disease in which one finds in the peripheral blood two cell populations: leukemic myeloblasts and normal granulocytes (Hiatus leucemicus, see p. 210).
This maturation arrest can take place in later stages, or may not exist at all in certain leukemic cell populations.
It may affect promyelocytes or metamyelocytes. It can also be found in lymphocytic leukemias. The disease may uniquely affect large lymphocytes or small lymphocytes or lymphoblasts. It is frequently possible to recognize a particular leukemic patient by the appearance of the cells one sees in the blood smears. In other words, the stage of maturation in which the cell is arrested varies from patient to patient.

Maturation Arrest and Autonomy of Leukemic Cells.
Autonomy is the failure of cancer cells to respond to normal regulatory factors of the organism. Leukemic cells are partly autonomous. In general, the earlier the stage of maturation at which the leukemic cell is arrested, the greater is the degree of autonomy. However, many exceptions to this rule exist. There are degrees of autonomy as there are degrees of maturation[3, 4].

Anarchy of Maturation[1-3]

As we have noted (p. 17), a synchrony in the maturation of different organelles exists in each cell line. A young nucleus corresponds to a young cytoplasm. A large nucleolus is present together with finely reticulated chromatin. This synchrony is profoundly affected in leukemia. In Giemsa-stained smears one can frequently see cells in which the nucleus is mature, but the cytoplasm remains basophilic and has undifferentiated or no granules or, conversely, a cell with an immature nucleus and fully differentiated granules.

The electron microscope often permits one to recognize that the anarchy of development involves additional organelles and different functions.

Examples of Anarchy of Maturation

Asynchrony between the Development of the Nucleus and the Granules.
Some cells have a myeloblastic nucleus but with granules characteristic of the myelocytic stage.

Asynchrony between the Development of the Nucleus and the Endoplasmic Reticulum.
In the mature granulocytes of the blood, it is rare to see remaining lamellae of the endoplasmic reticulum. In contrast, in some myeloblastic leukemias, one may see stacks of three to five sacs of RER. These zones stain blue with Giemsa and have long been recognized as Döhle bodies. They are not at all specific for leukemic cells.

Asynchrony between Cell Maturation and Phagocytosis.
Normally, the capacity of myeloblasts to phagocytize cells or microorganisms is minimal. In some leukemias the leukemic myeloblasts phagocytize and actively digest erythrocytes, both in vivo and in vitro.

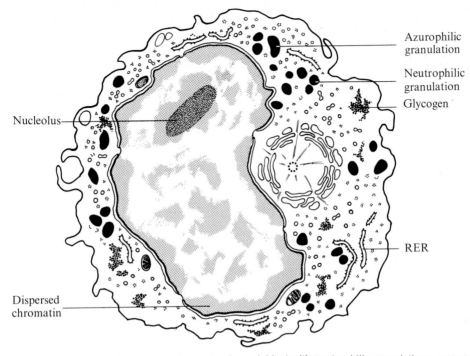

Azurophilic
granulation

Neutrophilic
granulation

Glycogen

Nucleolus

RER

Dispersed
chromatin

Fig. 1 – *Example of maturation arrest and anarchy*. A myeloblast with neutrophilic granulations, among other evidence of asynchrony in the development of cell organelles

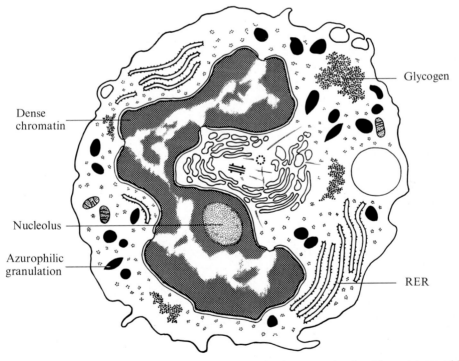

Glycogen

Dense
chromatin

Nucleolus

Azurophilic
granulation

RER

Fig. 2 – *Another example of maturation arrest and anarchy*. A metamyelocyte with persistence of RER and nucleolus, among other evidence of asynchrony in the development of cell organelles

3 — Mosaicism

Definition

Mosaicism is the existence of several populations of cells in the same organism. The mosaicism due to the presence of a population of leukemic cells in the presence of a normal cell line is further complicated by the mosaicism due to the emergence of subpopulations of the original leukemic cells*.

Hiatus Leucemicus[1]

This is the simplest case of mosaicism and was so named by the great hematologist Naegeli. In blood smears one finds only myeloblasts and mature granulocytes without the intermediate stages of maturation. This is due to the simultaneous presence of pathologic cells, which are arrested at the myeloblast stage, and mature granulocytes derived from the maturation of normal myeloblasts resident in the bone marrow.

With evolution of the disease, the blood smears contain only myeloblasts, indicating the progressive disappearance of the normal population of myeloid cells.

Emergence of Leukemic Subpopulations[2]

Very rarely a single population of leukemic cells exists from the very beginning to the final stages of the disease. In general, new subtypes of the original population continuously emerge. Among them those least responsive to the regulatory factors of the organism, be they immunologic or others, remain, undergo proliferation and outdistance all others (see p. 212).

In almost every human leukemia, periods exist in which one may note the mosaicism, be it by Giemsa stains, the electron microscope, chromosome studies, enzymatic, kinetic, or other even more refined techniques.

Numerous examples of such events can be seen in myeloid leukemia, in which one may see the simultaneous presence of several populations of myeloblasts characterized by different complement of granules or enzymes.

In some instances these subpopulations cannot be demonstrated, even if they actually exist. For instance, it is quite likely that the cells which are no longer responsive to treatment are actually a different subpopulation, even though they appear identical to the earlier cells by any test that has yet been devised.

Reinterpretation of the Leukemic Cell as Seen in Smears[3, 4]

In smears, the cells derived from different populations are intermingled. When the leukemic cells present have clear-cut malformations, it is easy to separate them. At other times, mosaicism can be the source of serious errors in cytologic diagnosis.

Let us take, for example, a cell, which on a smear has the characteristics of a myeloblast (M_1). We are frequently unable to distinguish whether we are dealing with 1) a normal myeloblast, 2) a leukemic myeloblast with maturation arrest at this stage, 3) a leukemic myeloblast which may, however, be able to mature to a later stage, which may vary from promyelocyte to mature granulocyte, 4) a cell belonging to another cell line, which has been pathologically altered, so that it gives the appearance of a myeloblast.

The examination by other techniques may or may not permit one to choose among these various possibilities in the individual cases.

---

* A population or subpopulation of leukemic cells is derived from a single modified cell and can, therefore, be called a clone. Unfortunately, this term is now applied in a precise (and unjustified) sense: one calls "monoclonal" a population of plasmacytes which produces a specific abnormal globulin. It is quite clear that such a clone can comprise several subpopulations characterized by differences which do not influence the synthesis of the globulin.

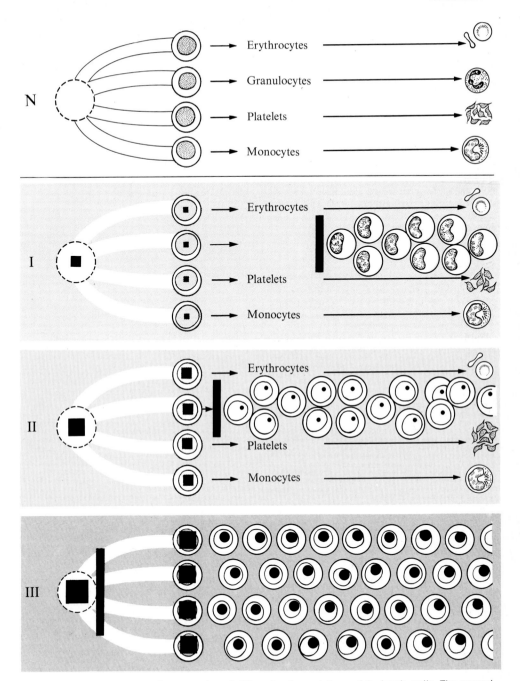

A schematic representation of progression of different subpopulations of leukemic cells. The normal hemopoietic stem cell gives rise to the four myeloid lines. A first abnormal cell (*I*) gives rise to erythrocytic, thrombocytic, and monocytic lines of normal appearance, and to myeloid cells with abnormal features and arrested maturation. A second abnormal stem cell (*II*) gives rise to more abnormal myeloid cells but still produces normal cells of the other lines. A third myeloid stem cell is so pathologic, that it can give rise to nothing but pathologic cells (*III*). In blood smears, all these populations are intermingled

4 — Progression

Definition[1—4]

Progression is characterized by the successive spread and takeover by one of the subpopulations.

Expansion of One of the Subpopulations

As we have noted, the mosaicism is due to a continual emergence of cells forming new populations. Among these cells, thousands are nonviable. Others remain stationary at least for a cetain time. These cells are rare in smears and may escape notice altogether, perhaps because they are not looked for too thoroughly.

Among all the populations that have emerged, however, one develops, replaces the other leukemic cells, and soon all of the cells of the bone marrow.

Causes of Progression[5—8]

One may hypothesize that the progression is due to a selection among the leukemic populations. This might be due to the fact that the cell has modified antigens leading to immunselection, or that they have not synthesized receptors of normal regulators. Mention should also be made of the selection of populations of leukemic cells by drugs. Only the cells resistant to treatment can continue to proliferate. In general, but not always, these are cells which from a morphologic point of view appear simplified or as used to be said "dedifferentiated."* In this way a myeloid leukemia, which progresses, may change from a promyelocytic to myeloblastic type. The progression is linked, but not always, to modifications of the karyotype (clonal evolution).

It is interesting to consider from this point of view those cases of leukemia or other malignancies of the hemopoietic system, in which the final evolution is represented by a "reticulum sarcoma" or a "lymphoma." Perhaps in some of these cases we are dealing with the progression of one of the subpopulations of the original cells.

Disappearance of Other Populations

Simultaneously with the preponderance of one of the leukemic populations, one often notes the progressive disappearance of all the others. This phenomenon is of considerable clinical importance, since it leads to the fatal outcome. Its mechanism is unknown. Several hypotheses may be advanced:

1) The leukemic cell may secrete a product which is toxic to the other marrow cells.
2) The mechanism of regulation of all marrow cells has become disordered.
3) The same mechanism, which has led to the appearance of new populations of leukemic cells, also leads to the development of a stem cell, which is so abnormal that it can no longer respond to the normal factors that lead to its commitment to the red cell and megakaryocytic series. Since this particular stem cell has a metabolic advantage in proliferation, it outgrows progressively all other normal cells.

---

* This term implies a return to an earlier normal stage and hence is inapplicable to leukemic cells.

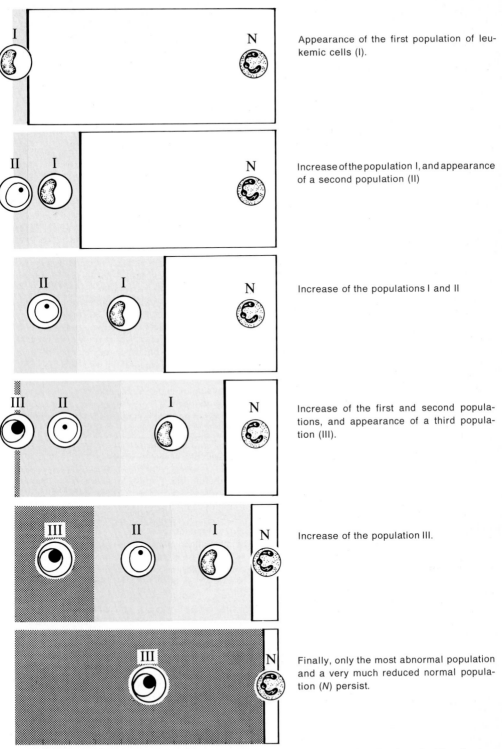

Appearance of the first population of leukemic cells (I).

Increase of the population I, and appearance of a second population (II)

Increase of the populations I and II

Increase of the first and second populations, and appearance of a third population (III).

Increase of the population III.

Finally, only the most abnormal population and a very much reduced normal population (N) persist.

A schematic representation of progression of different subpopulations of leukemic cells. The diagram indicates the percentages of different cell populations

5 – Provisional Conclusions on the Pathophysiology of Leukemic Cells

The preceding analysis of the principal features of leukemic cells calls for a synthesis. Such attempts at synthesis are notoriously hazardous. The reflections and conclusions on the nature of leukemia reached today are liable to be discarded tomorrow when the causes and treatment of the disease will be known.

I have, nevertheless, attempted to derive from the preceding analysis some general laws that appear to govern the behavior of leukemic cells, of which only secondary and variable effects are seen in smears.

Three phenomena appear to explain all symptoms of leukemia.

1) Autonomy[1, 2]

The leukemic cells obey the laws of the organism imperfectly. They have varying degrees of autonomy, though they are never entirely beyond control by the organism.

This autonomy is reflected both in the extent of proliferation and the degree of maturation. Proliferation and maturation, however, are not affected in parallel.

Possibly, the autonomy depends on lesions of the cell organelles that make them incapable of responding completely to the regulatory forces of the organism (see pp. 206 and 208).

2) Mosaicism[3]

New types of leukemic cells emerge constantly, leading to the simultaneous existence of multiple populations of leukemic cells (see p. 210). At times, these populations differ morphologically. At times they differ only physiologically, for example, in their resistance to treatment.

3) Progression[4]

The most autonomous cells replace all others, be they normal or pathologic (see p. 212). This is the foremost cause of the fatal outcome of the disease.

The first leukemic cells to emerge may already be so autonomous that they rapidly replace all others and lead to death without repeated progression and intervening mosaicism.

Causes of Cytologic Symptoms

Proliferation

The number of cells produced by the leukemic proliferation is not an essential characteristic of a given leukemia.

The number of cells in the circulation or in the hemopoietic centers may be large or small. The number depends on the degree of autonomy of the different populations of leukemic cells and the individual environment provided by each patient. Studies of these phenomena have recently received a new impetus from concepts of cytodemography and cytoecology (see p. 24). These disciplines are still in their infancy. They are faced with many technical difficulties, particularly as a result of the large number of different cell populations present in most leukemias or the small number of abnormal cells present in hemopoietic dysplasias (preleukemic states: see p. 203).

Pathology of Organelles

The morphologic abnormalities of the leukemic cell are not an essential feature of the disease. If abnormalities are present, they are more or less diagnostic, depending on their number or character. However, cells entirely normal on examination with the light and electron microscopes may nevertheless be leukemic.

Possibly, an analysis of different types of anarchy of maturation could lead, if not to the discovery of the nature of leukemia, at least to a better understanding of the relationship between the pathology of organelles and the leukemic process. I indicate here some working hypotheses:

1) A pathologic program of maturation (presumably resident in the chromosomes) cannot give appropriate orders for the normal development of organelles.

2) Pathologic organelles cannot obey the program of maturation.

3) Different parts of a single cell differ. Some organelles mature and function normally, others do not. This is to say, a sort of autonomy of organelles exists in the interior of the leukemic cell: they no longer respond to the governing program. The partial autonomy of organelles may thus lead to the partial autonomy of the entire cell.

In general, the pathology of organelles is not the cause of the leukemic change but merely a visible manifestation, which may not be present. The same may be said of the chemical, enzymologic, immunologic and genetic changes one may observe. All of them may be only epiphenomena more or less constantly associated with certain types of leukemia. This may apply even to karyotypic changes. It is conceivable, however, that a primary alteration in an organelle may lead to partial autonomy of the cell. In that case, the pathologic change of an organelle could conceivably be the first event in the development of a leukemia, provided that the pathologic change is retained during reproduction of the cell.

Causes of Clinical Symptoms

The clinical symptoms of the disease depend on the type of the proliferating cell and the degree of its proliferation. Thus, a promyelocytic leukemia produces specific clinical symptoms. So does a lymphoblastic leukemia with secretion of macroglobulins. Theoretically, an infinite number of clinical forms of leukemia can exist, each one depending on the proliferation of the abnormal cell populations in a given patient.

In practice, however, some types of leukemia occur more commonly than others or differences between some individual leukemias are so slight that one can distinguish a limited number of categories. One may then classify most leukemias based on their abnormalities (e.g., hairy cell leukemia).

The clinical malignancy of a leukemia does not depend on the severity of the lesions of the proliferating cell. It depends on the number and/or proportion of the affected cells. It depends most of all on the bone marrow aplasia of the normal marrow (see progression, p. 212), which eventually kills the patient.

Reflections on Future Research

It is impossible to predict the scientific discipline that will provide the observation or the reasoning to illuminate the etiology and indicate the eventual cure of leukemia. Virology and immunology are the two disciplines most extensively pursued at this moment, June 1977. In contrast, the study of the changes in the program of leukemic cells leading to their abnormal behavior which I have stressed in this book, and the effect of the microenvironment on the regulation of normal and leukemic cells appear to find little favor. Other approaches have even fewer advocates at present.

Yet, a study of the history of discoveries provides little hope that the solution will come from further refinements of our present means of diagnosis and treatment. Possibly we lack the appropriate concept or the appropriate technique to find the cause and cure of leukemia.

I like to recall two examples of great discoveries: one in which the lack of concept, the other in which the lack of a technique prevented progress for a long time:

In 1740, scurvy[5] killed more men in the British Navy than the combined armies of France and Spain. Many great minds tirelessly pursued research into the cause of that disease and were led a long way from the truth by their logical reasoning. Several circumstances contributed to this failure of some of the greatest physicians of the time. Many other conditions affected the sailors sick of scurvy, so that the disease remained ill defined until its cause was recognized. When ships carrying sick sailors reached port, the milder cases were taken ashore and recovered. This led to the conclusion that it was the contact with land which cured the scurvy. However, the worst cases, unfit to be released were left aboard and some still recovered. The "air of port" was, therefore, held responsible for the recoveries.

The solution of the problem required an entirely novel, though now trivial idea: the lack of a vitamin.

In 1938[6], erythroblastosis fetalis was a mysterious disease (it was even considered as a form of congenital leukemia by some pediatricians). By dint of extraordinarily cogent reasoning, Ruth Darrow concluded:

If the possible mechanisms giving rise to the destruction of eythrocytes are reviewed, it is found that all may be eliminated from consideration save one, the destruction of red cells by some form of immune reaction... The mother is actively immunized against fetal red cells or some component of them.... The antibodies formed in the maternal organism then pass to the child through the placenta...

Darrow was unable to prove her thesis. Immunization of the mother by fetal cells was only demonstrated when appropriate techniques became available as a result of the independent discovery of the blood groups of rhesus monkeys and their relationship to human transfusion reactions.

These examples illustrate how unrewarding speculations are apt to be when appropriate concepts are lacking or when hypotheses cannot be verified for lack of appropriate and reliable techniques. The preceding conclusions and hypotheses on leukemic cells are, therefore, presented here only as an invitation to stimulate efforts to prove or disprove them.

Chapter 10

Technique

*"I come now to describe the Manner of bringing the Blood
to a strict Examination before the Microscope, and shall offer
such Ways as I have myself experienced, not doubting but the
ingenious will contrive others, as they may find Occasion".*
Henry Baker, 1754[1].

General Remarks

The examination of smears stained with May-Grünwald-Giemsa or Wright is so widely used that it is superfluous to describe in detail the different techniques of making smears and staining them. The briefest mention of the technique I recommend will suffice and I will only add some points that are often neglected. I have also added brief notes on cytochemistry, shadowing, and examination in the Soret band which may be useful.

A brief discussion of the examination of cells by phase microscopy and electron microscopy is intended primarily to aid the interpretation of the pictures in this book and to indicate to the student the direction of current research.

Good and Bad Smears[2—6]

I have been reproached for showing in the color plates of my books only cells which are too "beautiful" and far removed from the images one sees every day. It is true that I have chosen the best portion of a smear where the cells are optimally spread, where there is no dust or stain precipitate on the slide nor bubbles in the red cells, and where the staining is good. It is equally true that the smears which the hematologist sees every day (and some which one finds in books) contain leukocytes that are unrecognizable. One finds disappearance of the chromatin structure, partial staining of nuclei, rupture of cells, misleading modifications of the nuances of colors in the cytoplasm of leukocytes which play such an important role in the diagnosis of the different stages of maturation. The colors of red cells, instead of being pink, may be bright red, blue, violet or even green; artifactual alterations of their shape can simulate the presence in the circulating blood of torocytes (see p. 224), echinocytes (see p. 52), elliptocytes, etc.

Sometimes even bad smears still allow one to make the correct diagnosis, but in other instances they make the diagnosis impossible or lead to an erroneous diagnosis.

The quality of smears cannot always be perfect. However, routine preparations should be at least good and bad ones should be discarded. Preparations intended for research should be excellent. Much more can be seen in a perfect preparation. The student should glance at the books of Jolly, Maximov, Wright, Downey and Ferrata and note that their illustrations were based on preparations of exceptional quality: a prerequisite for the important discoveries these pioneers of hematology made. In this book I have partly complied with the demands of the antiperfectionists and have included photographs of smears which are far from perfect, but are acceptable.

1 – A Page of History

This page is devoted to readers who are curious to know who first conceived the idea of making blood smears; how the technique was perfected by trial and error to give the most vivid colors and to discriminate best between the cells seen in smears; how scientific explanation of the differential staining allowed reinterpretation of the different colors observed; and finally how the present perfection of smears may permit one to consider their future development.

Discovery of the Technique of Making Smears[1-4]

Andral appears to have been the first to use smears systematically for the study of blood cells. He prepared them with a blast of air. In 1843 he wrote:

It is very easy to understand what happens when we place a drop of blood on a plate of glass and blow strongly upon it, in such a way as to spread it in an extremely thin layer. If the operation be well done and with rapidity, the globules remain perfectly unchanged, their surfaces are smooth, their edges very clear, and left to themselves, they never become either raspberry-like or festooned. In these circumstances, the evaporation of the serum is almost instantaneous.

Possibly other microscopists of the time also used this technique, particularly Hewson and Gulliver, if one is to judge by their drawings. The earlier literature describes only living cells.

The technique was progressively improved and the spreading by a blast of air was replaced by the present technique described by Norris in 1882. The word "smear" is a translation of the German *Ausstrich*, used from 1879 on by Ehrlich and his school.

Discovery of Hematologic Stains[5-14]

In 1877 Ehrlich, still a student, conceived the idea of staining tissues and blood with the substances discovered in Germany by the newly founded dye industry. He attempted to understand and predict the action of dyes by classifying them as acid, basic, and neutral. He discovered that the different categories of dyes stained different structures. Chemical reactions were thus accomplished at the level of the microscope and Ehrlich became the founder of cytochemistry. Originally Ehrlich stained the smears with a single substance; later he used compound dyes which colored the nucleus, the cytoplasm and the granules differentially. He noted that certain blue aniline dyes colored certain granules not the blue one expected, but a brick red color. He thus discovered metachromasia.

These "compound dyes" made possible the classification of neutrophils, eosinophils, basophils and of cells without granules. The differential count as we still know it today was born.

At the end of the last century, when blood parasites were discovered, the efforts to study them in smears gave a tremendous impetus to the study of blood cells. Many of the great hematologists at the turn of the century started their studies on malaria.

Romanovsky, to whom the majority of hematologists improperly attribute the paternity of modern dye combinations for the staining of blood smears, described in 1891 an improved mixture of methylene blue and eosin which was useful for visualization of the parasites of red blood cells. The mixture gave vivid colors, particularly a vivid red, which was originally called "red of methylene blue": this is the red of the azure which stains the azurophilic granules. We now know that this happy result was achieved by chance; in aqueous solution methylene blue acquires the property of combining with eosin which results in the appearance of the eosinate of the methylene azure.

Leishmann (1901) conceived the idea of dissolving the dyes in methyl alcohol. May and Grünwald (1902) described a mixture which did not stain the nucleus, but differentiated the granules particularly well. Giemsa (1902–1904) explained the action of the azure and made generally available mixtures of dyes which were simple to use and gave reproducible results.

Pappenheim (1912) combined the dyes developed by May-Grünwald and Giemsa under the name "universal panoptic staining": this is the staining method used today.

Wright, who with Whitman studied under Leishmann in Germany, introduced a single compound dye which is still widely used in the United States. Wright's stain gives less vivid colors than Giemsa, because azurophilic structures stain blue rather than red. Some use Wright's followed by Giemsa which gives results very similar to May-Grünwald-Giemsa (see also p. 221).

Finally, certain cytochemical methods have been applied to smears (see p. 222). Few of them are truly useful when applied to smears: they are very likely to find their greatest utility in electronmicroscopy.

"The further one looks back, the more one can look forward"
(Winston Churchill)

Hematology Without Smears and Microscope[1−6]

The preceding page summarizes the gradual development of techniques which has led to present day hematologic practice. The introduction of different staining methods has been followed by an understanding of the mechanism of the staining reactions and the beginning of cytochemistry. The use of smears and of rational staining technics has given birth to new ideas that have revolutionized hematology in the first three-quarters of this century. One might think that the smears, which have arrived at such a degree of perfection, are bound to continue to render a signal service to hematology. This is by no means certain.

If one considers the number of blood counts and hemograms performed daily all over the world, the replacement of the human eye and expensive microscope by instruments which are cheap and easy to use becomes a challenging possibility. One such development, at the moment expensive and imperfect, aims at the automatic reading of smears (see p. 220). Another aims at dispensing entirely with the smears and the microscope. Let us assume for a moment that the microscope had not been invented and we would have at our disposal a number of instruments which allow us to characterize the cells by volume, density, surface electric charge, deformability, spectral absorption, speed of sedimentation, growth kinetics in culture and so on. Undoubtedly, we would have found correlations between these parameters and clinical diseases and arrived at the prognosis and treatment of some of them. We might have arrived at classifications that do not coincide exactly with those we have derived from examination of stained smears. Who is to say, however, that they might not have been superior classifications, at least in some instances?

It is thus possible that in a few years stained smears will have only historical interest. Already today, a hematology without smears is conceivable but not a "hematology without the microscope," a catchword that has achieved some notoriety. Present or future microscopic techniques will remain irreplaceable in the study of the physiology and sociology of living cells and their ultrastructure.

2 – Preparation of Smears

I will only indicate the technique I use. Others undoubtedly have their advantages. They can be found in the bibliography.

Smears on Slides[1,2]

Prior cleaning of slides and removal of grease is indispensable. Glass slides should be washed in a mixture of sulphuric acid and chromate and stored in a mixture of equal parts of alcohol and ether. They should be dried with lint-free cloth and touched only at the edges. Commercially available "precleaned" slides must be carefully checked before accepting them as meeting the same exacting standards.

A drop of blood or diluent containing the cells, approximately 2 mm in diameter, is placed near one end of the slide, about 1 cm from the edge. The edge of a coverslip is brought in contact with the slide at an angle of approximately 30° between the drop and the end of the slide, while the coverslip stays in light contact with the slide. One moves the coverslip against the fluid, which spreads by capillarity along the line of contact of slide and coverslip. Immediately, by a continuous and steady movement, the coverslip is moved toward the opposite end of the slide. In this fashion, a uniform smear will result. The use of a special "spreader" is often recommended, consisting of a slide with one side ground and trimmed to produce a smooth smear narrower than the slide. A suitable coverslip is as satisfactory and can be discarded after a single use.

It is essential that the slide or the coverslip which is used as a spreader should be narrower than the slide, so that the margins of the smear are at some distance from the edge of the slide and can be readily observed under the microscope. This is important because the white cells tend to accumulate at the edge of the smear (see p. 223). A smear prepared in this manner has the necessary prerequisites for an examination: it is thin, uniform, with almost rectangular borders, except for the round margin at the end of the smear.

The thickness of the smear is a function of the speed with which the coverslip is moved across the slide. It is thick when the movement is slow, thin when it is rapid.

The smears should be dried very rapidly in order to conserve the shape of the red cells. This is best accomplished by rapid movement through the air. The use of slides previously heated to 40° or the preparation of smears in front of an infrared lamp result in almost instantaneous drying.

The erythrocytes and leukocytes have a larger surface in the dried smears than in the living state. They are more or less flattened by surface tension during the preparation of the smear, their thickness is reduced to less than 0.5 μm. Consequently, the internal details can be seen more clearly. Nevertheless, it must be kept in mind that we are dealing with artifacts due to the preparation of the smear and that the appearances of cells in smears differ markedly from those in the circulation (see p. 233, Fig. 1).

Automation of the Preparation and Reading of Smears[3-6]

Several instruments have been proposed for the automatic spreading of blood on slides by either mechanical means or a particular type of centrifugation, followed by automatic fixation and staining. The recognition of the forms and colors of cells can also be accomplished automatically by analysis of the analog signals by suitable computer programs. Instruments already available can recognize the common normal cell types and thus indicate which cells need to be examined by the hematologist for identification.

As already indicated (p. 219), automation may take routes other than the examination of blood smears. One approach is to measure electronically the absorbance and size of individual cells as they pass in multiple parallel streams by suitable sensors. The cells in each stream are either unstained or have been previously exposed to a dye or enzyme substrate which is more or less specific for a given cell type. An already available instrument reliably counts normal cells in this fashion and displays a two-dimensional pattern of absorbances and sizes on a screen, indicating various types of deviation from normality. Other approaches bypassing the use of smears are being explored.

Progress in pattern recognition may eventually allow the identification of some specific abnormal cells when present in appreciable numbers. Automation of the diagnosis of leukemic cells, however, would require enormously complex programs to match the diagnostic potential of the microscopist who takes into account the minute abnormalities of rare cells, judges the possible interrelation of different cells and supplements his examination by use of special stains, phase and electron microscopy.

First step: Fixation. Drying of the smears does not fix them. Leukocytes undergo changes and the red blood cells are distorted (torocytes) or hemolyze if exposed to the least trace of humidity. To fix them, a liquid fixative is needed. Nevertheless, very well dried preparations kept for several months or subjected to a high vacuum become fixed and water no longer hemolyzes the cells. Methanol applied for 3 min is a good fixative. The May-Grünwald dye itself, since it is dissolved in methanol, can serve as fixative. Slides may be put face upward on a staining rack consisting of two parallel rods 2 cm apart or on match sticks in a Petri dish. The number of drops of May-Grünwald necessary to cover the entire slide is applied. The stain is left in place for 3 min. The smears are now fixed, but are not stained, because the dyes are not ionized in methanol.

Second step: Staining with May-Grünwald[1,2]**.** As many drops of buffered water are added as the original number of drops of dye. Adequate mixing is assured by rocking the rack slightly. After 3 min, the slides are removed from the rack and washed. The smears are now of a reddish color and are put in a solution of Giemsa.

Third Step: Staining with Giemsa. It is convenient to put the slides into staining jars in a vertical position or face downward on match sticks in a Petri dish with the stain allowed to fill the space between the bottom of the Petri dish and slide by capillarity. In both cases, and particularly in the Petri dish, precipitation of dye is prevented. It is essential to use water of a neutral pH for both the preparation of the solution and the wash in order to obtain consistent results.*

The smears are put in a fresh solution of Giemsa and buffered water in proportions prescribed by the manufacturer for 10 to 30 min. After removal from the stain, the slides are washed in a stream of water and dried. They are now ready for examination.

It is unnecessary to cover the smears with a cover-slip. This "protection" does not enhance their preservation. A drop of immersion oil is put on the slide for examination with a high power objective. After examination, it is best to remove the oil by immersing the slide for a few minutes in a jar of xylol.

Appearances of Cells after Staining with May-Grünwald-Giemsa

The basic chromatin is reddish-purple and the oxychromatin pink.
The basophilic cytoplasm is light to dark blue.
The acidophilic cytoplasm is pink.
The neutrophilic granules are beige.
The acidophilic granules of eosinophils are orange.

The metachromatic granules of basophils are a dark violet.
The azurophilic granules are purple or violet-purple.
The basophilic stippling of erythrocytes is a cobalt blue.
The mitochondria and centrioles are not stained. Neither is the centrosome. This "negative staining" permits its localization in the midst of the stained cytoplasm if the cell is well-spread (see fig. 1, p. 177).

It must be realized that the colors described may be altered by age or by the use of a bad stain or of too acid or too alkaline solutions. One must judge whether the stain is adequate by the appearance of familiar cells.

If the smear is excessively blue, this may be the result of inadequate washing, prolonged staining of the smear, too high an alkalinity of the dye or the buffer. If the smear is too thick, the red cells tend to be blue or green. The cytoplasm of the lymphocytes becomes grey or lavender, the neutrophilic granules appear large and dark. The eosinophilic granules may appear grey or blue. If, in contrast, the smear is too red, it is likely that the dye, buffer, or water are too acid. The nuclear chromatin becomes reddish instead of violet; the erythrocytes are red or orange instead of pink, and the eosinophilic granules are a very brilliant red.

Staining with Wright's

The nuclei are a dark blue, the neutrophilic granules lilac, the eosinophilic granules red, and the azurophilic granules a dark blue. The other colors are identical to those described for Giemsa stain. This, however, is only true of the original Wright's stain. The stain used to-day in the United States under the label of Wright's gives colors quite similar to the May-Grünwald-Giemsa.

---

* Commercially available stains vary from manufacturer to manufacturer and sometimes from lot to lot, even if they carry the same designation.

4 — Cytochemical Stains

Cytochemistry attempts to identify chemical substances in their natural location. Apart from dyes, cytochemistry uses polarized light, X-ray diffraction, fluorescence, spectroscopy in the visible light and ultraviolet, micro-incineration, digestion by enzymes, specific solvents, etc. In some circumstances it is possible to measure the amount of certain substances in a single cell (for example nucleic acid or hemoglobin) and to study in this manner the quantitative variation of individual cells in diverse physiologic and pathologic states.

A description of cytochemical techniques and critical analysis of their value is beyond the scope of this book. I will only mention the techniques currently used for diagnosis. It should be noted that in leukemias the normal synthetic pathways may be so altered that cytochemical reactions are not always reliable for correct identification of the cell line of origin (see p. 206).

Perls' stain (Prussian Blue reaction)[1]

Perls' reaction stains hemosiderin, ferritin and iron containing micelles blue. It identifies normal sideroblasts and permits the identification of pathologic sideroblasts in which iron-containing micelles are present in the mitochondria (see pp. 34 and 54). It reveals the presence of hemosiderin and ferritin in leukocytes.

Benzidine Reaction for Hemoglobin[2]

Benzidine reaction stains hemoglobin-containing cells a yellowish green. It is useful when examination in the Soret band (see p. 228) cannot be used.

Peroxidase Reaction for Leukocyte Granulations[3, 4]

The peroxidase reaction stains the portion of the cell containing the enzyme. It is positive in some cells in which the granules are not readily seen and permits the identification of a cell as belonging to the granulocytic series (see p. 190). Cells of the monocytic series are irregularly and faintly positive.

NBT Reaction[5]

This reaction colors blue cells capable of reducing nitroso-blue to formozan. The reaction is negative in granulomatous disease. It has been thought to be useful in bacterial infections, but its utility is in question.

Alkaline Phosphatase Reaction[6]

This reaction stains the portion of the cell which contains the enzyme a blue-black or reddish color, depending on the particular dye used. It is almost always absent in untreated chronic myelogenous leukemia and is often, but not always, increased in agnogenic myeloid metaplasia and myelofibrosis. It is increased in infectious leukocytosis.

Nonspecific Esterase Reactions[7]

There are many varieties of esterases and many substrates to demonstrate them. They are used by some authors to distinguish pathologic myelocytic cells which give a faint reaction, not inhibited by sodium fluoride, from pathologic monocytic cells which give a strong reaction, inhibited by sodium fluoride. For this purpose α-naphtyl-butyrate appears the most useful substrate at present.

The naphtol-ASD-chloracetate preferentially stains granulocytes and thus is a "specific esterase" though the term is seldom used.

Lysozyme Reaction[8]

Immunofluorescence or the lysis of adjacent bacteria may be used. This reaction is positive in granulocytes and monocytes, the amount of lysozyme being greater in monocytic cells.

Feulgen Reaction[9]

The Feulgen reaction stains DNA. It is useful in combination with spectrophotometry for the measurement of the ploidy of the nucleus.

Methyl-Green Pyronin[10]

It stains the nucleus green and the ribonucleic acid red. It stains red what the Giemsa stains blue, and it is now seldom used.

Sudan Black[11]

Sudan black stains inclusions or granules that contain lipids. Sudan-positive granules suggest myelocytic or monocytic origin of the cells, but their absence does not exclude that derivation. In leukemias, granules of cell populations that are normally positive may fail to stain (see pp. 122 and 198).

PAS Reaction[12]

PAS stains mucopolysaccharides and glycogen red. In some cases of lymphosarcoma and acute lymphocytic leukemia the reaction may be strongly positive in pathologic lymphoblasts. In numerous cases of dyserythropoiesis, the reaction is strongly positive in erythroblasts. These include erythremias, leukemias, sideroblastic anemias, thalassemias, and hemopoietic dysplasias.

One must make sure that the smear is properly prepared and stained. It is mandatory to refuse to examine bad smears: this may result in errors of diagnosis (see p. 217, Good and Bad Smears).

The smears should be examined first under low power, in order to appreciate the distribution of the cells. The preparation may be satisfactory only in certain parts. It is essential to choose the best portion for examination under oil immersion. The erythrocytes must be in close proximity to each other but should not touch. If the preparation is too thick, the leukocytes are small and details cannot be recognized. If the preparation is too thin, the erythrocytes are flattened and the distribution of hemoglobin is artificially uniform, the red cells lacking the central pallor (see pp. 64 to 100).

Distribution of Cells in the Smears[1-5]

Cells are not uniformly distributed in either blood smears or bone marrow smears. Nonuniformity depends partly on the technique used, and partly on the combined influence of the viscosity of the blood, the specific gravity and size of the cells, the surface tension and other factors still unknown. In the center of a peripheral blood smear are the red cells with the smallest diameter and lymphocytes. At the edges one finds the large red cells, some granulocytes and lymphocytes. At the feather edge all cells are markedly flattened.

In bone marrow smears the nonuniformity is much more marked, because of the variable mixture of blood and marrow, depending on the technique used. The admixture of blood is minimized in techniques which separate the marrow particles from the blood and prepare smears of particles. Particles do not spread evenly as a rule, and individual cells may be unrecognizable in parts of the smear while other areas may contain well-preserved cells.

Megakaryocytes are present in the thicker portions of the smear. In some areas naked nuclei may be seen (which must not be confused with lymphocytes or normoblasts).

Cells which do not belong to the hemopoietic system should be noted whether they are normally present, such as fibroblasts, fat cells, osteoblasts and osteoclasts, or abnormal invaders of the bone marrow. Metastatic cells if present are generally easily recognized by their arrangement in clumps or sheets and by their individual monstrosities.

Unclassifiable cells may be due to lack of cell differentiation or artifactual alteration in the preparation of the smear. Squashed and disrupted cells are generally artifacts, but may be due to an unusual fragility of a particular cell type.

Differential Counts

Given the uneven distribution of cells in the smears, the safest way to insure a reliable count is to prepare a smear from a minute drop of blood, so that all the white cells in the entire preparation can be counted. A simpler method is to count all cells in longitudinal bands from one end of the smear to the other (at least where the preparation is not too thick or too thin). One or two hundred nucleated cells are counted.

Bone Marrow Differential Counts

It is recommended that at least 1000 cells be counted. Even then widely differing figures may be found in normals, either because different observers use slightly differing criteria for separating different classes of red and white cell precursors or because minor differences in technique lead to different admixtures of mature cells from the circulating blood aspirated with the marrow or for other reasons that are not well understood.

The differential count is still of little use in many instances without an estimate of the total cellularity*. For example, the interpretation of a high percentage of lymphocytes as due to overproduction of lymphocytes or due to failure of granulocyte production depends on the total cell count. Wide variations exist in the bone marrow cellularity and distribution of cell types from area to area. Consequently, time is better spent in careful examination of the individual features of a bone marrow aspirate than in attempting differential counts.

Splenic and Lymphnode Puncture

Most hematologic centers have rejected splenic puncture as too hazardous and lymphnode aspiration as providing too little information without simultaneous node biopsy. Careful examination of all available slides is necessary: pathologic cells, such as metastases or Reed-Sternberg cells, may be found in only a single slide.

---

* An estimate of total cellularity is obtained by examination of sections of marrow particles or of needle biopsies which are now readily obtained by use of the Jamshidi needle and performed routinely by many hematologists. Even bone marrow sections, however, give at best a semiquantitative estimate of cell number. Markedly hypoplastic areas are often found along with cellular areas in the same needle biopsy. Occasionally they may occupy an entire biopsy, although the total output of the marrow may be quite normal as indicated by a stable, normal level of red cells, white cells, and platelets in the peripheral blood. Not only metastatic cells but also myeloma and leukemic cells may have a very patchy distribution. Sections should always be taken for the identification of metastatic tumors; they often produce a connective tissue reaction and hence may be absent from the aspirate. Needle biopsy is also essential in the diagnosis of Hodgkin's disease and myelofibrosis and is mandatory whenever a "dry tap" is obtained by an experienced operator.

6 – Artifacts

Any preparation of blood cells in smears is of necessity an artifact. It distorts the appearance of cells. However, in well-prepared and stained smears one may safely interpret certain forms and colors as corresponding to actual features of the living cells and this is what I have attempted to do in this book. Unfortunately, all smears are not perfect (see note on p. 217). Moreover, some cells are so fragile that they are easily bruised or destroyed even in a well-made smear. Such artifacts have given rise to frequent errors. I will note the most common ones.

Distortion of the Nucleus Due to Adherence to Other Cells

In a number of leukocytes, particularly monocytes, the nucleus is stretched out as a result of adherence of the cells to neighboring erythrocytes during the preparation of the smears (Fig. 1). The chromatin may become rearranged in parallel strands oriented in the direction of the sheared nucleus. In some instances, the nucleus adheres to the periphery of the cytoplasm so as to divide it into two more or less equal fields.

Explosion of the Cell

This is due to mechanical trauma in the course of the preparation of the smear. The cytoplasm spreads and is squeezed into the space available between the neighboring cells (Fig. 4). Granules and the "negative images" of the centrosomes and mitochondria may be particularly well seen in these flattened and stretched-out portions of the cytoplasm (Figs. 3 and 4).

Basket Cells. At times the cytoplasm is no longer visible and the nucleus assumes the appearance of "nuclear ghosts" (basket cells). The chromatin network becomes distended. If the chromatin is thick, it becomes fenestrated; if it is thin, it assumes a finely spongy appearance. Nucleoli appear enlarged, generally pale blue and surrounded by a ring of chromatin. The disruption or explosion of cells may affect an already necrotic, but still circulating cell. Most of the time, however, there is no correlation between the number of basket cells in the smear and the number of dead leukocytes demonstrable with vital dyes (see p. 22). Hence, basket cells appear to be derived from living, but fragile cells. Many of the so-called Ferrata cells are nothing else than damaged myeloblasts and promyelocytes.

Red Cell Artifacts[1]

I have discussed the most important artifacts in chapter II (pp. 66–101). I mention here only those which occur in poorly made smears.

Selenocytes (selenoid bodies). These are crescent-shaped forms, much larger than the normal erythrocytes. They are derived from fragile red cells, which have burst in the process of preparation of the smear. They can be found in all hemolytic states. They are due to two factors:

1) The mechanical friction between the erythrocytes and the surface of the glass, which occurs as the smear is being made.
2) The quantity of lipids in the blood. Hyperlipidemia increases the number of selenocytes.

The first stage of the formation of selenocytes is characterized by the development of a hole in the membrane. As a result of the distention in the process of spreading, this hole becomes very large, and the remainder of the cell takes the form of a lightly colored crescent, which is much larger than the normal red cell.

Torocytes (doughnut cells). These are erythrocytes which have a sharply defined, thick peripheral ring of hemoglobin and a colorless center. They are produced when the preparation is dried too slowly or when traces of water have been mixed with the fixative solution (see p. 221). They must not be confused with leptocytes (see p. 90) in which, like in normal cells, the transition from the area of central pallor to the hemoglobinized periphery is always gradual.

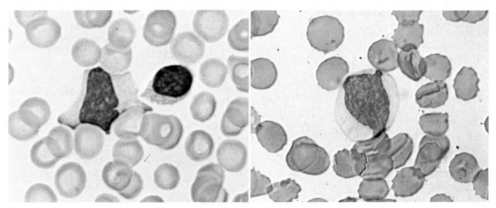

Figs. 1 and 2 — *Distorted nuclei*. The adhesion to red blood cells is responsible for a peculiar stretched-out appearance of the nuclei of the white blood cells

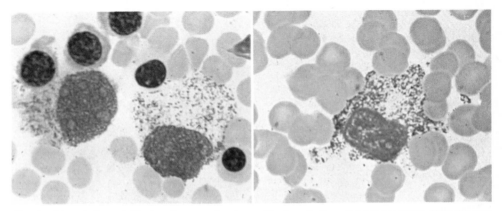

Fig. 3 — *Disrupted myeloblast and promyelocyte*, with dispersion of their granules

Fig. 4 — *A disrupted promyelocyte*. Note the centrosome surrounded by neutrophilic granules

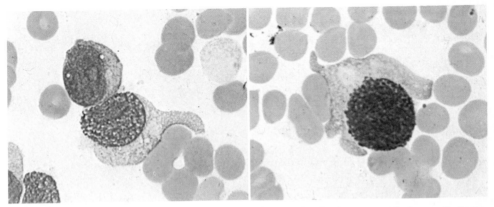

Figs. 5 and 6 — *Three leukemic cells*. Two cells, particularly fragile, have spread abnormally in the smear. Note the "reticular" structure of the nucleus, the nucleoli, the negative images of the unstained mitochondria and the centrosome

7 – Examination of Smears with Phase and Interference Microscopes

Phase Contrast Microscopy[1, 2]

The phase microscope has been the source of important advances in the study of living cells. Before the advent of the phase microscope, living cells could only be studied by maximal closure of the diaphragm of the condensor and slight defocusing of the objective. The resulting picture was inadequate or blurred. With phase contrast microscopy, the chromatin, the mitochondria, the centrosome, and specific granulations are clearly visible, often more distinctly than in the stained preparations. Combined with time lapse cinematography, detailed analysis of intracellular and cellular movements becomes possible. The phase microscope is now an indispensable part of the armamentarium of the cytologist.

Interference Microscopy[3, 4]

Interference microscopy has the advantage of permitting the measurement of the thickness of biological objects or, to be exact, their optical density. In practice this constitutes a measure of the dry weight of the cell. In general, the quality of images obtained with interference microscopes is less than optimal. However, the Nomarski system gives very satisfactory pictures and occasionally images superior to phase contrast microscopy. The cells appear in relief and resemble results of oblique shadowing with metals (see p. 228), except that we are dealing with living cells and only a semblance of shadowing.

The shadow effect gives an erroneous impression. Organelles which are actually in the interior of the cell, but which have a different refractive index, appear as if they were located on the surface of the cell or as depressions of the surface. This must be kept in mind in the interpretation of the results.

Halos around the cells or organelles (disturbing in phase contrast microscopy) are absent. Red cells, refractile granulations such as those of eosinophils, and intracellular highly refringent crystals such as hemoglobin crystals are seen sharply defined. However, mitochondria in a reticulocyte or a Heinz body in a red cell will appear as an invagination or as a granule on the surface of the red cell, depending on the adjustment of the microscope.

Techniques of Preparation[1]

The preparation of cells for the study in the living state with phase microscopy or interference microscopy requires a certain number of precautions. The leukocytes, being spherical, are surrounded by a pronounced halo which obscures all internal details. For an optimal image it is necessary that the cells be separated from each other and their thickness be decreased. This can be done by several methods which fall into two categories: 1) methods of compression, ranging from very slight reduction of the space occupied by the cell with only slight cellular changes to virtual crushing of the cell with serious alteration in its behavior; 2) methods utilizing spontaneous spreading of cells and applicable only to cells that possess that capability.

Examination of Stained Smears

In general, the examination of stained smears in phase contrast offers no advantages. The colors are altered or disappear and the outlines of cells and organelles are indistinct. Nevertheless, in very thin and faintly stained smears, one can sometimes see new details and define the outlines of cells depending on the change in color (Figs. 3 to 6), appreciate filiform pseudopods or very small granulations.

The procedure may be useful in cytochemistry when the colors, such as Feulgen's or methylgreen pyronin, are faint.

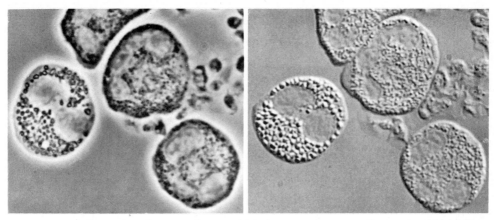

Figs. 1 and 2 — *Comparison between phase and interference microscopy.* One eosinophil and three PMN (living cells)

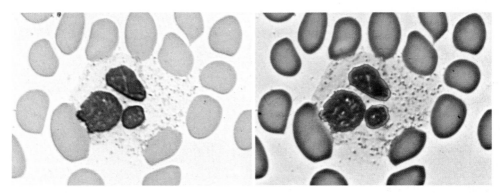

Figs. 3 and 4 — *Comparison of a smear viewed with the ordinary microscope and the phase contrast microscope.* A PMN disrupted during preparation of the smear is shown. Note that the colors have changed and that the granules appear particularly distinct with phase contrast (*right*)

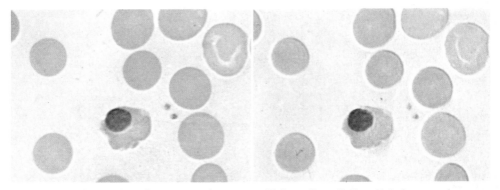

Figs. 5 and 6 — *Comparison of appearance in a smear with the ordinary* (left) *and interference microscope* (right). A normoblast, reticulocytes and erythrocytes are shown. Note the changed colors and the ''pseudo-shadowing''

Examination in the Soret Band[1,2]

With sunlight I made some observations on the spectrum of oxygenated blood. I did not find an absorption band in the so-called ultraviolet, but I did observe one in the violet, which to my knowledge has not previously been reported. It is easily distinguished with an ordinary eye piece provided one places a blue glass in front of the slit of the spectroscope.

In these words Soret described in 1878 for the first time the absorption band which carries his name and which has been used in a large number of investigations of hemoglobin. Examination of red cells in the Soret band allows one to identify and localize hemoglobin with high precision. It requires no special equipment, except a powerful light source such as a mercury vapor lamp or xenon arc and commercially available interference filters. These filters are convenient for photography and microcinematography because photographic films are very sensitive to wavelengths of approximately 400 nm. The filters cannot be used by themselves for direct visual observation, because the eye is very sensitive to parasitic green light which passes the interference filters. For this reason, it is necessary to combine an interference filter of 414 nm with a violet or dark blue filter (e.g. BG12) which excludes all light above 450 nm and effectively reduces the undesirable light to which the eye is particularly sensitive to less than 1%.

At a glance one can in this way distinguish the different stages of maturation of erythroblasts, recognize phagocytized erythrocytes, and certain pathologic alterations of red cells (see pp. 66 to 100).

Shadowing of Smears[3,4]

The shadowing technique consists of aiming a stream of anatomized metal, generally gold, at a cell. The stream disperses over the cell surface like a ray of light. The preparation appears to present a white shadow, which is reversed in the photographic print. The thickness of the cell can now be deduced in a precise fashion by the length of the shadow it casts.

Special apparatuses for shadowing are commercially available, but a simple vacuum bell jar connected to a good pump providing a vacuum of 10^{-5} mm Hg can give good results. The bell jar should be equipped with a tungsten crucible into which one introduces a few milligrams of gold or other metal. Approximately 10 cm below the crucible, the smear to be shadowed is placed under a variable angle, usually at 15° from the vertical. After evacuating the bell jar, one heats the crucible slowly until the gold sublimates and becomes deposited on the smear.

The thickness of the layer of gold deposited decreases in the smear with increasing distance from the source of its volatilization. One can, therefore, find in the same preparation cells which are more or less shadowed and can select an area where the details are most clearly seen. Certain objects which cannot be stained in smears, such as myelin forms (see Fig. 4), appear clearly in relief.

One may shadow preparations stained with Giemsa with excellent results. The cells then appear in both color and relief.

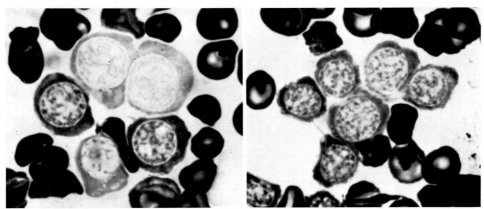

Figs. 1 and 2 — *Examination of red cell precursors and erythrocytes in the Soret band,* reflecting the increased concentration of hemoglobin with progressive maturation. Note (Fig. 2) a few leukocytes (white areas which deform the erythroblasts and erythrocytes)

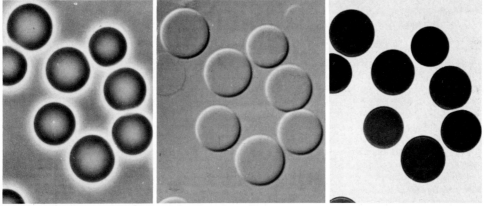

Fig. 3 — *Comparison between phase microscopy* (left), *interference microscopy* (middle) *and examination in the Soret band* (right). Normal erythrocytes slightly compressed between slide and coverslip

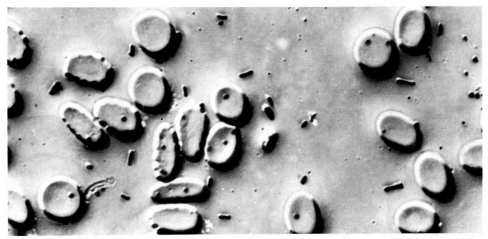

Fig. 4 — *Red cells on a smear shadowed with gold and examined with the light microscope.* The smear was made from stored blood. Note the presence of myelin figures, invisible in stained smears

9 — Examination with the Transmission Electron Microscope

A Word of History

It was only 30 years ago that the first electron microscopes were made available to cytologists.

These first instruments did not have the capabilities of those of today. On the whole, however, they were useful from the very beginning. They underwent a rapid evolution and within a few years, the basic type of electron microscope was developed from which current models differ only in details.

Originally accepted favorably and even enthusiastically by industry and bacteriologists, the new instrument was received with strong reservation by pathologists and hematologists. From 1950 to 1960 work with electron microscopes had the marks of a desperate battle: the instruments were much better than the techniques of specimen preparation. Eventually, however, the pains of the pioneers were amply repaid by the discoveries of the inframicroscopic universe. A new cytology was born which revolutionized biology and pathology.

A research instrument yesterday, the electron microscope has become today a valuable adjunct to clinical diagnosis. It has undergone an evolution similar to that of virtually all scientific instruments applied to the practice of medicine. In 1880, the optical microscope appeared to have no relation to clinical medicine; X-rays were long considered by the physician as a curiosity; electrocardiography and electroencephalography were received with the same skepticism. Again and again one finds the same phenomenon of mistrust of new instruments. Why the expense? What do they bring to the clinic? What does an electron microscope add to a proper examination of smears? Then a few years later one is astounded that so useful a technique has only been introduced with such delay in hospital services and in routine clinical pathology. Recently several papers have been published, expressing the same reservations about the scanning electron microscope ...

A Word on Technique[1, 2]

Preparative methods have recently made great progress, thanks to the introduction of new fixatives, new embedding substances, new dyes, new microtomes, and new methods of preparation.

Today one can routinely apply chemical, autoradiographic and immunochemical techniques to the study of the structure of the organelles. This can be done as readily in situ as after isolation of organelles by cell fractionation. The number of techniques applicable to blood cells is thus very large and they are being continuously improved.

General Remarks on Fixation

Today one may cut a cell into a thousand slices so thin that 50 are contained within the thickness of 1.0 µm and one can clearly see the cell organelles (see pp. 231–235). One cannot yet see the molecules and macromolecules of which they are formed. The reason is not in the inadequate power of resolution. Present instruments have a resolving power of 2 nm and hemoglobin actually measures 6×3 nm, ferritin 10×10 nm, and albumin 15×3 nm. The reason is the inadequacy of preparative techniques and particularly of fixation.

1) For examination with the optical microscope, standard fixation methods which involve coagulation of proteins are quite satisfactory. They maintain anatomical relationship and the shapes of individual cell structures down to the limit of resolution of approximately 0.3 µm. Hence no change is visible in the light microscope.

2) Between 0.3 µm and 3 nm (the size of cell organelles), the concept of fixation changes considerably. The requirement for the fixative is more demanding at this level; it should immobilize macromolecules while preserving the relationships that obtain in the living cell. Both the separation of macromolecules by rupture of bonds between them and agglomeration resulting from new intermolecular associations must be avoided.

3) Finally, a third size level, that of molecules less than 3 nm, must be considered in using modern microscopes that may have a resolving power down to 0.2 nm. Under these circumstances, ideal fixation ought to immobilize the molecules without any alteration so that they might eventually resume their original activity, as for example, enzymatic activity. Change in the molecule ought not to be irreversible. There is only one method known to accomplish this at the present time, the combination of freezing and dehydration, but freeze-drying poses considerable problems.

There is no freeze-drying technique perfectly suitable and applicable for blood cells which, it should be recalled, contain 70–80% water. As yet the results cannot be considered reliable, and the techniques are still being developed.

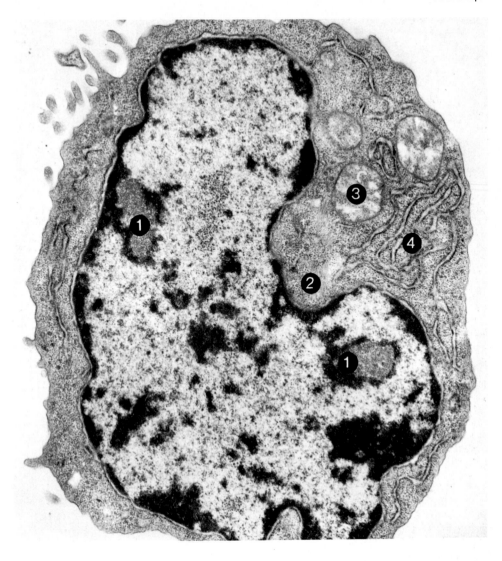

Section of a proplasmocyte. (1) Nucleoli surrounded by perinucleolar chromatin; (2) a bundle of micro-filaments; (3) mitochondria (slightly swollen); (4) RER. Note the microvilli of the cell surface. (Approximately × 20,000)

In conclusion, ideal fixation which would immobilize each molecule in situ is, for the moment, difficult to come by and applicable only to a few particular situations. Consequently, the usual fixatives must be retained for current work. As has been discussed, the criteria for a good fixative differ depending upon the magnification to be used. This basic fact should be kept in mind in interpreting the results when one uses the various electron-microscopic techniques. Moreover, it is essential to recognize artifacts caused by the various procedures, such as extraction of various materials during fixation itself or alterations resulting from washing, dehydration or embedding. Fixation of material for cytochemical and enzymatic study must be subjected to this type of critical review.

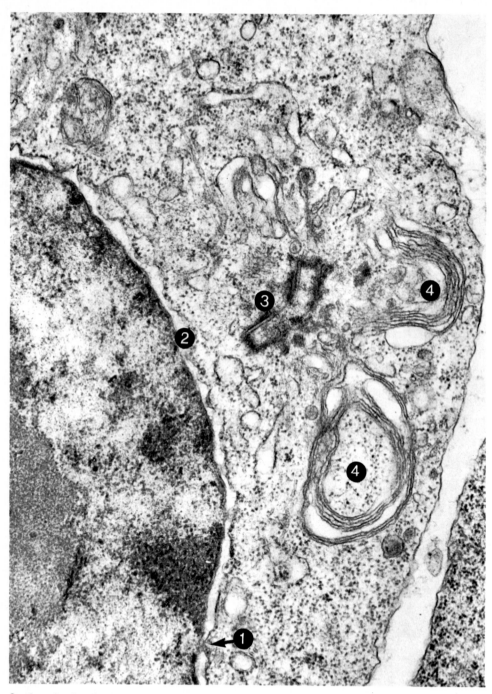

Section showing the nucleus, the nuclear membrane (the perinuclear cisterna is slightly dilated), *the centrioles and the Golgi complex.* (1) Nuclear pore; (2) perinuclear cisterna; (3) the two centrioles, which are in the same plane (an unusual arrangement); (4) the dispersed Golgi lamellae and vesicles (cf. figures, pp. 7 and 8). (Approximately ×30,000)

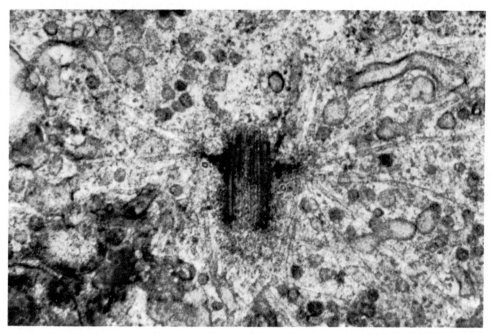

Fig. 1 – *Centriole, sectioned longitudinally.* The section is slightly oblique and only one set of satellites are seen. Note the attachment of the microtubules. (Approximately ×50,000)

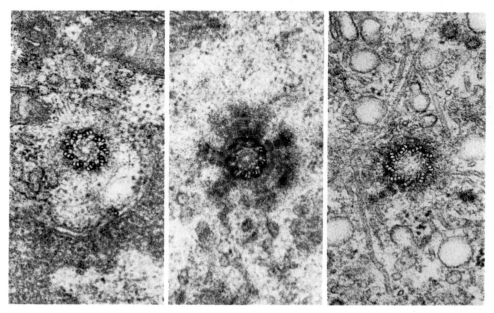

Figs. 2, 3 and 4 – *Transverse sections through centrioles at different levels.* The middle section passes through a crown of satellites (cf. figures, p. 8). (Approximately ×50,000)

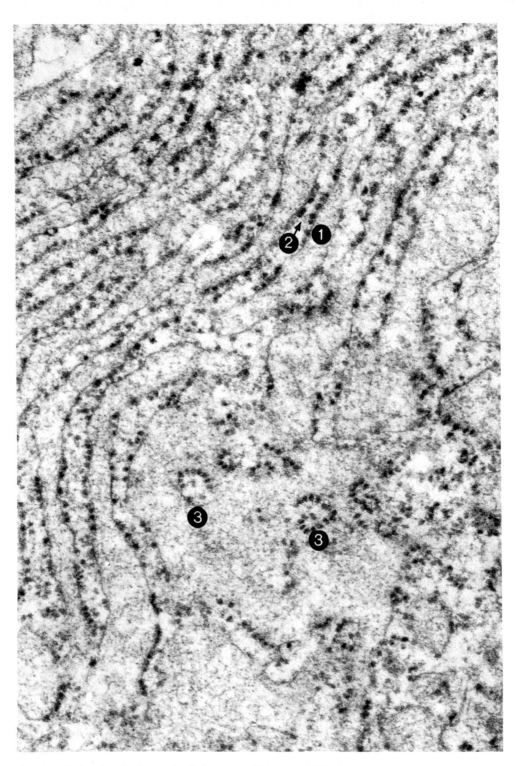

Rough endoplasmic reticulum and polyribosomes. (1) Lumen of RER, filled with proteins; (2) ribosomes, attached to the external surface of the RER; (3) polyribosomes. (Approximately ×100,000)

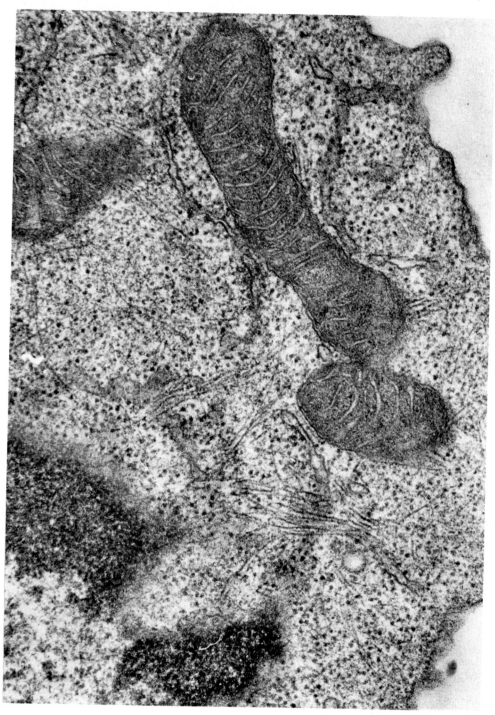

Mitochondria, microfibrils and monoribosomes. (Cf. figure, p. 10). (Approximately ×40,000)

10 – Examination with the Scanning Electron Microscope

The scanning electron microscope (SEM) permits the visualization of the surface of cells. One may turn a cell while observing it and examine it full face or in three-quarter or full profile. One thus receives in an instant a quantity of information that the transmission electron microscope could provide only by reconstruction from a thousand separate sections. True, the present SEM has a resolving power of only 20 nm, which places it between the light microscope with a resolving power of 200 nm and the transmission electron microscope with a resolving power of 0.2 nm. Even so, it can provide valuable insights, provided one asks appropriate questions. Among these are studies of the shape of normal and pathologic blood cells at rest and in motion. A look at this book, particularly the figures on the opposite page, proves this better than a long discussion.

The SEM demonstrates in a startling manner all the artifacts that occur in smears (see pp. 71 and 73). It shows (if the fixation is adequate) the true shape of the cells. Nevertheless, one must not lose sight of the fact that the SEM has its own artifacts. To be able to interpret these artifacts, a brief review of the principles of current techniques and the alterations which the shape of the cells can undergo in the different stages of preparation is necessary.

Techniques[1, 2]

Careful examination of living cells with the light microscope should precede preparation of cells for examination with the SEM in order to become thoroughly familiar with their shape under different experimental conditions and to appreciate artifactual changes induced by techniques of preparation. For fixation, a drop of the blood or cell suspension is allowed to fall into a fixative designed to harden the cell while maintaining its original shape. Commonly one uses glutaraldehyde (0.3%), osmic acid (2%) or both, either in sequence or in combination.

After fixation, the cells must be washed several times in distilled water in order to remove the protein molecules, salt crystals, and other substances that may adhere to the cellular membrane during fixation (and dessication) and which would interfer with visualization of the surface structure of the cell, the main objective of SEM examination.

Desiccation is indispensable as the SEM examination takes place in vacuum. Before desiccation, cells must be dehydrated by passing them through ascending concentrations of alcohol, then through graded solutions of propylene oxide up to and including pure propylene oxide. A small drop of the cell suspension is next deposited on a grid fixed to the stub for introduction into the microscope. Desiccation occurs almost instantaneously in air.

The "critical point" drying method needs to be used with very fragile cells. The method provides for cells to pass rapidly from the frozen to the dried state during evaporation, avoiding an intervening liquid phase which obliterates details of the cell surface. Before SEM examination, the cells must be covered by a thin layer of gold-palladium some 10 nm thick in a vacuum bell jar. This makes their surface conductive and thus allows rapid evacuation of electrons. Otherwise, electrons accumulate at some points and blur the picture.

Artifacts[3, 4]

The careful evaluation of the shape, the surface area and the volume of red cells, white cells, and platelets are of major importance for the study of the physiology of blood cells. It is, therefore, essential that the fixation and other stages of the technique do not produce changes of these parameters. The images provided by the scanning electron microscope are of value only to the extent that one remains constantly aware of the different stages of cell preparation and the artifacts each step may produce.

The environment in which the cells are kept may change the shape of living cells and the fixation may in turn modify these changes. For instance, echinocytes found in stored blood become discs again during the first stage of glutaraldehyde fixation.

The fixation can cause other artifacts: discocytes may become stomatocytes, or the volume may change. Leukocytes may lose a portion of their veils or their microvillosities.

Dehydration may affect many cellular structures. Red blood cells become torocytes, the pseudopods and microvillosities of leukocytes are effaced. Critical point drying avoids these artifacts.

Not all hematologists need a SEM for diagnosis. Once seen, the results of observation with the SEM can be applied by one's imagination to interpret the appearance of cells seen in phase contrast or even in fixed and stained smears. Details which remained unnoticed or were observed and discarded as unimportant may assume a new significance.

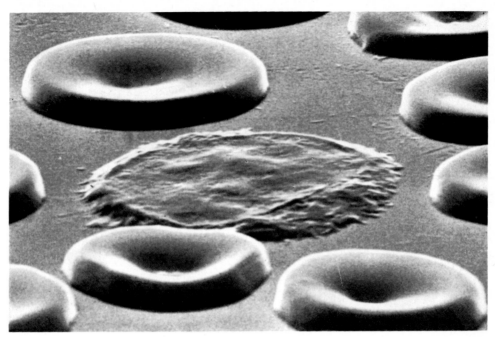

Fig. 1 — *A smear examined with the SEM.* The lymphocyte is reduced to a pancake (cf. Fig. 1, p. 151), the erythrocytes are flattened

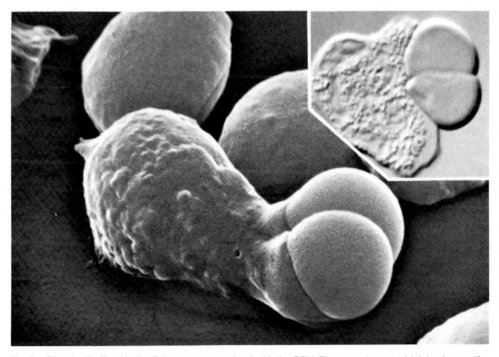

Fig. 2 — *Blood cells fixed in the living state, examined with the SEM.* The granulocyte, which is phagocytizing two spherocytes, has kept its original aspect.
In the insert, a similar cell is examined with the interference microscope. In this case, the impression of a three-dimensional presentation is an illusion (see p. 226)

Selected Bibliography

For an extensive bibliography with over 5000 references, I refer the interested reader to my book *Living Blood Cells and Their Ultrastructure**. In the present work I include only 'references to recent articles or to early work seldom cited in the hematologic literature. The numbers identifying the references are generally given immediately after the heading of each paragraph, even if they refer to a specific point. Since the entire title of each reference is given, the specific subject referred to can be easily identified.

The selection of references should not be taken as a recognition of priority. In fact, it is often difficult to decide who was the first to describe a new observation; it is even more difficult to determine who was the first to recognize its significance. At times, an observation is described by one author, but its significance is recognized by another, occasionally after a lapse of many years.

Often, several authors make the same discovery independently at virtually the same time. Others find again and describe as new an observation published years ago. Finally, some authors, in the words of Balthazar Gracian, have appreciated an explanation or hypothesis proposed by a colleague so much that they made it their own....

For any of these reasons I may have erred in my choice of references. I shall be grateful to any reader who will point our errors or omissions. They will be rectified in any second edition.

* Bessis, M. Living Blood Cells and Their Ultrastructure. Berlin-Heidelberg-New York: Springer-Verlag 1973.

Preface

On Morphology and Visual Images

ARNHEIM, R.: Visual thinking. In: *Education of Vision.* KEPES, G. (ed.). New York: Braziller **1965**, p. 1

BESSIS, M., MEL, H.: Hematology without the microscope. Blood Cells **1**, 401 (**1975**)

GOMBRICH, E.H.: *Art and Illusion: a Study in the Psychology of Pictorial Representation.* New Jersey: Princeton Univ. Pr. **1966**

VALERY, P.: Une vue de Descartes. In: *Variétés.* Paris: Gallimard **1944**, Vol. V

WALD, H.: L'homme et les signes. Int. Logic Rev. **4**, 102 (**1973**)

WEISS, P.: Within the Gates of the Science and Beyond. New York: Hafner **1971**, p. 197

Chapter 1
General Anatomy and Physiology of Blood Cells

p. 1–10 – Inventory of Cell Organelles

1 – BESSIS, M.: *Living Blood Cells and their Ultrastructure.* Berlin-Heidelberg-New York: Springer-Verlag 1973 Transl.: R.I. Weed

2 – NOVIKOFF, A.B., HOLTZMAN, E.: *Cells and Organelles.* New York: Holt, Rinehart and Winston 1970

3 – BRACHET, J., MIRSKY, A.E. (ed.): *The Cell. Biochemistry, Physiology, Morphology.* New York: Acad. Pr. 1959 to 1964

p. 12–Endocytosis

1 – SIMSON, J.V., SPICER, S.S.: Activities of specific cell constituents in phagocytosis (Endocytosis). Int. Rev. Exp. Pathol. **12**, 79 (1973)

2 – STOSSEL, T.P.: Phagocytosis. N. Engl. J. Med. **290**, 717, 774, 833 (1974)

3 – EDELSON, P.J., ZWIEBEL, R., COHN, Z.A.: The pinocytic rate of activated macrophages. J. Exp. Med. **142**, 1150 (1975)

4 – POLICARD, A., BESSIS, M.: Micropinocytosis and rhopheocytosis. Nature **194**, 110 (1962)

p. 14 – Digestion and Exocytosis

1 – DINGLE, J.T., FELL, H.B. (ed.): *Lysosomes in Biology and Pathology.* Amsterdam: North-Holland, Elsevier 1969

2 – HIRSCH, J.G., COHN, Z.A.: Digestive and autolytic functions of lysosomes in phagocytic cells. Fed. Proc. **23**, 1023 (1964)

3 – POLICARD, A.: La clasmatose. Nouv. Rev. Fr. Hématol. **2**, 897 (1962)

4 – SHIELDS, J.W.: Mononuclear cells, hyaline bodies and the plasma: an analytical review. Blood **17**, 235 (1961)

p. 16 – Cell Movements

1 – *Locomotion of tissue cells.* CIBA Foundation Symposium, n° 14. Amsterdam: North-Holland 1973

2 – JAHN, T.L., BOVEE, E.C.: Protoplasmic movements within cells. Physiol. Rev. **49**, 793 (1969)

3 – BESSIS, M., BOISFLEURY, A. DE: A catalogue of white blood cell movements (normal and pathologic). Blood Cells **2**, 365 (1976)

4 – BESSIS, M., LOCQUIN, M.: Sur la présence de mouvements propres de l'Aster et de vacuoles contractiles dans les granulocytes. C.R. Soc. Biol. **I44**, 483 (1950)

5 – BESSIS, M.: Cytological aspects of hemoglobin production. Harvey Lectures, series **58**, 125. New York: Acad. Press 1963

p. 17 – Hemopoiesis

1 – WEISS, P.A.: *Dynamics of Development: Experiments and Interferences.* New York: Acad. Pr. 1968

p. 18 – Stem Cells

1 – STOHLMAN, F. (ed.): *Hemopoietic Cellular Proliferation.* New York: Grune and Stratton 1970

2 – LAJTHA, L.G., SCHOFIELD, R.: On the problem of differentiation in haemopoiesis. Differentiation **2**, 313 (1974)

3 – NOWELL, P.C., WILSON, D.B.: Lymphocytes and hemic stem cells. Am. J. Pathol. **65**, 641 (1971)

4 – HOELZER, D., KURRLE, E., HARRISS, E.B., FLIEDNER, T.M., HAAS, R.J.: Evidence for stem cell function of resting bone marrow lymphocytes identified by the complete ^{3}H-thymidine labelling method. Biomedicine **22**, 285 (1975)

5 – ZUCKER-FRANKLIN, D., GRUSKY, G., L'ESPERANCE, P.: Granulocyte colonies derived from lymphocyte fractions of normal human peripheral blood. Proc. Natl. Acad. Sci. U.S.A. **71**, 2711 (1974)

6 – BARR, R.D., WHANG-PENG, J., PERRY, S.: Hemopoietic stem cells in human peripheral blood. Science **190**, 284 (1975)
7 – CRONKITE, E.P., FEINENDEGEN, L.E.: Notions about human stem cells Blood Cells **2**, 269 (1976)

p. 20 – Amplification and Maturation

1 – BESSIS, M., BRECHER, G.: A second look at stress erythropoiesis. Unanswered questions. Blood Cells **1**, 409 (1975)
2 – PAULUS, J.M.: *Production et Destruction des Plaquettes Sanguines*. Paris: Masson 1974, p. 136
3 – BESSIS, M.: Contribution à l'étude de la cytologie sanguine. I. Le synchronisme d'évolution nucléo-cytoplasmique. Rev. Hématol. **1**, 45 (1946)
4 – BRECHER, G., HALEY, J.E., PRENANT, M., BESSIS, M.: Macronormoblasts, macroreticulocytes and macrocytes. Blood Cells **1**, 547 (1975)

p. 22 – Agony, Death and Necrosis of Blood Cells

1 – BESSIS, M.: Studies on cell agony and death: an attempt at classification. In: CIBA Foundation Symp., *"Cellular injury"*. (DE REUCK, A.V.S., KNIGHT, J. (eds.). London: Churchill 1964, p. 287

p. 24 – Sociology of Blood Cells

1 – BASERGA, A.: Cyto-démographie et cyto-écologie. Bilan quotidien des mitoses hémopoiétiques. Nouv. Rev. Fr. Hématol. **4**, 469 (1964)
2 – ALEXANDROV, V.YA.: The problem of behaviour at the cellular level (Cytoethology). J. Theor. Biol. **35**, 1 (1972)
3 – WEISS, P., ANDRES, G.: Experiments on the fate of embryonic cells (chick) disseminated by the vascular route. J. Exp. Zool. **121**, 449 (1952)
4 – BESSIS, M.: Necrotaxis. Chemotaxis towards an injured cell. In: Antibiotics and Chemotherapy: *Chemotaxis, Its Biology and Biochemistry*. SORKIN, E. (ed.). Basel: Karger 1974, Vol. XIX, p. 369

Chapter 2 – Erythrocytic Series

A–NORMAL CELLS

p. 25 – General Comments

1 – HARRIS, J.W., KELLERMEYER, R.W.: *The Red Cell*. Cambridge: Harvard Univ. Pr. 1970
2 – BESSIS, M.: Cellular mechanism for the destruction of erythrocytes. Scand. J. Haematol. **2**, 59 (1965)

p. 32 – Polychromatophilic Normoblasts I and II (E$_4$ and E$_5$)

1 – BRETON-GORIUS, J.: Utilisation de la diaminobenzidine pour la mise en évidence, au microscope électronique, de l'hémoglobine intracellulaire. La réactivité des différents organelles des érythroblastes. Nouv. Rev. Fr. Hématol. **10**, 243 (1970)
2 – BRETON-GORIUS, J., GUICHARD, J.: Fine structural and cytochemical identification of microperoxisomes in developing human erythrocytic cells. Am. J. Pathol. **79**, 523 (1975)

p. 34 – Normal Sideroblasts

1 – MOURIQUAND, C.: Le sidéroblaste: étude morphologique et essai d'interprétation. Rev. Hématol. **13**, 79 (1958)
2 – GRUNEBERG, H.: Siderocytes: a new kind of erythrocytes. Nature **148**, 114 (1941)

3 – PAPPENHEIMER, A.M., THOMPSON, W.P., PARKER, D.D., SMITH, K.E.: Anaemia associated with unidentified erythrocytic inclusions. Q. J. Med. Sci. **14**, 75 (1945)
4 – BESSIS, M., BRETON-GORIUS, J.: Ferritin and ferruginous micelles in normal erythroblasts and hypochromic hypersideremic anemias. Blood **14**, 423 (1959)
5 – BESSIS, M., BRETON-GORIUS, J.: Différences entre sidéroblastes normaux et pathologiques. Etude au microscope électronique. Nouv. Rev. Fr. Hématol., **2**, 629 (1962)
6 – TANAKA, Y., BRECHER, G., BULL, B.: Ferritin localization on the erythroblast cell membrane and ropheocytosis in hypersiderotic human bone marrows. Blood **28**, 758 (1966)
7 – ZAIL, S.S., CHARLTON, R.W., TORRANCE, J.D., BOTHWELL, T.H.: Studies on the formation of ferritin in red cell precursors. J. clin. Invest. **43**, 670 (1964)
8 – CROSBY, W.H.: Siderocytes and the spleen. Blood **12**, 165 (1957)
9 – KOYAMA, S., AOKI, S., DEGUCHI, K.: Electron microscopic observations of the splenic red pulp with special reference to the pitting function. Mie Med. J. **14**, 143 (1964)

p. 36 – Extrusion of the Nucleus

1 – BESSIS, M., BRICKA, M.: Aspect dynamique des cellules du sang. Son étude par microcinématographie en contraste de phase. Rev. Hématol. **7**, 407 (1952)

p. 38—40 The Erythroblastic Island

1 – BESSIS, M.: L'ilôt érythroblastique, unité fonctionnelle de la moelle osseuse. Rev. Hématol. **13**, 8 (1958)

2 – BESSIS, M., BRETON-GORIUS, J.: Nouvelles observations sur l'ilôt érythroblastique et la rhophéocytose de la ferritine. Rev. Hématol. **14**, 165 (1959)

3 – BESSIS, M., BRETON-GORIUS, J.: Iron metabolism in the bone marrow as seen by electron microscopy. A critical review. Blood **19**, 635 (1962)

4 – LE CHARPENTIER, Y., PRENANT, M.: Isolement de l'ilôt érythroblastique. Etude en microscopie optique et électronique à balayage. Nouv. Rev. Fr. Hématol. **15**, 119 (1975)

5 – HERBEUVAL, H., HERBEUVAL, R.: *Cellules Cancéreuses et Insolites du Sang Circulant.* Paris: Masson, 1965, p. 62

p. 42 – The Erythroblastic Island: Phagocytic Functions

1 – MIESCHER, P.: Le mécanisme de l'érythroclasie à l'état normal. Rev. Hématol. **11**, 248 (1956)

2 – BESSIS, M.: Cellular mechanisms for the destruction of erythrocytes. Scand. J. Haematol. **2**, 59 (1965)

3 – EHRENSTEIN, G., LOCKNER, D. VON: Sites of the physiological breakdown of the red blood corpuscles. Nature **181**, 911 (1958)

4 – LONDON, I.M., WEST, R., SHEMIN, D., RITTENBERG, D.: On origin of bile pigment in normal man. J. Biol. Chem. **184**, 351 (1950)

5 – GLASS, J., YANNONI, S.H., ROBINSON, S.H.: Rapidly synthesized heme: relationship to erythropoiesis and hemoglobin production. Blood Cells **1**, 557 (1975)

6 – CROSBY, W.H.: Siderocytes and the spleen. Blood **12**, 165 (1957)

7 – KOYAMA, S., AOKI, S., DEGUCHI, K.: Electron microscopic observations of the splenic red pulp with special reference to the pitting function. Mie Med. J. **14**, 143 (1964)

p. 44 – Reticulocytes-Supravital Staining

1 – BESSIS, M., BRETON-GORIUS, J.: Le réticulocyte. Coloration vitale et microscopie électronique. Nouv. Rev. Fr. Hématol. **4**, 77 (1964)

2 – BRECHER, G.: New methylene blue as a reticulocyte stain. Am. J. Clin. Pathol. **19**, 895 (1949)

3 – SIMPSON, C.F., KLING, J.M.: The mechanism of mitochondrial extrusion from phenylhydrazine-induced reticulocytes in the circulating blood. J. Cell Biol. **36**, 103 (1968)

4 – MATSUMOTO, N., ISHIHARA, T., MIWA(S)UCHINO(F): The mechanism of mitochondrial extrusion from reticulocytes in the spleen from patients with erythrocyte pyruvate kinase (PK) deficiency. Acta Haematol. Jpn. **37**, 25 (1974)

p. 46 – Maturation of Reticulocytes

1 – BRECHER, G., HALEY, J.E., PRENANT, M., BESSIS, M.: Macronormoblasts, macroreticulocytes and macrocytes. Blood Cells **1**, 547 (1975)

2 – BESSIS, M., BRECHER, G.: A second look at stress erythropoiesis. Unanswered questions. Blood Cells **1**, 409 (1975)

3 – MEL, H.C., PRENANT, M., MOHANDAS, N.: Reticulocyte motility and form: studies on maturation and classification. Blood **49**, 1001 (1977)

4 – CROSBY, W.H.: Siderocytes and the spleen. Blood **12**, 165 (1957)

5 – REIFF, R.H., NUTTER, J.Y., DONOHUE, D.M., FINCH, C.A.: The relative number of marrow reticulocytes. Am. J. Clin. Pathol. **30**, 199 (1958)

6 – PAPAYANNOPOULOU, T., FINCH, C.A.: Radio-iron measurements of red cell maturation. Blood Cells **1**, 535 (1975)

p. 48 – Normal Erythrocytes: Discocytes

1 – WEINSTEIN, R.S.: The morphology of adult red cells. In: *The Red Blood Cell.* SURGENOR, D.M. (ed.). New York: Acad. Pr. 1974, Vol. I, p. 213

2 – WEED, R.I.: Membrane structure and its relation to haemolysis. Clin. Haematol. **4**, 3 (1975)

3 – BESSIS, M., WEED, R.I.: The structure of normal and pathologic erythrocytes. Adv. Biol. Med. Physics **14**, 35, 1973

p. 50 – Reversible Discocyte, Echinocyte and Stomatocyte Transformation

1 – DEUTICKE, B.: Transformation and restoration of biconcave shape of human erythrocytes induced by amphiphilic agents and changes of ionic environment. Biochem. Biophys. Acta **163**, 494 (1968)

2 – BRECHER, G., BESSIS, M.: Present status of spiculed red cells and their relationship to the discocyte-echinocyte transformation. A critical review. Blood **40**, 333 (1972)

3 – WEED, R.I., BESSIS, M.: The discocyte-stomatocyte equilibrium of normal and pathological red cells. Blood **41**, 471 (1973)

4 – BESSIS, M.: Red cell shapes: an illustrated classification and its rationale. Nouv. Rev. Fr. Hématol. **12**, 721 (1972)

5 – MOHANDAS, N., FEO, C.: A quantitative study of the red cell shape changes produced by anionic and cationic derivatives of phenothiazine. Blood Cells **1**, 375 (1975)

p. 52 – Echinocytes

1 – BRECHER, G., BESSIS, M.: Present status of spiculed red cells and their relationship to the discocyte-echinocyte transformation. A critical review. Blood **40**, 333 (1972)

2 – BESSIS, M., PRENANT, M.: Topographie de l'apparition des spicules dans les érythrocytes crénelés (échinocytes). Nouv. Rev. Fr. Hématol. **12**, 351 (1972)

3 – WEED, R.I., LACELLE, P.L., MERRILL, E.W.: Metabolic dependence of red cell deformability. J. Clin. Invest. **48**, 795 (1969)

4 – BESSIS, M., BRECHER, G.: Action du plasma conservé sur la forme des globules rouges. (Transformation discocyte-échinocyte). Nouv. Rev. Fr. Hématol. **11**, 305 (1971)

B–PATHOLOGIC CELLS

p. 54 – Pathologic Sideroblasts and Siderocytes (Sidero-mitochondria)

1 – BESSIS, M., BRETON-GORIUS, J.: Ferritin and ferruginous micelles in normal erythroblasts and hypochromic hypersideremic anemias. Blood **14**, 423 (1959)

2 – BESSIS, M., JENSEN, W.N.: Sideroblastic anaemia, mitochondria and erythroblastic iron. Brit. J. Haematol. **11**, 49 (1965)

3 – TANAKA, Y., BRECHER, G., BULL, B.: Ferritin localization on the erythroblast cell membrane and ropheocytosis in hypersiderotic human bone marrows. Blood **28**, 758 (1966)

4 – BOWMAN, JR., W.D.: Abnormal ("ringed") sideroblasts in various hematologic and non-hematologic disorders. Blood **18**, 662 (1961)

5 – BESSIS, M., ALAGILLE, D., BRETON-GORIUS, J.: Particularités des érythroblastes et des érythrocytes dans la maladie de Cooley. Etude au microscope électronique. Rev. Hématol. **13**, 538 (1958)

p. 56 – Iron in Inflammation

1 – BESSIS, M., BRETON-GORIUS, J.: L'îlot érythroblastique et la rhophéocytose de la ferritine dans l'inflammation. Nouv. Rev. Fr. Hématol. **1**, 569 (1961)

2 – MOURIQUAND, C., LEJEUNE, E., GERMAIN, D., MAITREPIERRE, J.: Le fer médullaire dans les maladies rhumatismales. Nouv. Rev. Fr. Hémat. **2**, 55 (1962)

3 – BESSIS, M., BRETON-GORIUS, J.: L'îlot érythroblastique et la rhophéocytose dans l'anémie ferriprive. Rev. Hématol. **15**, 233 (1960)

p. 58 – Reticulocyte and Erythrocyte Inclusions

1 – JENSEN, W.N., MORENO, G.D., BESSIS, M.: An electron microscopic description of basophilic stippling in red cells. Blood **25**, 933 (1965)

2 – VALENTINE, W.N., FINK, K., PAGLIA, D.E., HARRIS, S.R., ADAMS, W.S.: Hereditary hemolytic anemia with human erythrocyte pyrimidine 5′-nucleotidase deficiency. J. Clin. Invest. **54**, 866 (1974)

3 – VALENTINE, W.N., PAGLIA, D.E., FINK, K., MADOKORO, G.: Lead poisoning. Association with hemolytic anemia, basophilic stippling, erythrocyte pyrimidine 5′-nucleotidase deficiency and intra erythrocytic accumulation of pyrimidines. J. clin. Invest. **58**, 926 (1976)

p. 60 – Howell-Jolly Bodies. Cabot Rings

1 – DISCOMBE, G.: L'origine des corps de Howell-Jolly et des anneaux de Cabot. Sang **29**, 262 (1948)

2 – KOYAMA, S.: Studies on Howell-Jolly body. Acta Haematol. Jpn. **23**, 20 (1960)

3 – PICARD, D.: Nature et signification des anneaux de Cabot des hématies. C.R. Soc. Biol. (Paris) **147**, 1451 (1953)

4 – VAN OYE, E.: L'origine des anneaux de Cabot. Rev. Hématol. **9**, 173 (1954)

5 – FESSAS, P.: Inclusions of hemoglobin in erythroblasts and erythrocytes of thalassemia. Blood **21**, 21 (1963)

p. 62–63 – General Remarks on Pathologic Erythrocytes

1 – DISCOMBE, G., WATKINSON, G.: Atypical anemia, with spherocytes and target cells coexisting in the blood. Am. J. Med. Sci. **213**, 153 (1974)

2 – COHEN, F., ZUELZER, W.W., NEEL, J.V., ROBINSON, A.R.: Multiple inherited erythrocyte abnormalities in an American Negro family: hereditary spherocytosis, sickling and thalassemia. Blood **14**, 816 (1959)

3 – THOMPSON, R.B., ROBERTSON, M.G.: Three inherited intra-erythrocytic defects: hereditary spherocytosis, Hb S and Hb C. Acta Haematol. **32**, 233 (1964)

4 – BEUTLER, E.: Cellular mosaicism and heterogeneity in red cell disorders. Am. J. Med. **41**, 724 (1966)

5 – MAURER, H.S., VIDA, L.N., HONIG, G.R.: Homozygous sickle cell disease with coexistent hereditary spherocytosis in three siblings. J. Pediatr. **80**, 235 (1972)

p. 64–66 – General Remarks on Spiculated Red Blood Cells: Acanthocytes

1 – BRECHER, G., BESSIS, M.: Present status of spiculed red cells and their relationship to the discocyte-echinocyte transformation: a critical review. Blood **40**, 333 (1972)

2 – COOPER, R.A.: Anemia with spur cells: a red cell defect acquired in serum and modified in the circulation. J. Clin. Invest. **48**, 1820 (1969)

3 – TANAKA, K.R., VALENTINE, W.N., MIWA, S.: Pyruvate kinase (PK) deficiency hereditary non-spherocytic hemolytic anemia. Blood **19**, 267 (1962)

4 – ZARKOWSKY, H.S., OSKI, F.A., SHAAFI, R., SHOHET, S.B., NATHAN, D.G.: Congenital hemolytic anemia with high sodium, low potassium red cells. N. Engl. J. Med. **278**, 573 (1968)
5 – FEO, C.J., TCHERNIA, C., SUBTIL, E., LEBLOND, P.F.: Temporarily irreversible echinocytosis in 8 patients. Blood (in press) (1977)
6 – VRIES, A. DE, CONDREA, E., GITTER, S., DANON, D., KATCHALSKY, R., KESSLER, J., KOCHWA, S.: Crénelure des globules rouges et inhibition de la formation des rouleaux dans le sang humain en corrélation avec l'activité de la lipoprotéine-lipase. Sang **31**, 289 (1960)
7 – SINGER, K., FISHER, B., PERLSTEIN, M.A.: Acantocytosis: a genetic erythrocytic malformation. Blood **7**, 577 (1952)
8 – KAYDEN, H.J., BESSIS, M.: Morphology of normal erythrocyte and acanthocyte using Nomarski optics and the scanning electron microscope. Blood **35**, 427 (1970)
9 – WEED, R.I., BESSIS, M.: The discocyte-stomatocyte equilibrium of normal and pathologic red cells. Blood **41**, 471 (1973)
10 – BRECHER, G., HALEY, J.E., WALLERSTEIN, R.O.: Spiculed erythrocytes after splenectomy. Acanthocytes or non-specific poikilocytes? Nouv. Rev. Fr. Hématol. **12**, 751 (1972)
11 – FEO, C., MAIGNE, J.: Transformation acanthocyte-discocyte. Etude in vitro de l'action hémolysante de l'azide de sodium vis à vis des érythrocytes de sujets atteints d' abêtalipoprotéinémie. Nouv. Rev. Fr. Hématol. **14**, 119 (1974)

p. 68–72 – Codocytes and Target Cells

1 – BESSIS, M.: Red cell shapes: an illustrated classification and its rationale. Nouv. Rev. Fr. Hématol. **12**, 721 (1972)
2 – GJONE, E., TORSVIK, H., NORUM, K.R.: Familial plasma cholesterol ester deficiency. A study of the erythrocytes. Scand. J. Clin. Lab. Invest. **21**, 327 (1968)

p. 74 – Dacryocytes

1 – SLATER, L.M., MUIR, W.A., WEED, R.I.: Influence of splenectomy on insoluble hemoglobin inclusion bodies in beta thalassemic erythrocytes. Blood **31**, 766 (1968)

Monographs:

DACIE, J.V.: *The Haemolytic Anaemias, Congenital and Acquired. New York, Grune and Stratton 1960 to 1967*
Part 1 – The Congenital Anaemias. 2nd ed., 1960
Part 2 – The Auto-Immune Haemolytic Anaemias. 2nd ed., 1962

Part 3 – Secondary and Symptomatic Haemolytic Anaemias. 2nd ed. 1967
Part 4 – Drug-Induced Haemolytic Anaemias, Paroxysmal Noctural Hemoglobinuria, Haemolytic Disease of the Newborn. 2nd ed. 1967

p. 76 – Drepanocytes (Sickle Cells) and Intracellular Crystals

1 – WHITE, J.G.: Ultrastructural features of erythrocyte and hemoglobin sickling. Arch. Int. Med. **133**, 545 (1974)
2 – BESSIS, M., NOMARSKI, G., THIERY, J.P., BRETON-GORIUS, J.: Etudes sur la falciformation des globules rouges au microscope polarisant et au microscope électronique. II. L'intérieur du globule: comparaison avec les cristaux intra-globulaires. Rev. Hématol. **13**, 249 (1958)
3 – WHITE, J.G., HEAGAN, B.: Gels of normal and sickled hemoglobin. Comparative study. J. Exp. Med. **131**, 1079 (1970)
4 – WHITE, J.G., HEAGAN, B.: Tubular polymers of normal human hemoglobin. Am. J. Pathol. **59**, 101 (1970)
5 – DRUPT, F., POILLOT, M.H., LECLERC, M., LAVOLLAY, B., ALLARD, C., BACH, C.: Hémoglobine instable (mutant α) se singularisant par la formation de pseudo-drépanocytes in vitro. Nouv. Presse Méd. **5**, 1066 (1976)

p. 78 – Drepanocytes (Sickle Cells)–In vitro Formation

1 – JENSEN, W.N.: Fragmentation and the "freakish poikilocyte". Am. J. Med. Sci. **257**, 355 (1969)
2 – JENSEN, W.N., LESSIN, L.S.: Membrane alterations associated with hemoglobinopathies. Semin. Hematol. **7**, 409 (1970)

p. 80 – Drepano-echinocytes and Drepano-stomatocytes

1 – BESSIS, M., LESSIN, L.S.: The discocyte-echinocyte equilibrium of the normal and pathologic red cell. Blood **36**, 399 (1970)
2 – WEED, R.I., BESSIS, M.: The discocyte-stomatocyte equilibrium of normal and pathological red cells. Blood **41**, 471 (1973)
3 – JENSEN, W.N., BROMBERG, P.A., BESSIS, M.: Microincision of sickled erythrocytes by a Laser beam. Science, **155**, 704 (1967)
4 – DRUPT, F., POILLOT, M.H., LECLERC, M., LAVOLLAY, B., ALLARD, C., BACH, C.: Hémoglobine instable (mutant α) se singularisant par la formation de pseudo-drépanocytes in vitro. Nouv. Presse Méd. **5**, 1066 (1976)

p. 84 – Keratocytes

1 – BELL, R.E.: The origin of "burr" erythrocytes. Br. J. Haematol. **9**, 552 (1963)

2 – BRAIN, M.C., DACIE, J.V., HOURIHANE, D.: Microangiopathic haemolysis anaemia: the possible role of vascular lesions in pathogenesis. Br. J. Haematol. **8**, 358 (1962)

3 – BULL, B.S., KUHN, I.N.: The production of schistocytes by fibrin strands. (A scanning electron microscope study). Blood **35**, 104 (1970)

4 – BESSIS, M., BOISFLEURY, A. DE: Etude sur les poikilocytes au microscope à balayage, en particulier dans la thalassémie. Nouv. Rev. Fr. Hématol. **10**, 515 (1970)

p. 86 – Knizocytes

1 – BESSIS, M.: Red cell shapes: an illustrated classification and its rationale. Nouv. Rev. Fr. Hématol. **12**, 721 (1972)

2 – BULL, B.: Red cell biconcavity and deformability. A macromodel based on flow chamber observations. In: *Red Cell Shape. Physiology, Pathology, Ultrastructure.* BESSIS, M., WEED, R.I., LEBLOND, P.F. (eds.). Heidelberg: Springer-Verlag 1973, p. 115

p. 88 – Macrocytes and Megalocytes

1 – BRECHER, G., STOHLMAN, F.: Reticulocyte size and erythropoietic stimulation. Proc. Soc. Exp. Biol. Med. **107**, 887 (1961)

2 – BRECHER, G., PRENANT, M., HALEY, J., BESSIS, M.: Origin of stress macroreticulocytes from macronormoblasts. Nouv. Rev. Fr. Hématol. **15**, 13 (1975)

3 – BRECHER, G., HALEY, J.E., PRENANT, M.: Macronormoblasts, macroreticulocytes and macrocytes. Blood Cells **1**, 547 (1975)

4 – NATHAN, D.G., SHOHET, S.B.: Erythrocyte ion transport defects and hemolytic anemia: "hydrocytosis" and "desiccytosis". Semin. Hematol. **7**, 381 (1970)

5–6 – DACIE, J.V.: The *Haemolytic Anaemias, Congenital and Acquired.* New York: Grune and Stratton 1960 to 1967. Part 2–The Auto-Immune Haemolytic Anaemias. 2nd ed. 1962, pp. 345–357. Part 4–Drug-Induced Haemolytic Anaemias, Paroxysmal Nocturnal Haemoglobinuria, Haemolytic Disease of the Newborn. 2nd ed. 1967, p. 1271

p. 90 – Microcytes and Leptocytes

1 – NATHAN, D.G., SHOHET, S.B.: Erythrocyte ion transport defects and hemolytic anemia: "hydrocytosis" and "desiccytosis". Semin. Hematol. **7**, 381 (1970)

p. 92–94 – Schizocytes

1 – BRAIN, M.C., DACIE, J.V., HOURIHANE, D.: Microangiopathic haemolysis anaemia: the possible role of vascular lesions in pathogenesis. Br. J. Haematol. **8**, 358 (1962)

2 – BULL, B.S., RUBENBERG, M.L., DACIE, J.V., BRAIN, M.C.: Microangiopathic haemolytic anaemia: mechanism of red-cell fragmentation: in vitro studies. Br. J. Haematol. **14**, 643 (1968)

3 – RUBENBERG, M.L., REGOECZI, E., BULL, B.S., DACIE, J.V., BRAIN, M.C.: Microangiopathic haemolytic anaemia: the experimental production of haemolysis and red-cell fragmentation by defibrination in vivo. Br. J. Haematol. **14**, 627 (1968)

4 – BULL, B.S., KUHN, I.N.: The production of schistocytes by fibrin strands. (A scanning electron microscope study). Blood **35**, 104 (1970)

5 – POLICARD, A., BESSIS, M.: Fractionnement d'hématies par les leucocytes au cours de la phagocytose, C.R. Soc. Biol. (Paris) **147**, 982 (1953)

Monographs:

DACIE, J.V.: *The haemolytic anaemias, congenital and acquired.* New York, Grune and Stratton 1960 à 1967. Part I—The congenital anaemias (2nd ed.), 1960, Part II—The auto-immune haemolytic anaemias (2nd ed.), 1962, Part III—Secondary and symptomatic haemolytic anaemias (2nd ed.), 1967, Part IV—Drug-induced haemolytic anaemias, paroxysmal nocturnal haemoglobinuria, haemolytic disease of the newborn (2nd ed.), 1967

p. 98 – Hereditary Spherocytosis

1 – LEBLOND, P.F., BOISFLEURY, A. DE, BESSIS, M.: La forme des érythrocytes dans la sphérocytose héréditaire. Etude au microscope à balayage, Relation avec leur déformabilité. Nouv. Rev. Fr. Hématol. **13**, 873 (1973)

2 – LEBLOND, P.F., LACELLE, P.L., WEED, R.I.: Rhéologie des érythroblastes et des érythrocytes dans la sphérocytose congénitale. Nouv. Rev. Fr. Hématol. **11**, 537 (1971)

3 – COOPER, R.A., JANDL, J.H.: Bile salts and cholesterol in the pathogenesis of target cells in obstructive jaundice. J. Clin. Invest. **47**, 809 (1968)

4 – COOPER, R.A., JANDL, J.H.: The role of membrane lipids in the survival of red cells in hereditary spherocytosis. J. Clin. Invest. **48**, 736 (1969)

Monographs:

BELLINGHAM, A.J., PRANKERD, T.A.J.: Hereditary sphérocytosis. Clin. Haematol. **4**, 139 (1975)

DACIE, J.V.: *The Haemolytic Anaemias, Congenital and Acquired.* Part 1– The Congenital Anaemias. 2nd ed., 1960. Part 2– The Auto-Immune Haemolytic Anaemias. 2nd ed., 1962

p. 100 – Stomatocytes

1 – LOCK, P.S., SEPHTON SMITH, R., HARDISTY, R.M.: Stomatocytosis: a hereditary red cell anomaly associated with haemolytic anemia. Br. J. Haematol. **7**, 303 (1961)
2 – MILLER, D.R., RICKLES, R.F., LICHTMAN, M.A., LACELLE, P.L., BATES, J., WEED, R.I.: A new variant of hereditary hemolytic anemia with stomatocytosis and erythrocyte cation abnormality. Blood **38**, 184 (1971)
3 – STURGEON, P.: Hematological observations on the anemia associated with blood type Rh null. Blood **36**, 310 (1970)
4 – LEVINE, P., TRIPODI, D., STRUCK, J., ZMIJEWSKI, C.M., POLLACK, S.: Hemolytic anemia associated with Rh$_{null}$ but not with Bombay blood. A hypothesis based on differing antigenic structures. Vox Sang **24**, 417 (1973)

Monograph:

DACIE, J.V.: *The Haemolytic Anaemias, Congenital and Acquired.* Part 1– The Congenital Anaemias. 2nd ed., 1960

p. 102 – Mean Cell Diameter and Deformability of Erythrocytes

1 – YOUNG, TH.: Remarks on the measurement of minute particles, especially those of the blood and of pus. In: *An Introduction to Medical Literature Including a System of Practical Nosology, Intended as a Guide to Students, and an Assistant to Practitioners.* YOUNG, TH. London: Underwood and Blacks 1813, p. 545
2 – PIJPER, A.: The diffraction method of measuring red blood cells. J. Lab. Clin. Med. **32**, 857 (1947)
3 – TZANCK, A., BESSIS, M.: Un nouvel hémo-diffractomètre. Sang **18**, 71 (1947)
4 – WEED, R.I.: The importance of erythrocyte deformability. Am. J. Med. **49**, 147 (1970)
5 – BESSIS, M., MOHANDAS, N.: Mesure de la déformabilité cellulaire par une méthode diffractométrique. C.R. Acad. Sci. (Paris) **278**, 3263 (1974)
6 – BESSIS, M., MOHANDAS, N.: A diffractometric method for the measurement of cellular deformability. Blood Cells **1**, 307 (1975)
7 – BESSIS, M., MOHANDAS, N.: Red cell structure, shapes and deformability. Br. J. Haemat. **30**, (Suppl.), 1 (1975)
8 – BESSIS, M., MOHANDAS, N.: Déformation et orientation des globules rouges soumis à des forces de cisaillement. C.R. Acad. Sci. (Paris) **282**, 1567 (1976)

Monograph:

"Red cell rheology" 1977 Blood Cells, **3**, n° 1–2

Chapter 3 – Granulocytic Series

p. 105 – General comments

1 – CRONKITE, E.P., VINCENT, P.C.: Granulocytopoiesis. In: *Hemopoietic Cellular Proliferation.* STOHLMAN, F. (ed.). New York: Grune and Stratton 1970, p. 211
2 – VINCENT, P.C., CHANANA, A.D., CRONKITE, E.P., JOEL, D.D.: The intravascular survival of neutrophils labeled in vivo. Blood **43**, 371 (1974)

p. 106 – Granulocytopoiesis

1 – CRONKITE, E.P., VINCENT, P.C.: Granulocytopoiesis. In: *Hemopoietic Cellular Proliferation.* STOHLMAN, F. (ed.). New York: Grune and Stratton 1970, p. 211
2 – CRADDOCK, C.G.: Techniques for studying granulocytes kinetics. In: *Hematology.* WILLIAMS, W.J., BEUTLER, E., ERSLEV, A.J., RUNDLES, R.W. (eds.). New York: McGraw-Hill, 1972, p. 593
3 – CRONKITE, E.P., FEINENDEGEN, L.E.: Notions about human stem cells. Blood Cells **2**, 269 (1976)

4 – GRANT, L.: The sticking and emigration of white blood cells in inflammation. In: *The Inflammatory Process.* 2nd ed. ZWEIFACH, B.W., GRANT, L., MCCLUSKEY, R.T. (eds.). New York: Acad. Pr. 1973, Vol. II, p. 205

p. 108 – M$_1$ (Myeloblast)

1 – TCHERNIA, G., GRANGE, M.J., GOMBAUD-SAINTONGE, G.: Locomotion, necrotaxis and phagocytosis in normal and leukemic myeloblasts. Blood Cells **2**, 431 (1976)

p. 110 – M$_2$ (Promyelocyte)

1 – BAINTON, D.F., ULLYIOT, J.L., FARQUHAR, M.G.: The development of neutrophilic polymorphonuclear leukocytes in human bone marrow. Origin and content of azurophil and specific granules. J. Exp. Med. **134**, 907 (1971)
2 – BAINTON, D.F.: Neutrophil granules. Br. J. Haematol. **29**, 17 (1975)
3 – BRETON-GORIUS, J., REYES, F.: Ultrastructure of human bone marrow cell maturation. Int. Rev. Cytol. **46**, 251 (1976)

p. 112 – M₃, M₄ and M₅ (Myelocytes and Metamyelocyte)

1 – BAINTON, D.F.: Neutrophil granules. Br. J. Haematol. **29**, 17 (1975)
2 – CATINI, C., GHERI, G., MILIANI, A., DI GUGLIELMO, R.: Démonstration et caractérisation cytochimique des mucopolysaccharides acides des séries granulocytaires normales. Nouv. Rev. Fr. Hématol. **15**, 307 (1975)
3 – BRETZ, U., BAGGIOLINI, M.: Biochemical and morphological characterisation of azurophil and specific granules of human neutrophilic polymorphonuclear leukocytes. J. Cell. Biol. **63**, 251 (1974)

p. 114 – M₆ and M₇ (Polymorphonuclear Neutrophils)

1 – HADEN, R.L.: Qualitative changes in neutrophilic leukocytes. Am. J. Clin. Pathol. **5**, 354 (1935)
2 – MARSH, J.C., BOGGS, D.R., CARTWRIGHT, G.E., WINTROBE, M.M.: Neutrophil kinetics in acute infection. J. Clin. Invest. **46**, 1943 (1967)
3 – UNDRITZ, E.: Eine neue Sippe mit erblich-konstitutioneller Hochsegmentierung der Neutrophilenkerne. Schweiz. Med. Wochenschr. **88**, 1000 (1958)
4 – NORBERG, B.: Neutrophil segmentation and radial segmentation. Effects of cooling and demecolcine on the lobulation of neutrophil nuclei. Scand. J. Haematol. **6**, 274 (1969)

p. 116 – Movements of PMN

1 – PENNY, R., GALTON, D.A.G., SCOTT, J.T., EISEN, V.: Studies on neutrophil function. I. Physiological and pharmacological aspects. Br. J. Haematol. **12**, 623 (1966)
2 – BESSIS, M., BRICKA, M.: Etude sur l'étalement des leucocytes du sang humain au microscope à contraste de phase et par la méthode de l'ombrage. Rev. Hématol. **4**, 350 (1949)
3 – BESSIS, M., BOISFLEURY, A. DE: Les mouvements des leucocytes étudiés au microscope électronique à balayage. Nouv. Rev. Fr. Hématol. **11**, 377 (1971)
4 – BESSIS, M., LOCQUIN, M.: Sur la présence de mouvements propres de l'aster et de vacuoles contractiles dans les granulocytes. C.R. Soc. Biol. (Paris) **144**, 483 (1950)
5 – WILKINSON, P.C.: *Chemotaxis and Inflammation*. London: Churchill Livingstone 1974
6 – KELLER, H.U., HESS, M.W., COTTIER, H.: Physiology of chemotaxis and random motility. Semin. Hematol. **12**, 47 (1975)
7 – MILLER, M.E.: Pathology of chemotaxis and random motility. Semin. Hematol. **12**, 59 (1975)

Monograph:

"White cell movements". 1976 Blood Cells **2**, n° 3

p. 118 – Phagocytosis

1 – BESSIS, M., BOISFLEURY, A. DE: Etude des différentes étapes de l'érythrophagocytose par microcinématographie et microscopie électronique à balayage. Nouv. Rev. Fr. Hématol. **10**, 223 (1970)
2 – BESSIS, M., TABUIS, J.: Action cytologique des sérums anti-leucocytes et anti-plaquettes. Rev. Hématol. **9**, 127 (1954)
3 – HARGRAVES, M.M.: Discovery of the L.E. cell and its morphology. Proc. Mayo Clin. **44**, 579 (1969)
4 – ROBINEAUX, R., KOURILSKY, R., BUFFE, D.: Recherches sur la formation de la cellule de Hargraves. Ann. Inst. Pasteur (Paris) **91**, 109 (1956)
5 – BAINTON, D.F.: Sequential degranulation of the two types of polymorphonuclear leukocyte granules during phagocytosis of microorganisms. J. Cell Biol. **58**, 249 (1973)
6 – SIMSON, J.V., SPICER, S.S.: Activities of specific cell constituents in phagocytosis (Endocytosis). Int. Rev. Exp. Pathol. **12**, 79 (1973)
7 – HIRSCH, J.G.: Neutrophil leukocytes. In: *The Inflammatory Process*. 2nd ed. ZWEIFACH, B.W., GRANT, L., MCCLUSKEY, R.T. (eds.). New York: Acad. Pr. 1974, Vol. I, p. 411
8 – LEHRER, R.I., CLINE, M.J.: Leukocyte myeloperoxidase deficiency and disseminated candidiasis; the role of myeloperoxidase in resistance to Candida infection. J. Clin. Invest. **48**, 1478 (1969)
9 – COOPER, M.R., CHATELET, L.R. DE, MCCALL, C.E., LA VIA, M.F., SPURR, C.L., BAEHNER, R.L.: Complete deficiency of leukocyte glucose-6-phosphate dehydrogenase with defective bactericidal activity. J. Clin. Invest. **51**, 769 (1972)
10 – STOSSEL, T.P., ROOT, R.K., VAUGHAN, M.: Phagocytosis in chronic granulomatous disease and the Chediak-Higashi syndrome. N. Engl. J. Med. **286**, 120 (1972)

p. 120 – Nuclear Appendages

1 – DAVIDSON, W.M., SMITH, D.R.: A morphological sex difference in the polymorphonuclear neutrophil leucocytes. Br. Med. J. **2**, 6 (1954)
2 – LAMBOROT-MANZUR, M., TISHLER, P.V., ATKINS, L.: Fluorescent drumsticks in male polymorphs. Lancet **1971/I**, 973
3 – RICCI, N., CASTOLDI, G.L., DALLAPICCOLA, B., BASERGA, A.: Small drumsticks and long Y chromosomes. Br. Med. J. **1**, 346 (1971)

4 – BESSIS, M., BRETON-GORIUS, J.: Rapports entre noyau et centrioles dans les granulocytes étalés. Role des microtubules. Nouv. Rev. Fr. Hématol. **7**, 601 (1967)

5 – ACHONG, B.G., EPSTEIN, M.A.: Fine structure of the Burkitt tumor. J. Nat. Cancer Inst. **36**, 877 (1966)

6 – MCDUFFIE, M.C.: Nuclear blebs in human leukaemic cells. Nature **214**, 1341 (1967)

7 – AHEARN, M.J., LEWIS, C.W., CAMPBELL, L.A.: Nuclear bleb formation in human bone marrow cells during cytosine arabinosine therapy. Nature **215**, 196 (1967)

8 – DAVIES, H.G., TOOZE, J.: Electron and light microscope observations on the spleen of the newt Triturus cristatus: the surface topography of the mitotic chromosomes. J. Cell. Sci. **1**, 331 (1966)

9 – DUPLAN, J.F., BESSIS, M., BRETON-GORIUS, J.: Les appendices nucléaires (caryoschizes) des granulocytes après irradiation générale. Etude au microscope électronique. Nouv. Rev. Fr. Hématol. **9**, 205 (1969)

10 – SMITH, G.F., O'HARA, P.T.: Nuclear pockets in normal leucocytes. Nature **215**, 773 (1967)

11 – MEHES, K.: Nuclear projections in neutrophils. Blood **28**, 598 (1966)

p. 122 – Abnormalities of Cytoplasm and Granules

1 – LEHRER, R.I., CLINE, M.J.: Leukocyte myeloperoxidase deficiency and disseminated Candidiadis: the role of myeloperoxidase in resistance to Candida infection. J. clin. Invest. **48**, 1478 (1969)

2 – UNDRITZ, E.: Die Alius-Grignashi-Anomalie: der erblich-konstitutionelle Peroxydasedefekt der Neutrophilen und Monozyten. Blut **14**, 129 (1968)

3 – STRAUSS, R.G., BOVE, K.E., JONES, J.F., MAUER, A.M., FULGINITI, V.A.: An anomaly of neutrophil morphology with impaired function. N. Engl. J. Med. **290**, 478 (1974)

4 – ZUCKER-FRANKLIN, D.: The phagosomes in rhumatoid synovial fluid leukocytes: a light, fluorescent and electron microscope study. Arthritis Rheum. **9**, 24 (1966)

5 – MARMONT, A., ROSSI, F., DAMASÌO, E., GORI, E.: Recenti aspetti patogenetici dell' attaco acuto di gotta. Reumatismo **4**, 152 (1969)

6 – JENIS, E.H., TAKEUCHI, A., DILLON, D.E., RUYMANN, F.E.: The May-Hegglin anomaly: ultrastructure of the granulocyte inclusion. Am. J. Clin. Pathol. **55**, 187 (1971)

7 – CAWLEY, J.C., HAYHOE, F.G.: The inclusions of the May-Hegglin anomalie and Döhle bodies of infection: an ultrastructural comparison. Br. J. Haematol. **22**, 491 (1972)

8 – PONDER, E., PONDER, R.: The cytology of the polymorphonuclear leucocyte in toxic conditions. J. Lab. Clin. Med., **28**, 316 (1942)

9 – MCCALL, C.E., KATAYAMA, I., COTRAN, R.S., FINLAND, M.: Lysosomal and ultrastructural changes in human "toxic" neutrophils during bacterial infection. J. Exp. Med. **129**, 267 (1969)

p. 124 – Eosinophils

Monograph:

ZUCKER-FRANKLIN, D.: Eosinophil function and disorders. Adv. Int. Med. **19**, 1 (1974)

p. 126 – Basophils

1 – BESSIS, M., TABUIS, J.: Formation de cristaux à partir de leucocytes basophiles. C.R. Soc. Biol. (Paris) **149**, 873 (1955)

2 – BESSIS, M., BRETON-GORIUS, J.: Structure périodique observée au microscope électronique dans des cristaux formés à partir de leucocytes basophiles. Nouv. Rev. Fr. Hématol. **153**, 720 (1959)

3 – KEYHANI, E., BRETON-GORIUS, J.: Etude au microscope électronique de la formation et de la maturation des granulations des leucocytes basophiles de la moelle osseuse de l'homme et du cobaye. Localisation cytochimique du glycogène. Acta Anat. **82**, 337 (1972)

Monograph:

PARWARESCH, M.R.: The *Human Blood Basophil*. Berlin: Springer-Verlag 1976

Chapter 4 – Thrombocytic Series

p. 129 – General remarks

1 – PAULUS, J.M. (ed.): *Platelet Kinetics.* Amsterdam: North-Holland 1971

2 – PAULUS, J.M.: *Production et Destruction des Plaquettes Sanguines.* Paris: Masson 1974

3 – BALDINI, M.G., EBBE, S. (eds.): *Platelets. Production Function, Transfusion and Storage.* New York: Grune and Stratton 1974

4 – PENINGTON, D.G., STREATFIELD, K.: Heterogeneity of megakaryocytes and platelets. Ser. Haematol. **8**, 22 (1975)

p. 130 – Thrombocytopoiesis

1 – PAULUS, J.M.: *Production et Destruction des Plaquette Sanguines.* Paris: Masson 1974

2 – PENINGTON, D.G., STREATFIELD, K.: Heterogeneity of megakaryocytes and platelets. Ser. Haematol. **8**, 22 (1975)

3 – LEVAL, M. DE, PAULUS, J.M.: Megakaryocytes: uninucleated polyploid or multinucleate cells? In: *Platelet Kinetics.* PAULUS, J.M. (ed.). Amsterdam: North-Holland 1971, p. 190

4 – JACKSON, C.W.: Some characteristics of rat megakaryocyte precursors identified using cholinesterase as a marker. In: *Platelets. Production, Function, Transfusion and Storage.* BALDINI, M.G., EBBE, S. (eds.). New York: Grune and Stratton 1974, p. 33

5 – COOPER, G.W.: Suppression of mouse platelet production by hypertransfusion and stimulation by thrombopoietin. In: *Platelets. Production, Function, Transfusion and Storage.* BALDINI, M.G., EBBE, S. (eds.). New York: Grune and Stratton, 1974, p. 73

6 – ODELL, T.T., MURPHY, J.R., JACKSON, C.W.: Stimulation of megakaryocytopoiesis by acute thrombocytopenia in rats. Blood **48**, 765 (1976)

7 – PENINGTON, D.G., STREATFIELD, K., WESTE, S.M.: Megakaryocyte ploidy and ultrastructure in stimulated thrombopoiesis. In: *Platelets. Production, Function, Transfusion and Storage.* BALDINI, M.G., EBBE, S. (eds.). New York: Grune and Stratton, 1974, p. 115

8 – EBBE, S., STOHLMAN, F., OVERCASH, J., DONOVAN, J., HOWARD, D.: Megakaryocyte size in thrombocytopenic and normal rats. Blood **32**, 775 (1968)

p. 132 and 134 – Megakaryocytes

1 – PAULUS, J.M.: *Production et Destruction des Plaquettes Sanguines.* Paris: Masson 1974

2 – BALDINI, M.G., EBBE, S. (eds.): *Platelets. Production, Function, Transfusion and Storage.* New York: Grune and Stratton 1974

3 – BRETON-GORIUS, J., GUICHARD, J.: Ultrastructural localization of peroxidase activity in human platelets and megakaryocytes. Am. J. Pathol. **66**, 277 (1972)

4 – BRETON-GORIUS, J., REYES, F.: Ultrastructure of human bone marrow cell maturation. Int. Rev. Cytol. **46**, 251 (1976)

p. 136 – Platelet Formation

1 – THIERY, J.P., BESSIS, M.: Mécanisme de la plaquettogenèse. Etude in vitro par la microcinématographie. Rev. Hématol. **11**, 162 (1956)

2 – PAULUS, J.M.: *Production et Destruction des Plaquettes Sanguines.* Paris: Masson 1974

3 – BESSIS, M., THIERY, J.P.: Los megacariocitos y las plaquetas examinados con el microscopio electronico. Sangre **1**, 123 (1956)

4 – BEHNKE, O., PEDERSEN, N.T.: Ultrastructural aspects of megakaryocyte maturation and platelet release. In: *Platelets. Production, Function, Transfusion and Storage.* BALDINI, M.G., EBBE, S. (eds.). New York: Grune and Stratton, 1974, p. 21

p. 138 – Platelets

1 – ZUCKER-FRANKLIN, D.: Microfibrils of blood platelets: their relationship to microtubules and the contractile protein. J. Clin. Invest. **48**, 165 (1969)

2 – WHITE, J.G.: Platelet morphology. In: *The Circulating Platelet.* JOHNSON, S.A., (ed.). New York: Acad. Pr. 1971

3 – TS'AO, CH.: Rough endoplasmic reticulum and ribosomes in blood platelets. Scand. J. Haematol. **8**, 134 (1971)

4 – WHITE, J.G.: Identification of platelet secretion in the electron microscope. Ser. Haematol. **6**, 429 (1973)

5 – HOVIG, T.: The ultrastructural basis of platelet function. In: *Platelets. Production, Function, Transfusion and Storage.* BALDINI, M.G., EBBE, S. (eds.). New York: Grune and Stratton 1974, p. 221

6 – ZUCKER, M.B.: Platelets. In: *The Inflammatory Process.* 2nd ed. ZWEIFACH, B.W., GRANT, L., MCCLUSKEY, R.T. (eds.). New York: Acad. Pr. 1974, Vol. I, p. 511

7 – WHITE, J.G., GERRARD, J.M.: Ultrastructural features of abnormal blood platelets. A review. Am. J. Pathol. **83**, 590 (1976)

8 – BENTFELD, M.E., BAINTON, D.F.: Cytochemical localization of lysosomal enzymes in rat megakaryocytes and platelets. J. Clin. Invest. **56**, 1635 (1975)

p. 140 – Movements of Platelets

1 – BESSIS, M., TABUIS, J.: Aspect dynamique des plaquettes sanguines à l'état normal, et pathologique. Analyse d'un film en contraste de phase. Rev. Hématol. **10**, 753 (1955)

2 – BESSIS, M., BURSTEIN, M.: Etudes sur les thrombocytes au microscope électronique. Rev. Hématol. **3**, 48 (1948)

3 – BESSIS, M., BRETON-GORIUS, J.: Les microtubules et les fibrilles dans les plaquettes étalées. Nouv. Rev. Fr. Hématol. **5**, 657 (1965)

4 – BETTEX-GALLAND, M., LUSCHER, E.F.: Thrombosthenin, the contractile protein from blood platelets and its relation to other contractile protein. Adv. Protein Chem. **20**, 1 (1965)

5 – WHITE, J.G.: Identification of platelet secretion in the electron microscope. Ser. Haematol. **6**, 429 (1973)

6 – HOVIG, T.: The ultrastructural basis of platelet function. In: *Platelets. Production, Function, Transfusion, and Storage.* BALDINI, M.G., EBBE, S. (eds.). New York: Grune and Stratton 1974, p. 221

7 – ZUCKER, M.B.: Platelets. In: *The Inflammatory Process.* 2nd ed. ZWEIFACH, B.W., GRANT, L., MCCLUSKEY, R.T. (eds.). New York: Acad. Pr. 1974, Vol. I, p. 511

8 – BALDINI, M.G., EBBE, S. (eds.): *Platelets. Production, Function, Transfusion and Storage.* New York: Grune and Stratton 1974

p. 142–144 – General Pathology of Megakaryocytes and Platelets

General comments:

WHITE, J.G., GERRARD, J.M.: Ultrastructural features of abnormal blood platelet. Am. J. Pathol. **83**, 590 (1976)

1 – BRETON-GORIUS, J., DREYFUS, B., SULTAN, C., BASCH, A., D'OLIVEIRA, J.G.: Identification of circulating micromegakaryocytes in a case of refractory anemia: an electron microscopic-cytochemical study. Blood **40**, 453 (1972)

2 – SMITH, W.B., ABLIN, A., GOODMAN, J.R., BRECHER, G.: Atypical megakaryocytes in preleukemic phase of acute myeloid leukemia. Blood **42**, 535 (1973)

3 – BRECHER, G.: Atypical megakaryocytes. A reflection of a stem cell disorder. In: *Platelets. Production, Function, Transfusion and Storage.* BALDINI, M.G., EBBE, S. (eds.). New York: Grune and Stratton 1974, p. 93

4 – MALDONADO, J.E.: Dysplastic platelets and circulating megakaryocytes in chronic myeloproliferative diseases. II. Ultrastructure of circulating megakaryocytes. Blood **43**, 811 (1974)

5 – BRETON-GORIUS, J.: Development of two distinct membrane systems associated in giant complexes in pathological megakaryocytes. Ser. Haematol. **8**, 49 (1975)

6 – BALDINI, M.G., EBBE, S. (eds.): *Platelets. Production, Function, Transfusion and Storage,* New York: Grune and Stratton 1974

7 – PAULUS, J.M., BRETON-GORIUS, J., KINET-DENOEL, C., BONIVER, J.: Megakaryocyte ultrastructure and ploidy in human macrothrombocytosis. In: *Platelets. Production, Function, Transfusion and Storage.* BALDINI, M.G., EBBE, S. (eds.). New York: Grune and Stratton 1974, p. 131

8 – PENINGTON, D.G., STREATFIELD, K.: Heterogeneity of megakaryocytes and platelets. Ser. Haematol. **8**, 22 (1975)

9 – WHITE, J.G.: Ultrastructural defects in congenital disorders of platelet function. Ann. N.Y. Acad. Sci. **201**, 205 (1972)

10 – BEHRENS, W.E. VON: Mediterranean macrothrombocytopenia. Blood **46**, 199 (1975)

Chapter 5 – Lymphocytic Series

p. 145 – General Remarks

1 – GOWANS, J.L.: Lymphocytes. In: *The Harvey Lectures* 1968/1969. Ser. 64, New York: Acad. Pr. 1970, p, 87

2 – CRONKITE, E.P., CHANANA, A.D.: Lymphocytopoiesis. In: *Formation and Destruction of Blood Cells.* GREENWALT, T.J., JAMIESON, G.A. (eds.). Philadelphia: Lippincott 1970, p. 284

p. 146 – The Lymphoblast and Large Lymphocyte

1 – SEMAN, G.: La mise en évidence des nucléoles sur frottis par le bleu de méthylène boraté. Rev. Et. Clin. Biol. **5**, 196 (1960)

p. 148 – The small lymphocyte

1 – ANTEUNIS, A.: Origin and fate of the multivesicular bodies in PHA stimulated lymphocytes. Cell Tissue Res. **149**, 497 (1974)

2 – LAZARUS, S.S., VETHAMANY, V.G., SCHNECK, L., VOLK, B.W.: Fine structure and histochemistry of peripheral blood cells in Niemann-Pick disease. Lab. Invest. **17**, 155 (1967)

3 – NOONAN, S.M., WEISS, L., RIDDLE, J.M.: A new cytoplasmic inclusion in Tay-Sachs lymphocytes. J. Cell Biol. **63**, 246a (1974)

p. 150 – Movements of Lymphocytes

1 – BARAT, N., AVRAMEAS, S.: Surface and intracellular localization of concanavalin A in human lymphocytes. Exp. Cell Res. **76**, 451 (1973)

2 – BIBERFELD, P., BIBERFELD, G., MOLNAR, Z., FAGRAEUS, A.: Fixation of cellbound antibody in the membrane immunofluorescent test. J. Immunol. Methods **4**, 135 (1974)

3 – ROSSE, W. (in press)

p. 152 – Antigenic Stimulation – I. General Remarks

1 – MURPHY, M.J., HAY, J.B., MORRIS, B., BESSIS, M.: Ultrastructural analysis of antibody synthesis in cells from lymph and lymph nodes. Am. J. Pathol. **66**, 25 (1972)

2 – HAY, J.B., MURPHY, M.J., MORRIS, B., BESSIS, M.: Quantitative studies on the proliferation and differentiation of antibody-forming cells in lymph. Am. J. Pathol. **66**, 1 (1972)

3 – BUERKI, H., COTTIER, H., HESS, M.W., LAISSUE, J., STONER, R.D.: Distinctive medullary and germinal center proliferative patterns in mouse lymph nodes after regional primary and secondary stimulation with tetanus toxoid. J. Immunol. **112**, 1961 (1974)

4 – JOEL, D.D., CHANANA, A.D., CRONKITE, E.P.C., SCHIFFER, L.M.: Modification of skin allograft immunity by extracorporeal radiation of lymph. Transplantation **5**, 1192 (1967)

p. 154 – Antigenic Stimulation – II. Transformation of Lymphocytes

1 – NOWELL, P.C.: Phytohemagglutinin: an initiator of mitosis in cultures of normal human leucocytes. Cancer Res., **20**, 462 (1960)

2 – DOUGLAS, S.D.: Human lymphocyte growth in vitro: morphologic, biochemical and immunologic significance. Int. Rev. Exp. Pathol. **10**, 41 (1971)

3 – ANTEUNIS, A.: Origin and fate of the multivesicular bodies in PHA stimulated lymphocytes. Cell Tissue Res. **149**, 497 (1974)

p. 156 – B- and T-Lymphocytes and Others

1 – BENTWICH, Z., KUNKEL, H.G.: Specific properties of human B and T lymphocytes and alterations in disease. Transplant. Rev. **16**, 29 (1973)

2 – "Separation of T and lymphocyte subpopulations". Transplant. Rev. **25**, 1975

3 – DWYER, J.M.: Identifying and enumerating human T and B lymphocytes. Prog. Allergy **21**, 178 (1976)

4 – SMITH, G.F., PENROSE, L.S., O'HARA, P.T.: Crystalline bodies in lymphocytes of patients with Down's syndrome. Lancet **1967/II,** 452

Chapter 6 – Monohistiocytic Series

p. 157 – General Remarks

1 – METCHNIKOFF, E.: *Leçons sur la Pathologie Comparée de l'Inflammation.* Paris: Masson 1892
– METCHNIKOFF, E.: *Lectures on the comparative pathology of inflammation.* Delivered at the Pasteur Institute, 1891. London: Kegan, Trench, Trubner 1896 (translation, New York: Dove 1968)

2 – METCHNIKOFF, E.: *L'immunité dans les Maladies Infectieuses.* Paris: Masson 1901

3 – KIYONO, K.: *Die vitale Carminspeicherung.* Jena: Fischer 1914

4 – ASCHOFF, L.: Das reticulo-endotheliale System. Ergeb. Inn. Med. Kinderheilkd. **26**, 1 (1924)

5 – VAN FURTH, R.: *Mononuclear Phagocytes in Immunity, Infection, and Pathology.* Oxford: Blackwell 1975

p. 158 – Monocytes

1 – YAM, L.T., LI, C.Y., CROSBY, W.H.: Cytochemical identification of monocytes and granulocytes. Am. J. Clin. Pathol. **55**, 283 (1971)

2 – LI, C.Y., YAM, L.T., CROSBY, W.H.: Histochemical characterization of cellular and structural elements of the human spleen. J. Histochem. Cytochem. **20**, 1049 (1972)

3 – KAPLOW, L.S.: Cytochemical heterogeneity of human circulating monocytes. Acta Cytologica **19**, 358 (1975)

4 – DANIEL, M.TH., FLANDRIN, G., LEJEUNE, F., LISO, P., LORTHOLARY, P.: Les estérases spécifiques monocytaires. Utilisation dans la classification des leucémies aigües. Nouv. Rev. Fr. Hématol. **11**, 233 (1971)

5 – NICHOLS, B.A., BAINTON, D.F.: Differentiation of human monocytes in bone marrow and blood. Sequential formation of two granule populations. Lab. Invest. **29**, 27 (1973)

p. 162 – Histiocytes (Macrophages). Phagocytosis of Lymphocytes

1 – LEWIS, W.M., LEWIS, M.R.: The transformation of white blood cells into clasmatocytes (macrophages), epithelioid cells, and giant cells. J. Amer. Med. Assoc. **84**, 798 (1925)

2 – GORDON, A.S., COHN, Z.A.: The macrophage. Int. Rev. Cytol. **36**, 171 (1973)

Monographs:

CARR, I.: *The Macrophage. A Review of Ultrastructure and Function.* New York: Acad. Pr. 1973

GORDON, S., COHN, Z.A.: The macrophage. Int. Rev. Cytol. **36**, 171 (1973)

STEINMAN, R.M., COHN, Z.A.: The metabolism and physiology of the mononuclear phagocytes. In: *The Inflammatory Process.* 2nd ed. ZWEIFACH, B.W., GRANT, L., MCCLUSKEY, R.T. (eds.). New York: Acad. Pr. 1974, Vol. I, p. 449

p. 164 – Histiocytes (Macrophages). Phagocytosis of Erythrocytes

1 – ROUS, P.: Destruction of the red blood corpuscles in health and disease. Physiol. Rev. **3**, 75 (1923)

2 – HARRIS, J.W.: *The Red Cell.* Cambridge: Harvard Univ. Pr. 1963, Vol. I

3 – BESSIS, M.: Cellular mechanisms for the destruction of erythrocytes Scand. J. Haematol. Ser. Haematol. **2**, 59 (1965)

4 – MIESCHER, P.: Le mécanisme de l'érythroclasie à l'état normal. Rev. Hematol. **11**, 248 (1956)

5 – EHRENSTEIN, G. VON, LOCKNER, D.: Sites of the physiological breakdown of the blood corpuscles. Nature **181**, 911 (1958)

6 – WAGNER, H.N., RAZZAK, M.A., GAERTNER, R.A.: Removal of erythrocytes from the circulation. Arch. Int. Med. **110**, 90 (1962)

p. 166 – Histiocytes (Macrophages). Formation of Hemosiderin

1 – NEUMANN, E.: Beiträge zur Kenntnis der pathologischen Pigmente. Virchows Arch. Pathol. Anat. **111**, 25 (1888)

2 – LAUFBERGER, V.: Sur la cristallisation de la ferritine. Bull. Soc. Chim. Biol. **19**, 1575 (1937)

3 – BESSIS, M., BRETON-GORIUS, J.: Trois aspects du fer dans des coupes d'organes examinées au microscope électronique (ferritine et dérivé dans les cellules intestinales, les érythroblastes et les cellules réticulaires.) C.R. Acad. Sci. (Paris) **245**, 1271 (1957)

4 – RICHTER, G.W.: The cellular transformation of injected colloidal iron complexes into ferritin and hemosiderin in experimental animals. A study with the aid of electron microscopy. J. Exp. Med. **109**, 197 (1959)

5 – THIERY, J.P.: Etudes des réactions cytochimiques du fer au microscope électronique. J. Microscopie **1**, 127 (1962)

6 – RICHTER, G.W., BESSIS, M.C.: Commentary on hemosiderin. Blood **25**, 370 (1965)

7 – RICHTER, G.W.: Electron microscopy of hemosiderin: presence of ferritin and occurence of crystalline lattices in hemosiderin deposits. J. Biophys. Biochem. Cytol. **4**, 55 (1958)

8 – BESSIS, M., BRETON-GORIUS, J.: Différents aspects du fer dans l'organisme. I. Ferritine et micelles ferrugineuses. II. Differentes formes de l'hémosidérine. J. Biophys. Biochem. Cytol. **6**, 231 (1959)

9 – BESSIS, M., CAROLI, J.: A comparative study of hemochromatosis by electron microscope. Gastroenterology **37**, 538 (1959)

p. 168 – Gaucher Disease

1 – BRADY, R.O.: Biochemical and metabolic basis of familial sphingolipidoses. Semin. Hematol. **9**, 273 (1972)

2 – LULLMANN, H., LULLMAN-RAUCH, R., WASSERMANN, O.: Drug-induced phospholipidosis. German Med. **3**, 128 (1973)

3 – JORDAN, S.W.: Electron microscopy of Gaucher cells. Exp. Mol. Pathol. **3**, 76 (1964)

4 – LORBER, M., NEMES, J.L.: Identification of ferritin within Gaucher cells. An electron microscopic and immunofluorescent study. Acta Haematol. **37**, 189 (1967)

5 –VAN SLYCK, E.J., WALDMANN, R., REBUCK, J.W.: Unavailability of iron in Gaucher's cells. N. Engl. J. Med. **291**, 261 (1974)

p. 170 – Niemann-Pick Disease and Other Storage Diseases

1 – TANAKA, Y., BRECHER, G., FREDRICKSON, D.S.: Cellules de la maladie de Niemann-Pick et de quelques autres lipoidoses. Nouv. Rev. Fr. Hématol., **3**, 5 (1963)

2 – MCKUSICK, V.A.: The nosology of mucopolysaccharidoses. Am. J. Med. **47**, 730 (1969)

3 – MAROTEAUX, P., LAMY, M.: Hurler's disease, Morquio's disease and related mucopolysaccharidoses. J. Pediatr. **67**, 312 (1965)

4 – FRICKER-ALDER, H.: Die Aldersche Granulationsanomalie; Nachuntersuchungen des erstbeschriebenen Falles und Überblick über den heutigen Stand der Kenntnisse. Schweiz. Med. Wochenschr. **88**, 989 (1958)

5 – SILVERSTEIN, M.N., ELLEFSON, R.D.: The syndrome of the sea-blue histiocytes. Semin. Hematol. **9**, 299 (1972)

6 – STAWISKY, A., ROSNER, F., CHODSKY, S.: The sea-blue histiocyte syndrome, A review: genetic and biochemical studies. Semin. Hematol. **9**, 285 (1972)

7 – RYWLIN, A.M., HERNANDEZ, J.A., CHASTAIN, D.E., PARDO, V.: Ceroid histiocytosis of spleen and bone marrow in idiopathic thrombopenic purpura (ITP): a contribution to the understanding of the sea-blue histiocyte. Blood **37**, 587 (1971)

8 – GOLDE, D.W., SCHNEIDER, E.L., BAINTON, D.F., PENTCHEV, P.G., BRADY, R.O., EPSTEIN, C.J., CLINE, M.J.: Pathogenesis of one variant of sea-blue histiocytosis. Lab. Invest. **33**, 371 (1975)

p. 172 – Storage Cells Due to Overload

1 – ALBRECHT, M.: ''Gaucher-Zellen'' bei chronisch myeloischer Leukämie. Blut **13**, 169 (1966)

2 – KATTLOVE, H.E., WILLIAMS, J.C., GAYNOR, E., SPIVAK, M., BRADLEY, R.M., BRADY, R.O.: Gaucher cells in chronic myelocytic leukemia: an acquired abnormality. Blood **33**, 379 (1969)

3 – LEE, R.E., ELLIS, L.D.: The storage cells of chronic myelogenous leukemia. Lab. Invest. **24**, 261 (1971)

4 – GELFAND, M., GRIBOFF, S.: Gaucher's disease and acute leukemia. J. Mount Sinai Hosp. N.Y. **28**, 278 (1961)

5 – ZAINO, E.C., ROSSI, M.B., PHAM, T.D., AZAR, H.A.: Gaucher's cells in thalassemia. Blood **38**, 457 (1971)

6 – BELTRAMI, C.A., BEARZI, I., FABRIS, G.: Storage cells of spleen and bone marrow in thalassemia: an ultrastructural study. Blood **41**, 901 (1973)

7 – VAN DORPE, A., BROECKAERT-VAN-ORSHOVEN, A., DESMET, V., VERWILGHEN, R.L.: Gaucher-like cells and congenital dyserythropoietic anemia, type II (HEMPAS). Br. J. Haematol. **25**, 165 (1973)

8 – CHANDRA, P., ROSNER, F., SAWITSKY, A.: Sea-blue histiocytes in thrombocytopenic purpura. Ann. Int. Med. **79**, 901 (1973)

9 – DOSIK, H., ROSNER, F., SAWITSKY, A.: Acquired lipidosis: Gaucher-like cells and "blue-cells" in chronic granulocytic leukemia. Semin. Hematol. **9**, 309 (1972)

10 – STEINBERG, M.H., DREILING, B.J.: Chronic granulocytic leukemia. Prolonged survival, muscle infiltration and sea-blue histiocytosis. Am. J. Med. **55**, 93 (1973)

11 – WHITE, J.G., WITKOP, C.J., GERRITSEN, S.M.: The Hermansky-Pudlak syndrome. Ultrastructure of bone marrow macrophages. Am. J. Pathol. **70**, 329 (1973)

Chapter 7 – Plasmocytic Series

p. 173 – General Remarks

1 – THIERY, J.P.: Etude au microscope électronique de l'ilôt plasmocytaire. J. Microscopie **1**, 275 (1962)

2 – HALL, J.G., MORRIS, B., MORENO, G.D., BESSIS, M.C.: The ultrastructure and function of the cells in lymph following antigenic stimulation. J. Exp. Med. **125**, 91 (1967)

p. 178 – Flaming plasmocytes

1 – AVRAMEAS, S., BOUTEILLE, M.: Ultrastructural localization of antibody by antigen label with peroxidase. Exp. Cell Res. **53**, 166 (1968)

2 – MURPHY, M.J., HAY, J.B., MORRIS, B., BESSIS, M.: An ultrastructural analysis of antibody synthesis in cells from lymph and lymph nodes. Am. J. Pathol. **66**, 25 (1972)

3 – THIERY, J.P.: Ultrastructure et fonction des cellules impliquees dans la réaction immunitaire. Bull. Soc. Chim. Biol. **50**, 1077 (1968)

p. 178 – Flaming Plasmocytes

1 – PARASKEVAS, F., HEREMANS, J., WALDENSTRÖM, J.: Cytology and electrophoretic pattern in φ_{1A} (β_{2A}) myeloma. Acta Med. Scand. **170**, 575 (1961)

2 – BRITTIN, G.M., TANAKA, Y., BRECHER, G.: Intranuclear inclusions in multiple myeloma and macroglobulinemia. Blood **21**, 335 (1963)

3 – BESSIS, M., BRETON-GORIUS, J., BINET, J.L.: Etude comparée du plasmocytome et du syndrome de Waldenström. Examen au microscope électronique. Nouv. Rev. Fr. Hématol. **3**, 159 (1963)

4 – THIERY, J.P.: Micro-cinematographic contributions to the study of plasma cells. In: *Cellular Aspects of Immunity*. WOLSTENHOLME, G.E.W., O'CONNOR, C.M. (eds.). CIBA Foundation Symposium, London: Churchill, 1960, p. 59

5 – BESSIS, M., THIERY, J.P.: Ultra-centrifugation de leucocytes normaux et leucémiques. Examen au microscope électronique. Rev. Hématol. **10**, 583 (1955)

p. 180–182 – Plasmocytes with Inclusions

1 – MOTT, F.W.: Observations on the brains of men and animals infected with various form of trypanosomes. Preliminary note. Proc. R. Soc. London [Biol.] **76**, 235 (1905)

2 – RUSSELL, W.: An address on a characteristic organism of cancer. Br. Med. J. **2**, 1356 (1890)

3 – THIERY, J.P.: Etude sur le plasmocyte en contraste de phase et en microscopie électronique. III. Plasmocytes à corps de Russell et à cristaux. Rev. Hématol. **13**, 61 (1958)

4 – LAMY, N., ROYER, P., NEZELOF, C.: Presence d'inclusions cellulaires dans la moelle osseuse chez des sujets atteints de gargoylisme. Presse Méd. **67**, 1058 (1959)

5 – SNAPPER, I., SCHNEID, B.: On the influence of stilbamidine upon myeloma cells. Blood **1**, 534 (1946)

6 – ZUCKER-FRANKLIN, D., FRANKLIN, E.C.: Intracellular localization of human amyloid by fluorescence and electron microscopy. Am. J. Pathol. **59**, 23 (1970)

Chapter 8 – Mastocytic Series

p. 183–184 – Mastoblasts and Mastocytes

1 – SELYE, H.: *The Mast Cells*. Washington: Butterworths 1965

2 – BLOOM, G.D.: Structure and biochemical characteristics of mast cells. In: *The Inflammatory Process*. 2nd ed. ZWEIFACH, B.W., GRANT, L., MCCLUSKEY, R.T. (eds.). New York: Acad. Pr. 1974, Vol. I, p. 545

3 – YAM, L.T., LI, C.Y., CROSBY, W.H.: Cytochemical identification of monocytes and granulocytes. Am. J. Clin. Pathol. **55**, 283 (1971)

4 – PADAWER, J.: Phagocytosis of particulate substances by mast cells. Lab. Invest. **25**, 320 (1971)

5 – KESSLER, S., KUHN, C.: Scanning electron microscopy of mast cell degranulation. Lab. Invest. **32**, 71 (1975)

6 – RÖHLICH, P., ANDERSON, P., UVNAS, B.: Electron microscope observations on compound 48/80-induced degranulation in rat mast cells. Evidence for sequential exocytosis of storage granules. J. Cell Biol. **51**, 465 (1971)

7 – BEAVEN, M.A.: Histamine. N. Engl. J. Med. **294**, 320 (1976)

Chapter 9 – Leukemias

p. 187 – Reinterpretation of "Classification" of Leukemias

1 – MEL, H.C.: On classification of leukemia. Blood Cells **1**, 259 (1975) (Discussion, p. 263)

2 – BESSIS, M.: Pathology of the leukemic cell or Reasons why some leukemias are unclassifiable. Blood Cells **1**, 183 (1975)

3 – GALTON, D.A.G., DACIE, J.V.: Classification of the acute leukaemias. Blood Cells **1**, 17 (1975)

4 – MATHE, G., BELPOMME, D., DANTCHEV, D., POUILLART, P., NAVARES, L., HAUSS, G., SCHLUMBERGER, J.R., LAFLEUR, M.: Search for correlations between cytological types and therapeutic sensitivity of acute leukaemias. Blood Cells **1**, 37 (1975)

5 – BENNETT, J.M., REED, C.E.: Acute leukemia cytochemical profile-diagnostic and clinical implications. Blood Cells **1**, 101 (1975)

6 – CATOVSKY, D., FRISCH, B., VAN NOORDEN, S.: B, T, and "Null" cell leukemias. Electron cytochemistry and surface morphology. Blood Cells **1**, 115 (1975)

7 – MOORE, M.A.S.: Marrow culture–a new approach to classification of leukemias. Blood Cells **1**, 149 (1975)

8 – GAVOSTO, F., MASERA, P.: Different cell proliferation models in myeloblastic and lymphoblastic leukemia. Contribution of cell kinetics to the classification of acute leukemias. Blood Cells **1**, 217 (1975)

9 – BROUET, J.C., PREUD'HOMME, J.L., SELIGMANN, M.: The use of B and T membrane markers in the classification of human leukemias, with special reference to acute lymphoblastic leukemia. Blood Cells **1**, 81 (1975)

10 – Chinese Encyclopedia "Celestial Emporium of Benevolent Knowledge" cited by Jorge Luis Borges in "Other Inquisitions" 1937–1952. Transl. by R.L. SIMMS, Univ. of Texas Press, Austin and London, 1975, p. 103

p. 190 – Pseudolymphoblastic Leukemias

1 – BESSIS, M.: Cytologic diagnosis of leukemias by electron microscopy. In: Recent Results in Cancer Research. MATHE, G., POUILLART, P., SCHWARZENBERG, L. (eds.). 1973, Vol. XLIII, p. 63

2 – BESSIS, M., MAIGNE, J.: Le diagnostic des variétés de leucémies aiguës par la réaction des peroxydases au microscope électronique. Son intérêt et ses limites. Rev. Eur. Clin. Biol. **15**, 691 (1970)

3 – SMITH, W.B., ABLIN, A., GOODMAN, J.R., BRECHER, G.: Atypical megakaryocytes in preleukemic phase of acute myeloid leukemia. Blood **42**, 535 (1973)

4 – LI, C.Y., YAM, L.T., CROSBY, W.H.: Histochemical characterization of cellular and structural elements of the human spleen. J. Histochem. Cytochem. **20**, 1049 (1972)

5 – BOGGS, D.R.: Hematopoietic stem cell theory in relation to possible lymphoblastic conversion of chronic myeloid leukemia. (Editorial). Blood **44**, 449 (1974)

6 – SARIN, P.S., ANDERSON, P.N., GALLO, R.C.: Terminal deoxynucleotidyl transferase activities in human blood leukocytes and lymphoblast cell lines: high levels in lymphoblast cell lines and in blast cells of some patients with chronic myelogenous leukemia in acute phase. Blood **47**, 11 (1976)

7 – MCCAFFREY, R., HARRISON, T.A., PARKMAN, R., BALTIMORE, D.: Terminal deoxynucleotidyl transferase activity in human leukemic cells and in normal human thymocytes. N. Engl. J. Med. **292**, 775 (1975)

p. 192 – Promyelocytic Leukemia (Syndrome of Jean BERNARD)

1 – HILLSTEAD, L.K.: Acute promyelocytic leukemia. Acta Med. Scand. **159**, 189 (1957)

2 – BERNARD, J., LASNERET, J., CHOME, J., LEVY, J.P., BOIRON, M.: A cytological and histological study of acute promyelocytic leukemia. J. Clin. Pathol. **16**, 319 (1963)

3 – BERNARD, J., FLANDRIN, G.: A propos de la leucémie à promyélocytes Nouv. Rev. Fr. Hématol. **13**, 755 (1973)

4 – GRALNICK, H.R., TAN, H.K.: Acute promyelocytic leukemia. A model for understanding the role of the maligant cell in homostasis. Human Pathol. **5**, 661 (1974)

5 – SULTAN, C., HEILMANN-GOUAULT, M., TULLIEZ, M.: Relationship between blast-cell morphology and occurence of a syndrome of dis-

seminated intravascular coagulation. Br. J. Haematol. **24**, 255 (1973)
6 – LISO, V., TROCCOLI, G., GRANDE, M.: Cytochemical study of acute promyelocytic leukaemia. Blut **30**, 261 (1975)
7 – KONDO, K., YOSHITAKE, J., TAKEMURA, K.: The fine structure of Auer bodies. J. Electron Microsc. (Tokyo) **15**, 237 (1966)
8 – BRETON GORIUS, J., HOUSSAY, D.: Auer bodies in acute promyelocytic leukemia. Demonstration of their fine structure and peroxidase localization. Lab. Invest. **28**, 135 (1973)
9 – MATSUOKA, M., HATTORI, A., MIZUSHINA, T., JINBO, C.: The ultrastructure of the cryofibrinogen in acute promyelocytic leukemia displaying the defibrination syndrome and fibrilar inclusions in promyelocytes. Acta Med. Biol. **17**, 49 (1969)
10 – QUATTRIN, N.: Leucémies aigües à basophiles. Nouv. Rev. Fr. Hématol. **13**, 745 (1973)

p. 194 – Chronic Lymphocytic Leukemias-Special Types

1 – BERNARD, J., BESSIS, M., SOULIER, J.P., THIERY, J.P.: Etude au microscope électronique d'une leucémie lymphoide avec cristaux intracytoplasmiques. Rev. Hématol. **14**, 227 (1959)
2 – BESSIS, M.: Ultrastructure of lymphoid and plasma cells in relation to globulin and antibody formation. Lab. Invest. **10**, 1040 (1961)
3 – HUREZ, D., FLANDRIN, G., PREUD'HOMME, L., SELIGMANN, M.: Unreleased intracellular monoclonal macroglobulin in chronic lymphocytic leukemia. Clinical exp. Immunol. **10**, 223 (1972)
4 – CLARK, C., RYDELL, R.E., KAPLAN, M.E.: Frequent association of IgM λ with crystalline inclusions in chronic lymphatic leukemic lymphocytes. N. Engl. J. Med. **289**, 113 (1973)
5 – BESSIS, M.: *Living Blood Cells and Their Ultrastructure*. Berlin-Heidelberg-New York: Springer-Verlag, 1973, p. 603
6 – BESSIS, M.: Etudes au microscope électronique des leucocytes normaux et leucémiques. Acta Union Int. Cancer **7**, 646 (1951)
7 – BESSIS, M., THIERY, J.P.: Etudes au microscope électronique sur les leucémies humaines. II. Les leucémies lymphocytaires. Comparison avec la leucémie de la souris de souche AK. Nouv. Rev. Fr. Hématol. **2**, 387 (1962)
8 – TERRASSE, J., CAVAROC, M., BESSIS, M.: Un cas de leucémie avec inclusions sphériques. Rev. Hématol. **13**, 331 (1958)
9 – VILLANT, G., FRANZINI, C.: Rare granulazioni atipiche di cellule linfoidi leucemiche. Haematol. Arch. **42**, 1541 (1957)

10 – LAGIOS, M.D., FRIEDLANDER, L.M., WALLERSTEIN, R.O., BOHANNON, R.A.: Atypical azurophilic crystals in chronic lymphocytic leukemia. A case report and comparison with other crystalline inclusions. Am. J. Clin. Pathol. **62**, 342 (1974)

p. 196 – Macroglobulinemias (WALDENSTRÖM's)

1 – WALDENSTRÖM, J.: Incipient myelematosis or essential hyperglobulinemy with fibrinogenopenia. A new syndrome. Acta Med. Scand. **117**, 216 (1944)
2 – WALDENSTRÖM, J.: Studies on conditions associated with disturbed gamma-globulin formation (gammapathies). Harvey Lectures **56**, 211 (1960/1961)
3 – WALDENSTRÖM, J.G.: *Monoclonal and Polyclonal Hypergammaglobulinemia*. Cambridge: Univ. Pr. 1968
4 – BESSIS, M., BRETON-GORIUS, J., BINET, J.L.: Etude comparée du plasmocytome et du syndrome de Waldenström. Examen au microscope électronique. Nouv. Rev. Fr. Hématol. **3**, 159 (1963)
5 – BRITTIN, G.M., TANAKA, Y., BRECHER, G.: Intranuclear inclusions in mutiple myeloma and macroglobulinemia. Blood **21**, 335 (1963)
6 – MARMONT, A., CHIAPPINO, G., DAMASIO, E., D'AMORE, E.: Démonstration par microscopie en immunofluorescence des éléments cellulaire macroglobulinopoiétiques dans la maladie de Waldenström. Schweiz. Med. Wochenschr. **93**, 1445 (1963)
7 – PREUD'HOMME, J.L., SELIGMANN, M.: Immunoglobulins on the surface of lymphoid cells in Waldenström's macroglobulinemia. J. Clin. Invest. **51**, 701 (1972)
8 – CLAUVEL, J.P., DANON, F., SELIGMANN, M.: Immunoglobulines monoclonales décelées en l'absence de myélome ou de macroglobulinémie de Waldenström. Evolution dans 54 observations. Nouv. Rev. Fr. Hématol. **11**, 677 (1971)

p. 198 – SEZARY's Syndrome

1 – SEZARY, A., BOUVRAIN, Y.: Erythrodermic avec présence de cellules monstrueuses dans le derme et le sang circulant. Bull. Soc. Fr. Dermatol. Syphiligr. **45**, 254 (1938)
2 – SEZARY, A.: Une nouvelle réticulose cutanée: la réticulose maligne leucémique à histiomonocytes monstrueux et à forme d'érythrodermie oedémateuse et pigmentée. Ann. Dermatol. Syphiligr. (Paris) **9**, 52 (1949)
3 – WINKELMANN, R.K.: History of the Sezary syndrome. Mayo Clin. Proc. **49**, 515 (1974)
4 – CLENDENNING, W.E., BRECHER, G., VAN SCOTT, E.J.: Mycosis fungoides. Relationship to malignant cutaneous reticulosis and the Sezary syndrome. Arch. Dermatol. **89**, 785 (1964)

5 – ROSAS-URIBE, A., VARIAKOJIS, D., MOLNAR, Z., RAPPAPORT, H.: Mycosis fungoides: an ultrastructural study. Cancer **34**, 634 (1974)

6 – CROSSEN, P.E., MELLOR, J.E.L., FINLEY, A.G., RAVICH, R.B.M., VINCENT, P.C., GUNZ, F.W.: The Sezary syndrome. Cytogenetic studies and identification of the Sezary cell as abnormal lymphocyte. Am. J. Med. **50**, 24

7 – FLANDRIN, G., BROUET, J.C.: The Sezary cell: cytologic, cytochemical and immunologic studies. Mayo Clin. Proc. **49**, 575 (1974)

8 – LUTZNER, M.A., JORDAN, H.W.: The ultrastructure of an abnormal cell in Sezary's syndrome. Blood **31**, 719 (1968)

9 – LUTZNER, M.A., EMERIT, I., LUREPAIRE, R., FLANDRIN, G., GRUPPER, CH., PRUNIERAS, M.: Cytogenetic, cytophotometric and ultrastructural study of large cerebriform cells of the Sezary syndrome and description of a small cell variant. J. Natl. Cancer Inst. **50**, 1145 (1973)

10 – PRUNIERAS, M.: DNA content and cytogenetics of the Sezary cell. Mayo Clin. Proc. **49**, 548 (1974)

11 – LITOVITZ, T.L., LUTZNER, M.A.: Quantitative measurements of blood lymphocytes from patients with chronic lymphocytic leukemia and the Sezary syndrome. J. Natl. Cancer Inst. **53**, 75 (1974)

12 – ZUCKER-FRANKLIN, D., MELTON, J.W., QUAGLIATA, F.: Ultrastructural, immunologic and functional studies on Sézary cells: a neoplastic variant of thymus-derived (T) lymphocytes. Proc. Natl. Acad. Sci. U.S.A. **71**, 1877

13 – BRAYLAN, R., VARIAKOJIS, D., YACHIN, S.: The Sezary syndrome lymphoid cell: abnormal surface properties and mitogen responsiveness. Br. J. Haematol. **31**, 553 (1975)

14 – BURG, G., BRAUN-FALCO, O.: T- und B-Lymphozyten in Hautveränderungen kutaner Lymphome. Dtsch. Med. Wochenschr. **100**, 2562 (1975)

General references:

Symposium on the Sezary cell. Mayo Clinic Proc. **49**, (8), 1974

PRUNIERAS, M.: The Sézary's syndrome. Trans. St. Johns Hosp. Dermatol. Soc. **61**, 1 (1975)

p. 200 – Tricholeukocytic Leukemia (Hairy Cell Leukemia)

1 – BOURONCLE, B.A., WISEMAN, B.K., DOAN, C.A.: Leukemic reticuloendotheliosis. Blood **13**, 609 (1958)

2 – SCHREK, R., DONNELLY, J.M.: "Hairy" cells in blood in lympho reticular neoplastic disease and "flagellated" cells of normal lymph nodes. Blood **27**, 199 (1966)

3 – LEE, S.L., ROSNER, F., ROSENTHAL, N., ROSENTHAL, R.L.: Reticulum cell leukemia: clinical and hematologic entity. N. Y. State J. Med. **69**, 422 (1969)

4 – PLENDERLEITH, I.H.: "Hairy cell" leukemia. Can. Med. Assoc. J. **102**, 1056 (1970)

5 – FLANDRIN, G., DANIEL, M.T., FOURCADE, M., CHELLOUL, N.: Leucémie à "tricholeucocyte" ("Hairy cell" leukemia). Etude clinique et cytologique de 55 observations. Nouv. Rev. Fr. Hématol. **13**, 609 (1973)

6 – TRUBOWITZ, S., MASEK, B., FRASCA, J.M.: Leukemic reticuloendotheliosis. Blood **38**, 288 (1971)

7 – GOLOMB, H.M., BRAYLAN, R., POLLIACK, A.: "Hairy cell" leukemia (leukemic reticuloendotheliosis): a scanning electron microscopic study of eight cases. Br. J. Haematol. **29**, 455 (1975)

8 – TCHERNIA, N.J. GRANGE, G., GOMBAUD-SAINTONGE, G.: Locomotion necrotaxis and phagocytosis in normal and leukemic myeloblasts. Blood Cells **2**, 431 (1977)

9 – MITUS, W.J., MEDNICOFF, I.B., WITELLS, B., DAMESHEK, W.: Neoplastic lymphoid reticulum cells in the peripheral blood: a histochemical study. Blood **17**, 206 (1961)

10 – YAM, L.T., CASTOLDI, G.L., GARVEY, M.B., FINKEL, H.E.: Functional cytogenetic and cytochemical study of the leukemic reticulum cells. Blood **32**, 90 (1968)

11 – YAM, L.T., LI, C.Y., LAM, K.W.: Tartrate-resistant acid phosphatase iso-enzyme in the reticulum cells of leukemic reticulo-endotheliosis N. Engl. J. Med. **284**, 357 (1974)

12 – ANDAY, G.J., GOODMAN, J.R., TISHKOFF, G.H.: An unusual cytoplasmic ribosomal structure in pathologic lymphocytes. Blood **41**, 439 (1973)

13 – KATAYAMA, I., LI, C.Y., YAM, L.T.: Ultrastructural characteristics of the "hairy cells" of leukemic reticuloendotheliosis. Am. J. Pathol. **67**, 361 (1972)

14 – DANIEL, M.T., FLANDRIN, G.: Fine structure of abnormal cells in Hairy cell (tricholeucocytic) leukemia with special reference to their in vitro phagocytic capacity. Lab. Invest. **30**, 1 (1974)

15 – HAAK, H.L., MAN, J.C.H. DE, HIJMANS, W., KNAPP, W., SPECK, B.: Further evidence for the lymphocytic nature of leukaemic reticuloendotheliosis ("Hairy cell" leukaemia) Br. J. Haematol. **27**, 31 (1974)

16 – SCHNITZER, B., HAMMACK, W.J.: B-lymphocyte nature of "Hairy cells" in "hairy-cell" leukaemia. Lancet **1974/II**, 649

17 – GORDAN, G.S., ROOF, B.S.: Humors from tumors: diagnostic potential of peptides. Ann. Int. Med. **76**, 501 (1972)

p. 202 – Unclassifiable Leukemias

1 – BESSIS, M.: Pathology of the leukemic cell or Reasons why some leukemias are unclassifiable. Blood Cells **1**, 183 (1975)

2 – FRIEND, C., PATULEIA, M.C., HARVEN, E. DE: Erythrocytic maturation in vitro of murine (Friend) virus-induced leukemic cells. Natl. Cancer Inst. Monogr. **22**, 505 (1966)

Monograph:

Unclassifiable leukemias. Blood Cells **1**, n° 1, (1975)

p. 203 – Hemopoietic Dysplasias (Preleukemic States)

1 – General discussion: I: *Is Preleukemic states an adequate designation?* In: Hemopoietic dysplasias. (Preleukemic states). Blood Cells **2**, 347 (1976)

2 – DREYFUS, B.: Preleukemic states. I. Definition and Classification. II. Refractory anemia with an excess of myeloblasts in the bone marrow (smoldering acute leukemia). Blood Cells **2**, 33 (1976)

3 – LINMAN, J.W., BAGBY, G.C.: The preleukemic syndrome: clinical and laboratory features, natural course, and management. Blood Cells **2**, 11 (1976)

4 – BERNARD, J.: Preleukemic states. Blood Cells **2**, 5 (1976)

5 – PIERRE, R.V.: Cytogenetic studies in preleukemia: studies before and after transition to acute leukemia in 17 subjects. Blood Cells **1**, 163 (1975)

6 – KILLMANN, S.A.: Peleukemia. Does-it exist? Blood Cells **2**, 81 (1976)

7 – DREYFUS, B., ROCHANT, M., SULTAN, C.: Anémies réfractaires: enzymopathies acquises des cellules-souches hématopoiétiques. Nouv. Rev. Fr. Hématol. **9**, 65 (1969)

8 – BOIVIN, P., GALAND, C., DREYFUS, B.: Activités enzymatiques érythrocytaires au cours des anémies réfractaires. Nouv. Rev. Fr. Hématol. **9**, 105 (1969)

9 – CAEN, J., SULTAN, Y., DREYFUS, B.: Etude des fonctions plaquettaires dans 7 cas d'anémie réfractaires. Nouv. Rev. Fr. Hématol. **9**, 123 (1969)

10 – SALMON, CH., ROCHANT, H., MANNONI, P., CARTRON, J.P., JACQUET, A., LIBERGE, G., DREYFUS, B.: Etude des modifications des antigenes de groupes sanguins dans 11 cas d'"anémies réfractaires". Nouv. Rev. Fr. Hématol. **9**, 113 (1969)

11 – BESSIS, M., DREYFUS, B., BRETON-GORIUS, J., SULTAN, C.: Etude au microscope électronique de onze cas d'anémies réfractaires avec enzymopathies multiples. Nouv. Rev. Fr. Hématol. **9**, 87 (1969)

12 – BRETON-GORIUS, J., COQUIN, J., VILDE, Y., DREYFUS, B.: Cytochemical and ultrastructural studies of aberrant granules in the neutrophils of two patients with myeloperoxidase deficiency during a preleukemic state. Relationship to abnormal bactericidal activity. Blood Cells **2**, 187 (1976)

13 – MALDONADO, J.E., MAIGNE, J., LECOQ, D.: Comparative electronmicroscopic study of the erythrocytic line in refractory anemia (preleukemia) and myelomonocytic leukemia. Blood Cells **2**, 157 (1976)

14 – KINET-DENOEL, C., BRETON-GORIUS, J.: Teneurs en ADN, ultrastructure et activité peroxidasique des mégacaryocytes médullaires dans un cas d'anémie réfractaire. Nouv. Rev. Fr. Hématol. **13**, 661 (1973)

15 – MALDONADO, J.E.: Giant platelet granules in refractory anemia (preleukemia) and myelomonocytic leukemia: a cell marker? Blood Cells **1**, 129 (1975)

Monograph:

"Hemopoietic dysplasias (Preleukemic states)", Blood Cells, **2**, (1/2) (1976)

p. 206 – Pathology of Cell Organelles

1 – BESSIS, M.: Cytologic diagnosis of leukemias by electron microscopy. In: Recent Results in Cancer Research. MATHE, G., POUILLART, P., SCHWARZENBERG, L. (eds.). Heidelberg: Springer-Verlag 1973, Vol. XLIII, p. 63

2 – BESSIS, M., BRETON-GORIUS, J.: Pathologie et asynchronisme de développement des organelles cellulaires au cours des leucémies aigües granulocytaires. Nouv. Rev. Fr. Hématol. **9**, 245 (1969)

3 – BESSIS, M.: Studies on cell agony and death. An attempt at classification. In: CIBA Foundation Symposium, "Cellular injury". London: Churchill 1964, p. 287

p. 208 – Anarchy and Arrest of Maturation

1 – BESSIS, M.: *Living Blood Cells and Their Ultrastructure.* Berlin-Heidelberg-New York: Springer-Verlag 1973, p. 558

2 – BESSIS, M.: Contribution à l'étude de la cytologie sanguine. I. Le synchronisme d'évolution nucléo-cytoplasmique. II. Interprétation des conceptions cytogénétiques. III. Etudes cytologique des leucémies et des réticulosarcomes. Rev. Hématol. **1**, 45 (1946)

3 – BESSIS, M.: L'autonomie des cellules leucémiques et des cellules cancéreuses en général. Rev. Hématol. **12**, 142 (1957)

4 – FURTH, J.: Conditioned and autonomous neoplasms: a review. Cancer Res. **13**, 477 (1953)

257

p. 210 – Mosaicism

1 – NAEGELI, O.: *Differentialdiagnose in der inneren Medizin.* Leipzig: Thieme 1937, p. 65
2 – BESSIS, M.: Pathology of the leukemic cell or Reasons why some leukemias are unclassifiable. Blood Cells **1**, 183 (1975)
3 – BACCARINI, M., KILLMANN, S.A.: Cytokinetic studies in chronic myeloid leukaemia: evidence for early presence of abnormal myeloblasts. Scand. J. Haematol. **9**, 283 (1972)
4 – BAINTON, D.F.: Abnormal neutrophils in acute myelogenous leukemia: identification of subpopulations based on analysis of azurophil and specific granules. Blood Cells **1**, 101 (1975)

p. 212 – Progression

1 – BESSIS, M.: Contribution à l'étude de la cytologie sanguine. Le synchronisme d'évolution nucléocytoplasmique. Interprétation des conceptions cytogénétiques. Etude cytologique des leucémies et des réticulosarcomes. Rev. Hématol. **1**, 45 (1946)
2 – FURTH, J.: Conditioned and autonomous neoplasms: a review. Cancer Res. **13**, 477 (1953)
3 – BESSIS, M.: L'autonomie des cellules leucémiques et des cellules cancéreuses en général. Rev. Hématol. **12**, 142 (1957)
4 – FOULDS, L.: Progression in carcinogenesis. Acta Union Int. Cancer **17**, 148 (1961)
5 – LEJEUNE, J., BERGER, R., HAINES, M., LAFOURCADE, J., VIALATTE, J., SATCÉ, P., TURPIN, R.: Constitution d'un clone à 54 chromosomes au cours d'une leucoblastose congénitale chez une enfant mongolienne. C.R. Acad. Sci. (Paris) **256**, 1195 (1963)
6 – GROUCHY, J. DE, NAVA, C. DE, CANTU, J.M., BILSKI-PASQUIER, G., BOUSSER, J.: Models for clonal evolutions. A study of chronic myelogenous leukemia. Am. J. Hum. Genet. **18**, 485 (1966)
7 – PEDERSEN, B.: Clonal evolution and progression in chronic myeloid leukemia. Blood Cells **1**, 227 (1975)
8 – PEDRINIS, E., ZIMMERMANN, A., BERTSCHMANN, M., HESS, M.W., COTTIER, H.: Problems relating to immunoselection of leukemias. Blood Cells **2**, 329 (1976)

p. 214 – Provisional Conclusions on the Pathophysiology of Leukemic Cells

1 – FURTH, J.: Conditioned and autonomous neoplasms: a review. Cancer Res. **13**, 477 (1953)
2 – BESSIS, M.: L'autonomie des cellules leucémiques et des cellules cancéreuses en général. Rev. Hématol. **12**, 142 (1957)
3 – BESSIS, M.: Pathology of the leukemic cell or Reasons why some leukemias are unclassifiable. Blood Cells **1**, 183 (1975)
4 – PEDRINIS, E., ZIMMERMANN, A., BERTSCHMANN, M., HESS, M.W., COTTIER, H.: Problems relating to immunoselection of leukemias. Blood Cells **2**, 329 (1976)
5 – LIND, J.: *A Treatise of the Scurvy.* Edinburgh: A. Kincaid 1753
6 – DARROW, R.: *Icterus gravis* (erythroblastosis) *neonatorum*: an examination of etiologic considerations. Arch. Pathol. **25**, 378 (1938)

Chapter 10 – Technique

p. 217 – General Remarks

1 – BAKER, H.: *Le Microscope à la Portée de Tout le Monde* (Translation from English, 1743). Paris: A. Jombert 1754, p. 131
2 – WRIGHT, J.H.: The origin and nature of the blood plates. Boston Med. Surgical J. **154**, 643 (1906)
3 – FERRATA, A.: Morphologia del Sangue Normale e Pathologico. Milano: Soc. Ed. Libraria 1912
4 – JOLLY, J.: *Traité Technique d'Hématologie.* Paris: Maloine 1923
5 – MAXIMOW, A.: Bindegewebe und blutbildende Gewebe. In: Hand. Mikrosk. Anat. Mensch. W. von Mollendorff (ed.) Berlin: Springer-Verlag 1927, Vol. II, part 1
6 – DOWNEY, H. (ed.): *Handbook of Hematology.* New York: Hoeber 1938

p. 218 – A Page of History

1 – ANDRAL, G.: *Essai d'Hematologie Pathologique.* Paris: Fortin, Masson 1843
2 – HEWSON, W.: On the figure and composition of the red particles commonly called the red globules. Philos. Trans. **63**, 303 (1773)
HEWSON, W.: In: *The Works of Williams HEWSON,* F.R.S. GULLIVER, G. (Ed.). London: Sydenham Soc. 1846 pl. V, p. 312
3 – GULLIVER, G.: Observations on the size and shapes of the red corpuscles of the blood of vertebrates, with drawings of them to a uniform scale, and extended and revised tables of measurements. Proc. Sci. Meet. Zool. Soc. London 1875, part 2 pp. 474–495, pl.LV
4 – NORRIS, R.: *The Physiology and Pathology of the Blood.* London: Smith Elder 1882

5 – EHRLICH, P.: Beiträge zur Kenntnis der Anilinfärbungen und ihrer Verwendung der mikroskopischen Technik. Arch. Mikr. Anat. **13**, 263 (1877)

6 – EHRLICH, P.: Methodologische Beiträge zur Physiologie und Pathologie der verschiedenen Formen der Leukocyten. Z. klin. Med. **1**, 553 (1879)

7 – ROMANOVSKY, D.: Zur Frage der Parasitologie und Therapie der Malaria. St. Petersb. Med. Wochenschr. **16**, 297, 307 (1891)

8 – LEISHMANN, W.B.: A simple and rapid method of producing Romanowsky staining in malarial and other blood films. Br. Med. J. **2**, 757 (1901)

9 – MAY, R., GRUNWALD, L.: Über Blutfärbungen. Zentralbl. Inn. Med. **23**, 265 (1902)

10 – GIEMSA, G.: Färbemethoden für Malariaparasiten. Zentralbl. Bakteriol. [Orig. A] **31**, 429, **32**, 307 (1902)

11 – GIEMSA, G.: Eine Vereinfachung und Vervollkommnung meiner Methylenazur-Methylenblau-Eosin-Färbemethode zur Erzielung der Romanowsky-Nochtschen Chromatinfärbung. Zentralbl. Bakteriol. [Orig. A] **37**, 308 (1904)

12 – PAPPENHEIM, A.: Zur Blutzellfärbung im klinischen Bluttrockenpräparat und zur histologischen Schnittpräparatfärbung der hämatopoetischen Gewebe nach meinen Methoden. Folia Haematol. (Leipzig) **13**, 337 (1912)

13 – WRIGHT, J.H.: A rapid method for the differential staining of blood films and malarial parasites. J. Med. Res. **7**, 138 (1902)

14 – WHITMAN, R.C.: The staining of leucocytes. Boston Med. Surg. J. **145**, 449 (1901)

Monograph:

CONN, H.J.: The History of Staining. Geneva (USA), Biological stain commission 1933

p. 219 – Hematology Without the Microscope

1 – BESSIS, M., MEL, H.: Hematology without the microscope. Blood Cells **1**, 269 (1975)

2 – BRITTIN, G.M., BRECHER, G.: Instrumentation and automation in clinical hematology. In: Progress in Hematology. BROWN, E., MOORE, C.V. (eds.). New York: Grune and Stratton 1971, Vol. VII, p. 299

3 – THORELL, B.: Automated cell population analysis. Blood Cells **1**, 71 (1975)

4 – HAANEN, C.A.M., HILLEN, H.F.P., WESSELS, J.M.C. (eds.): First International Symposium on Pulse-Cytophotometry. Ghent (Belgium), Eur. Pr. Medikon 1975

5 – MELAMED, M.R., KAMENTSKY, I.A.: Automated cytology. Int. Rev. Exp. Pathol. **14**, 205 (1975)

6 – Automated cytology. 4th Engineering Foundation Conference. J. Histochem. Cytochem. **24**, (1), (1976)

p. 220 – Preparation of Smears

1 – CARTWRIGHT, G.E.: *Diagnostic Laboratory Hematology* (4th ed). New York: Grune and Stratton 1968

2 – DACIE, J.V., LEWIS, S.M.: *Practical Haematology*. 5th ed. Edinburgh: Churchill, Livingstone 1975

3 – INGRAM, M., MINTER, F.M.: Semiautomatic preparation of coverglass blood smears using a centrifugal device. Am. J. Clin. Pathol. **51**, 214 (1969)

4 – BACUS, J.W.: An automated leukocyte classifier. Evaluation of performance in the clinical laboratory. Am. J. Clin. Pathol. **62**, 302 (1974)

5 – BACUS, J.W.: Erythrocyte morphology and centrifugal "spinner" blood film preparations. J. Histochem. Cytochem. **22**, 506 (1974)

6 – EGAN, J.J., BENZEL, J.E., HART, D.J., CHRISTOPHER, E.A.: Evaluation of an automated differential leukocyte counting system. III-Detection of abnormal cells. Am. J. Clin. Pathol. **62**, 537 (1974)

p. 221 – May-Grünwald-Giemsa Stain

Monographs:

1 – CARTWRIGHT, G.E.: *Diagnostic Laboratory Hematology*. (4th ed.) New York: Grune and Stratton 1968

2 – DACIE, J.V., LEWIS, S.M.: *Practical Haematology*. (5th ed.) Edinburg: Churchill, Livingstone 1975

p. 222 – Cytochemical Stains

1 – SUNDBERG, R.D., BROMAN, H.: The application of the Prussian blue stain to previously stained films of blood and bone marrow. Blood **10**, 160 (1955)

2 – LO BUE, J., DORNFEST, B.S., GORDON, A.S., HURST, J., QUASTLER, H.: Marrow distribution in rat femurs determined by cell enumeration and ^{59}Fe labeling. Proc. Soc. Exp. Biol. Med. **112**, 1058 (1963)

3 – STRAUS, W.: Factors affecting the cytochemical reaction of peroxidase with benzidine and the stability of the blue reaction product. J. Histochem. Cytochem. **14**, 462 (1964)

4 – KAPLOW, L.S.: Simplified myeloperoxidase stain using benzidine dihydrochloride. Blood **26**, 215 (1965)

5 – BAEHNER, R.L., NATHAN, D.G.: Leukocyte oxidase: defective activity in chronic granulomatous disease. Science **155**, 835 (1967)

6 – KAPLOW, L.S.: Leukocyte alkaline phosphatase cytochemistry: applications and methods. Ann. N.Y. Acad. Sci. **155**, 911 (1968)

7 – LI, C.Y., YAM, L.T., CROSBY, W.H.: Histochemical characterization of cellular and structural elements of the human spleen. J. Histochem. Cytochem. **20**, 1049 (1972)

8 – BRIGGS, R.S., PERILLIE, P.E., FINCH, S.C.: Lysozyme in bone marrow and peripheral blood cells. J. Histochem. Cytochem. **14**, 167 (1966)

9 – LISON, L.: *Histochimie et Cytochimie Animales. Principes et Méthodes.* (3me ed.), Paris: Gauthier-Villars 1960

10 – WHITE, W.C.: The cytoplasmic basophilia of bone marrow cells. J. Pathol. Bacteriol. **59**, 223 (1947)

11 – SHEEHAN, H.L., STOREY, G.W.: An improved method of staining leucocyte granules with Sudan black B. J. Pathol. Bacteriol. **59**, 336 (1947)

12 – HAYHOE, F.G., QUAGLINO, D., FLEMANS, R.S.: Consecutive use of Romanowsky and Periodic-Acid-Schiff techniques in the study of blood and bone-marrow cells. Br. J. Haematol. **6**, 23 (1960)

General reference

STUART, J., GORDON, P.A., LEE, T.R.: Enzyme cytochemistry of blood and marrow cells. Histochem. J. **7**, 471 (1975)

p. 223 – Examination of Smears

1 – BARNETT, C.W.: The unavoidable error in the differential count of the leukocytes of the blood. J. Clin. Invest. **12**, 77 (1933)

2 – MACGREGOR, R.G.S., RICHARDS, W., LOH, G.L.: The differential leucocyte count. J. Pathol. Bacteriol. **51**, 337 (1940)

3 – VAN DEN BERGHE, L., BLITSTEIN, I.: Critique des méthodes d'examen de la moelle osseuse. Rev. Pathol. Med. Exp. **23**, 289 (1954)

4 – STAVEM, P.E.R.: The distribution of erythrocytes and reticulocytes in blood smears. Acta Med. scand. **175**, Suppl. 409 (1964)

5 – BACUS, J.W.: The observer error in peripheral blood cell classification. Am. J. Clin. Pathol. **59**, 223 (1973)

p. 224 – Artifacts

1 – BETTIGOLDE, R.E.: Red-cell-staining artifacts and how to avoid them. N. Engl. J. Med. **271**, 1156 (1964)

p. 226 – Examination of Smears with Phase and Interference Microscopes

1 – BESSIS, M.: *Living Blood Cells and Their Ultrastructure.* Berlin: Springer-Verlag 1973, p. 686–699

2 – BESSIS, M., LOCQUIN, M.: Microscopie en contraste de phase appliquée à la cytologie. Examen comparé des techniques et des dispositifs. In: Le Contraste de Phase et le Contraste par Interférences. Paris: Rev. d'Optique 1952, p. 237

3 – BESSIS, M., THIERY, J.P.: Les cellules du sang vues au microscope à interférences (système Nomarski). Rev. Hématol. **12**, 518 (1957)

4 – PADAWER, J.: The Nomarski interference-contrast microscope. An experimental basis for image interpretation. J. R. Micr. Soc. **88**, 305 (1968)

p. 228 – Examination in the Soret Band and Shadowing of Smears

1 – SORET, J.L.: Recherches sur l'absorption des rayons ultra-violets par diverses substances. Arch. Sci. Phys. Naturelles **61**, 322 (1878)

2 – NOMARSKI, G., BESSIS, M.: Examen et microphotographie de cellules contenant de l'hémoglobine à l'aide de filtres interférentiels transmettant la bande de Soret. Rev. Hématol. **14**, 399 (1959)

3 – WILLIAMS, R.C., WYCKOFF, R.W.G.: The thickness of electron microscopic objects. J. Appl. Phys. **15**, 712 (1944)

4 – BESSIS, M.: La méthode de l'ombrage appliquée aux frottis de sang. Mise en évidence du relief cellulaire de certains organites intra-cellulaires et des mitochondries. Coloration des hématexodies. Rev. Hématol. **4**, 95 (1949)

p. 230 – Examination with a Transmission Electron Microscope

1 – KOEHLER, J.K.: *Advances Techniques in Biological Electron Microscopy.* Berlin-Heidelberg: Springer-Verlag 1973

2 – HAYAT, M.A. (ed.): *Principles and Techniques of Electron Microscopy: Biological Applications.* New York-London: Van Nostrand Reinhold 1974, Vol. III

p. 236 – Examination with the Scanning Electron Microscope

1 – JOHARI, O., CORVIN, I. (ed.): *Scanning electron microscopy.* Proc. 6th Annu. Scann. Electr. Micr. Symposium, IIT Res. Inst., Chicago, 1973

2 – JOHARI, O., CORVIN, I. (ed.): *Scanning Electron Microscopy,* IIT Research Institute, Chicago (Proc. Annu.), 1974ff

3 – BESSIS, M., BOISFLEURY, A. DE: Etude sur les poikilocytes au microscope à balayage, en particulier dans la thalassémie. Nouv. Rev. Fr. Hématol. **10**, 515 (1970)

4 – BESSIS, M., WEED, R.I.: Preparation of red blood cells (RBC) for SEM. Survey of various artifacts. Proc. Vth Annu. SEM Symp., Chicago, IIT Res. Inst., 1972, p. 289

Notes on the Illustrations

Color Photographs

A good photomicrograph, well reproduced, is almost identical with the image seen in the microscope. Hence, it can give the reader an exact view of a cell and its environment. The reader may observe details that are not mentioned in the legend or that may have escaped the notice of the author altogether.

Bad photomicrographs can be due to inappropriate illumination (particularly a faulty color temperature, an inappropriate choice of film or a poor quality and staining of the smear itself. The type of glass, the pH of the stain or of the washwater, the time elapsed between preparation and staining of the smear, and any other variable will affect the appearance of the smear (see pp. 217, 220).

Good photomicrographs may be badly reproduced. A shade of color, usually the red or blue, may appear to dominate the picture. In that case, the hematologist must reinterpret the reproduction to avoid errors of diagnosis.

All color photographs are reproduced at a uniform magnification of 1250× (except those on pp. 133 and 135 where the magnification is indicated in the legends). The diameter of a normal red cell happens to be 10 mm at that magnification.

Drawings in Color

Drawings indicate what one *ought to see* by integrating the impression received from observations on a large number of cells, often omitting some features and emphasizing others. Paradoxically, the drawing may be closer to the truth than a photograph.

The great hematologists of the turn of the century lavished particular care on the design of their illustrations. The colored drawings of Ferrata, Jolly, Naegeli, Wright and many others are astounding by the clarity with which each detail and delicate nuance are reproduced. The biographers of the great hematologists of yesterday never fail to mention the name of the artists who collaborated with them during their entire life. Those familiar with the old texts will recognize at a glance the drawings of Mrs. Sansoni or of Elizabeth Krause.

The magnification of the colored drawings differs depending on the details on which attention has been focused. The adjacent color photographs of known (and constant) magnification (see above) allow one to gauge the actual size of the cells.

Black and White Photographs

Apart from a few electron micrographs (pp. 230–237), all the black and white photographs are of red cells, and are taken in the Soret band for optimal definition (see p. 228).
Their magnification is uniformly 1250×.

Schematic Drawings of the Cell Ultrastructure

Any diagram is interpretative. The code used in the diagrams to identify the different organelles is given on pp. 4–11.

Notwithstanding the enormous progress made in recent years, one must keep in mind that we are still only in the middle ages of cytology. The diagrams of the electron-microscopic pictures must be likened to the drawings of the early anatomists. They depicted the aorta, the stomach, liver, heart and lungs without understanding their functions or interrelationships.

I have attempted to present diagrams which are not too abstract and which recall the appearance of the individual organelles in order to avoid gross errors and to promote scientific reflection.

Diagrams of Cell Movements

Some hematologists have never seen living cells, not even in movies which show (with time lapse acceleration) the movements and behavior of cells. Without such experience it may be difficult to fully comprehend the diagrams and brief descriptions dealing with that aspect of cell physiology: time is difficult to represent in diagrams.

Typography

The layout of paragraphs and legends is, in a fashion, part of the pictorial presentation: The amount of text and its relationship to each figure, and the typography have been designed to facilitate immediate comprehension and lasting retention.

List of Abbreviations

DNA	Desoxyribonucleic acid
E	Mature erythrocyte
E_1	Proerythroblast
E_2	Basophilic erythroblast I
E_3	Basophilic erythroblast II
E_4	Polychromatophilic erythroblast I
E_5	Polychromatophilic erythroblast II
Hb	Hemoglobin
M_1	Myeloblast
M_2	Promyelocyte
M_3	Myelocyte I
M_4	Myelocyte II
M_5	Metamyelocyte
M_6	Mature nonsegmented granulocyte (band cell)
M_7	Mature segmented granulocyte
MK_1	Basophilic megakaryocyte
MK_2	Granular megakaryocyte
MK_3	Platelet forming megakaryocyte
PMN	Polymorphonuclear neutrophil ($=M_7$)
RER	Rough endoplasmic reticulum
RNA	Ribonucleic acid
Rtc	Reticulocyte
SER	Smooth endoplasmic reticulum

Units of Measurements

$\mu m = micron = micrometer = 10^{-6}$ meter
$nm = nanometer = 10$ Angstroms $= 10^{-9}$ meter

Stains

SMEARS, when not further specified are smears stained with May-Grunwald-Giemsa or today's Wright's stain (see p. 221). For the history of Wright's stain, see p. 218.

Subject Index

Blood Cells

North America
1977: Volume 3 (3 issues):
US $ 46.60 including postage
and handling.
Subscriptions are entered
with prepayment only.
Please write to:
Springer-Verlag New York Inc.,
175 Fifth Avenue,
New York, NY 10010, USA

All countries
(except North America)
1977: Volume 3 (3 issues):
DM 108,– plus postage and
handling.
Please write to:
Springer-Verlag,
Wissenschaftliche Infor-
mation, Zeitschriften,
Postfach 105 280,
6900 Heidelberg, W.-Germany

Editor-in-Chief: Marcel Bessis, Institut de Pathologie Cellulaire, Hô-
pital de Bicêtre, 94270 Kremlin-Bicêtre, France
Associate Editor: Stephen B. Shohet, School of Medicine, Cancer
Research Institute, San Francisco, Calif. 94145, USA
Senior Consulting Editors: Jean Bernard, Paris, France; George
Brecher, San Francisco, Calif., USA
Editorial Board: D.F. Bainton, San Francisco, Calif., USA; E.P.
Cronkite, New York, N.Y., USA; T.M. Fliedner, Ulm, FRG; W.N.
Jensen, Washington, D.C., USA; S.-A. Killmann, Copenhagen, Den-
mark; L.G. Lajtha, Manchester, England; H.C. Mel, Berkeley, Calif.,
USA; B. Morris, Canberra, Australia; A.H.T. Robb-Smith, Oxford,
England; B. Thorell, Stockholm, Sweden; R.I. Weed, Rochester, N.Y.,
USA; J.G. White, Minneapolis, Minn., USA;

The journal is also staffed by an international group of Consulting
Editors.

Blood Cells publishes symposium proceedings that deal with selected
topics concerning the physiology or pathophysiology of the blood at
the cellular level. It also includes original papers and editorials, both
invited and unsolicited, submitted by distinguished international
scientists. The emphasis on alterations in structure and function at the
cellular level makes this new journal important to every researcher
concerned with the biochemistry, physiology, ultrastructure, immunology, and biophysics of blood cells in health and disease. Many of
the contributions have been presented at symposia organized three
times a year by the editors of Blood Cells. A unique feature is the
publication of general discussions which include questions by other
experts and the authors' replies.
Each issue focuses on a particular subject area within the field of
hematology.

Unclassifiable Leukemias

Editors: M. BESSIS, G. BRECHER
81 figures, 38 tables, 1 color-plate. VI, 270 pages. 1975
ISBN 3-540-07242-X
This monograph comprises the first issue of the international journal
Blood Cells.

In this monograph, a group of well known hematologists from both
sides of the Atlantic who have a theoretical interest in the differential
diagnosis of the leukemia and a practical concern with its differential
treatment discuss the power and limitations of currently available
methods for classification of "unclassifiable" leukemias. Starting out
from the classical distinction of lymphoid and granulocytic leukemias
in Giemsa stains, the papers cover common and rare subgroups
studied by phase, transmission and scanning electron microscopy,
determination of enzymes and immunologic surface markers.
Morphologic classification is related to clinical course and treatment.
Of particular value are the extensive debates between participants
which follow each paper and clarify controversial points. The incisive
discussions specify both agreements and disagreements between the
experts and lead beyond "authoritative" statements to speculation on
matters of current research interests such as the role of very small subpopulations in the evolution of leukemias and the eventual refrac-
toriness to treatment.

Springer-Verlag
Berlin
Heidelberg
New York

MARCEL BESSIS

Corpuscles

Atlas of Red Blood Cell Shapes

121 figures, 147 pages. 1974
ISBN 3-540-06375-7
Distribution rights for Japan: Maruzen Co. Ltd.,
Tokyo

The hematologist usually examines red blood
cells by looking at smears through a light microscope. This type of examination is likely to remain
the routine technique of blood cytology for a long
time to come. Still we should not forget that blood
smears are artifacts. The smearing flattens the cell
completely, obliterating many of their characteristics and distorting others.
This atlas reveals the shape of red blood cells as
displayed by scanning electron microscopy. As
with the transmission electron microscope, the
cells must be fixed, but they can be observed
in their three dimensions and revolved before
our very eyes, thus providing details of the surface
and the shape of cells never before visualized.
Having seen the results of studies with the new
techniques, hematologists can now distinguish
nuances in the appearance of red cells in routine
smears which previously could not have been
appreciated. Thus routine blood smears assume a
new significance.
Moreover, there is an inherent beauty in the
pictures displayed here, a beauty which is
enchanting to both the specialist and the layman.

Red Cell Shape

Physiology, Pathology, Ultrastructure

Edited by Marcel Bessis, Robert I. Weed,
Pierre F. Leblond

147 figures. VIII, 180 pages. 1973
ISBN 3-540-06257-2

This monograph is a collection of contributions
from a group of researchers who share a common
interest in the interrelationships between the
shape, structure, and functional characteristics
of normal and pathologic erythrocytes.
The text is divided into four sections.
Section I deals with red cell shape, including the
presentation of a rational, descriptive nomenclature and a discussion of postsplenectomy
changes.
Section II deals with biochemical factors which
underlie the discocyte-echinocyte (crenated) and
discocyte-stomatocyte (cup-shaped) transformations. This section includes discussions of

plasma such as cellular ATP, calcium, aging, and
various chemical agents, as determinants of shape.
Section III, which deals with biophysical
measurements, includes studies of the deformability of cells of different shapes and a model
of membrane structure which is proposed
to account for the dimensions of red cells which
undergo shape change.
The final section (Section IV) deals with ultrastructural studies of normal, sickled, and Heinz-
body-containing red cells. This section also
includes a study of cells injured by antibody and
complement.

MARCEL BESSIS

Living Blood Cells and their Ultrastructure

Translated by Robert I. Weed

521 figures, 2 color-plates. XXII, 767 pages. 1973
ISBN 3-540-05981-4
Distribution rights for Japan: Maruzen Co. Ltd.,
Tokyo

This timely reexamination of blood cell cytology
emphasizes the importance and uniqueness of
the information to be gained from studying cells
in the living state.
Chapter I deals with differentiation, hemopoiesis,
the structure of organelles, cell dynamics, and cell
death. Chapters II–VIII deal, respectively, with
each of the different blood cell lines and
Chapter IX discusses the morphologic changes of
hematologic malignancies and their significance.
Chapter X is concerned with techniques for the
study of living cells and electron microscopic
methodology as applied to blood cells. An
appendix to this chapter contains extensive
references to routine hematologic techniques and
normal values.
The text is abundantly illustrated with carefully
choosen photomicrographs. Throughout the
book the appearance of each cell is discussed as
it is seen in a Giemsa-stained preparation, in the
living state, and with both the scanning and
transmission electron microscopes.
There are approximately 5,000 selected references
to reviews and outstanding articles dealing with
the physiologic and biochemistry which underlie
the morphologic changes discussed in the book.

Springer-Verlag
Berlin Heidelberg New York